Also in the Varioum Collected Studies Series:

DAVID A. KING
Islamic Mathematical Astronomy

DAVID A. KING
Islamic Astronomical Instruments

P.M. HARMAN
After Newton: Essays on Natural Philosophy

ROBERT FOX
The Culture of Science in France, 1700-1900

A. RUPERT HALL
Newton, his Friends and his Foes

A.J. TURNER
Of Time and Measurment
Studies in the History of Horology and Fine Technology

R.W. HOME
Electricity and Experimental Physics in Eighteenth-Century Europe

ROSHDI RASHED
Optique et mathématiques
Recherches sur l'histoire de la pensée scientifique en arabe

JOHN M. RIDDLE
Quid pro quo: Studies in the History of Drugs

PAUL KUNITZSCH
The Arabs and the Stars
Texts and Traditions on the Fixed Stars and their Influence in Medieval Europe

CURTIS WILSON
Astronomy From Kepler to Newton

BRUCE S. EASTWOOD
Astronomy and Optics From Pliny to Descartes
Texts, Diagrams and Conceptual Structures

BERNARD R. GOLDSTEIN
Theory and Observation in Ancient and Medieval Astronomy

Astronomy in the Service of Islam

The author introducing his son Max to the delights of Euclid and Ptolemy
(Cairo, 1973)

David A. King

Astronomy in the Service of Islam

VARIORUM

Published by VARIORUM
Ashgate Publishing Limited
Gower House, Croft Road,
Aldershot, Hampshire GU11 3HR
Great Britain

Ashgate Publishing Company
Old Post Road,
Brookfield, Vermont 05036
USA

ISBN 0-86078-357-X

British Library CIP data
King, David A.
Astronomy in the Service of Islam. -
(Variorum Collected Studies Series; CS416)
I. Title II. Series
520.917671

Library of Congress CIP data
King, David A.
Astronomy in the service of Islam / David A. King.
p. cm. -- (Collected studies series; CS416)
ISBN 0-86078-357-X (alk. paper): $94.95 (est.)
1. Astronomy--Islamic countries--History. 2. Islam and science--History. 3. Calendar, Islamic--History. 4. Time (Islamic law)--History. 5. Qiblah--History. 6. New moon--Visibility--Forecasting--History. I. Title. II. Series: Collected studies; CS416.
QB23.K52 1993 93-25664
620' .917'671--dc20 CIP

The paper used in this publication meets the minimum requirements of American National Standard for Information Sciences - Permanence of Paper for Printed Library Materials, ANSI Z39.48-1984. ∞ ™

Printed by Galliard (Printers) Ltd
Great Yarmouth, Norfolk, Great Britain

COLLECTED STUDIES SERIES CS416

CONTENTS

This volume contains xiv+ 333 pages

PUBLISHER'S NOTE

The articles in this volume, as in all others in the Collected Studies Series, have not been given a new, continuous pagination. In order to avoid confusion, and to facilitate their use where these same studies have been referred to elsewhere, the original pagination has been maintained wherever possible.

Each article has been given a Roman number in order of appearance, as listed in the Contents. This number is repeated on each page and quoted in the index entries.

ACKNOWLEDGEMENTS

It is a pleasure to express my gratitude to my assistant Kurt Maier for proof-reading all of the articles in this volume as well as the additional text.

For permission to reprint the various articles I thank Dr. Howard J. Moore, editor of *impact of science on society* at U.N.E.S.C.O., Paris (I); Mr. Bill Boland, Director of the New York Academy of Sciences Press (II and XIII); Dr. Michael Hoskin of Churchill College, Cambridge, editor of the *Journal of the History of Astronomy*, and Science History Publications, Ltd., of Chalfont St. Giles, Bucks. (U.K.) (III and XII); the American University of Beirut Press (IV); Dr. Th. Dijkema and the editors of the *Encyclopaedia of Islam* at E. J. Brill, Leiden (V, VIII, IX, X and XI); Copenhagen University Library Press (VI); New York University Press (VII); my colleague Professor Fuat Sezgin of Frankfurt am Main, editor of the *Zeitschrift für Geschichte der arabisch-islamischen Wissenschaften* (XIV); as well as my co-author Dr. Gerald S. Hawkins of Washington, D.C. (XII).

Above all I am grateful to John Smedley of Variorum for his tolerance, forbearance, patience and encouragement.

To Max,
in gratitude for twenty years together

PREFACE

This volume supplements two earlier ones published by Variorum in their Collected Studies Series: *Islamic Mathematical Astronomy* (1st. ed., 1986; 2nd. revised ed., 1993) and *Islamic Astronomical Instruments* (1987). There is already a substantial amount of material in those volumes relating to the theme "Astronomy in the Service of Islam".

The studies reprinted here are intended to underline firstly that there were two, virtually independent traditions in medieval Islamic astronomy, and secondly that aspects of Islamic ritual involving astronomy were approached in two, almost mutually-exclusive ways. Thus the legal scholars advocated observation of the lunar crescent for regulating the calendar, simple arithmetical shadow schemes for regulating the times of prayer, and the use of astronomical horizon phenomena for determining the direction of prayer. The astronomers, on the other hand, developed sophisticated theories and mathematical procedures and tables for predicting lunar crescent visibility, for pinpointing the times of prayer to the nearest minute, and for determining the direction of Mecca to within a fraction of a degree. More information of a general nature on this topic is contained in Paper I, also published in various other languages, though, alas, not in Arabic.

Western readers will find here material that is new to both the history of science and to Islamic Studies. Muslim readers will find information which adds to our knowledge of the development and practice of Islamic ritual. The evidence for the generalizations made in Paper I is contained in the literature listed in the bibliography, none of which is in Arabic; the original sources identified in that literature are, however, all in Arabic, and they all are available for consultation. It is inevitable that such a wide range of previously-unstudied sources should lead to new understanding. Much has been written in recent decades by persons interested only in the scientific knowledge which Europe inherited from the Islamic world. Here, for the first time, is a sober account, written by a friend of Islam and based on countless hitherto-unstudied sources, of various aspects of the Islamic scientific tradition which were of no interest whatsoever to medieval Europe. They should be of interest to modern Muslims, for they attest to the way in which during some 1400 years the scientists of medieval Islam on the one hand, and the scholars of Islamic religious law on the other, came to terms with the practical implications of three of the five pillars of Islam. That these two groups - Muslim scientists and Muslim legal scholars - should handle these problems in different ways was inevitable (the same was the case in medieval Judaism and Christianity). There was, however, as far as I know, never any serious conflict between these two groups until the present century. The present conflicts could be partially resolved by studying the way in which the great legal scholars and the great astronomers of past centuries dealt with these problems.

Papers II-IV deal with lunar crescent visibility, but only from the point of view of the astronomers. There is now a substantial literature on the various visibility theories advocated by the astronomers, but a critical investigation of the writings of the legal scholars on this subject remains a task for the future.[1]

Papers V-VIII deal with astronomical time-keeping in general and the regulation of the times of prayer in particular. The reader may wish to consult *IMA* VIII-XII for a description of the corpuses of tables used in medieval Cairo, Damascus, Jerusalem, Tunis, Taiz and Istanbul. Papers VI-VII deal with universal solutions, mainly to problems of time-keeping, but not exclusively, and they illustrate the sophistication of Muslim developments in spherical astronomy.[2] Paper V provides an overview of the two approaches to time-keeping by the legal scholars and the astronomers. Paper VIII provides an introduction to Islamic gnomonics, one practical manifestation of Muslim concern for time-keeping. At first a subject treated by Muslim scientists, sundials later came to be a prominent feature in most major mosques.[3]

---

1 Numerous relevant sources have been published by Muḥammad ibn ʻAbd al-Wahhāb of Marrakesh in a very useful work entitled *al-ʻAdhb al-zulāl fī mabāḥith ru'yat al-hilāl* (Qatar: Maṭbaʻat Idārat al-Shu'ūn al-Dīniyya, 1977).

2 All known tables for astronomical time-keeping are analyzed in my long-promised *Studies in Astronomical Timekeeping in Medieval Islam*, and a detailed study of the Damascus corpus entitled *Shams al-Dīn al-Khalīlī and the Culmination of the Islamic Science of Astronomical Timekeeping* has been completed more recently. My paper "A Survey of Medieval Islamic Shadow Schemes for Simple Time-Reckoning," *Oriens* 32 (1990), pp. 191-249, not included here for reasons of space, shows that the legal scholars of medieval Islam had no time for any real tables and that they proposed quite different means for regulating the prayers.

3 Another paper on the institutions of the muezzin and the *muwaqqit* (the professional astronomer associated with mosques in the central lands of Islam after the thirteenth century), presented at a conference on Islamic intellectual history held at Harvard University in 1988 and not included here, is to be published in *Oriens*.

Since I started working on Islamic time-keeping some progress has been made by others in the history of astronomical time-keeping in medieval Europe. I refer mainly to the edition by S. Eisner (1980) of the tables for Oxford by Nicholas of Lynn. But there is still a great deal of work to be done on tables for time-keeping in European manuscripts and early printed works. In a forthcoming book on the corpus of tables for Damascus prepared by al-Khalīlī in the fourteenth century I point out that the auxiliary tables of Ḥabash and al-Khalīlī and other Muslim astronomers had their counterparts in tables of similar functions by Regiomontanus in the fifteenth century and Magini in the early seventeenth. As this book was in press I stumbled across a sixteenth-century German manuscript full of tables for time-keeping for various latitudes, including one serving all latitudes between 39° and 63°.[4] In the nineteenth and early-twentieth century, tables for solving the basic problem of the determination of time from solar or stellar altitudes for any latitude were still being prepared in Europe and the United States; this activity started in Baghdad in the ninth century.

Papers IX-XIV deal with the qibla, the sacred direction of Islam. My own development as a historian of Islamic science is well reflected in Papers IX and X from the *Encyclopaedia of Islam*. As the new edition of the *Encyclopaedia* progressed from K to M (from the mid-1970's to the late 1980's) I was able to provide the articles "Ḳibla" and "Makka: As Centre of the World". When writing the first article, based exclusively on mathematical treatments of the qibla problem by medieval scientists as researched mainly by K. Schoy and E. S. Kennedy, I had no inkling of the existence of the material on which the second article is based. Now a whole new corpus of material on the qibla is uncovered, and this is investigated in a forthcoming monograph,[5] summarized in Paper X.

At the end of the article "Ḳibla" I commented that some medieval mosques seemed to have been oriented by traditional procedures rather than calculation. That the alignment of the Ka'ba, discussed for the first time in Paper XII, and the primitive wind schemes of the Arabs of pre-

---

4 MS Schweinfurt Stadtsbibliothek 21. The tables are now described in a catalogue of an exhibition (item H18) to be inaugurated in Schweinfurt in November, 1993, edited by Dr. Uwe Müller. A first attempt at a comparison of Islamic and Christian applications of astronomy to religious practices is in my paper "Aspects of Applied Science in Mosques and Monasteries," presented at the Symposium "Science and Theology in Medieval Islam, Judaism, and Christianity," Madison, Wisconsin, April 15-17, 1993.

5 "The Sacred Geography of Islam", to appear under the auspices of Islamic Art Publications, S.p.A.

Islamic Arabia surveyed in Paper XI, had as much influence in the orientation of medieval Islamic religious architecture as did the qibla tables of the medieval scientists is now clear from Papers X and XIII and the more detailed study, as yet unpublished, on which they are based. In Paper XIV I describe the earliest mathematical methods and tables for finding the qibla from ninth-century Baghdad. The remarkable cartographic solution to the qibla problem from Isfahan, *ca.* 1700, which came to my attention only in 1989 and which is illustrated in Paper I, shows how Muslim initiative with regard to the determination of the qibla reached another peak. I had previously thought that the qibla table of al-Khalīlī, now known to be but one of several highly sophisticated solutions, marked the culmination of this activity.

At this time of Muslim reawakening in matters relating to the applications of astronomy to daily life in the Muslim community, attested by the valuable works of Dr. Muhammad Ilyas of Malaysia, it is important to remember that these applications have a history of over a thousand years. But we can appreciate the Muslim achievement in this area, as in others, only if we look at what the Muslim scientists actually wrote. By ignoring their writings and simply extolling their virtues, or by suppressing what the scholars of the sacred law had to say on the same subjects, we do them and ourselves a great disservice.

Four of the articles are dedicated to teachers, colleagues and friends, namely, two historians of science, E. S. Kennedy (II) and Asger Aaboe (VI), and two Arabists, the late Malcolm H. Kerr (IV) and the late R. Bayly Winder (VII). Any merits this book may have are to a great extent thanks to my contacts with these and other distinguished teachers and colleagues.

Frankfurt am Main,
July, 1993

DAVID A. KING

I

Science in the service of religion: the case of Islam

In Islam, as in no other religion in human history, the performance of various aspects of religious ritual has been assisted by scientific procedures. The organization of the lunar calendar, the regulation of the astronomically-defined times of prayer, and the determination of the sacred direction of the Kaaba in Mecca—these are topics of traditional Islamic science still of concern to Muslims today, and each has a history going back close to fourteen hundred years. But the techniques advocated by the scientists of medieval Islam on the one hand and by the scholars of religious law on the other were quite different, and our present knowledge of them is based mainly on research conducted during the past twenty years on one small fraction of the vast literary heritage of the Muslim peoples.

Most historians of Islamic science have concentrated on scientific knowledge which was transmitted to the West; by so doing they have tended to overlook the essence of Islamic science. Indeed, most modern accounts of science in the medieval Islamic world, whether by Western or Muslim writers, have ignored what may well be called the Islamic aspects of Islamic science. These have been researched recently, using the vast amount of relevant medieval Arabic manuscripts available in libraries around the world; most of the results of this research have appeared in scholarly journals not easily obtainable outside academic libraries. The time is therefore ripe for an overview. In fact, this article is the first attempt in the non-scholarly literature to survey the way in which science, particularly astronomy, has been used for purposes relating to Muslim religious life for well over a millennium. Even so, it is not an overview of Islamic astronomy in general, for it deals with only three of the many topics dealt with by the scholars of medieval Islam.

To understand Muslim activity in this domain we must realize that there were two main traditions of astronomy in the Islamic Near East, *folk astronomy* and *mathematical astronomy*. Folk astronomy, based on naked-eye observation of celestial phenomena and devoid of theory or computation, has generally been overlooked by historians of science with their predilection for hard-core scientific achievements. Yet, as we shall see, it was far more influential in Islamic society than mathematical astronomy, which as the name indicates was based on systematic observation, theory and mathematical procedures.

A historical investigation of the Islamic aspects of Islamic science provides answers to several questions. First, why is there so much confusion in the modern Islamic world about the determination of the beginning of Ramadan, the sacred month of fasting? Second, why are there five prayers in Islam? These are not specifically prescribed in either the Koran, the ultimate source of Islamic sacred law, or the hadith, the literature dealing with the sayings and practice of the Prophet Muhammad, the second main authority for the sacred law of Islam. Third, why are medieval mosques invariably not oriented properly towards Mecca? This problem has often puzzled those few historians of Islamic architecture who have taken the trouble to measure mosque orientations, but, as we shall see, it is now largely resolved, thanks to the evidence of newly discovered medieval texts. Furthermore these texts cast new light on the significance of the Kaaba itself and on its original function.

Folk astronomy and mathematical astronomy

The Arabs of the Arabian peninsula before Islam had an intimate acquaintance with the sun, moon and fixed stars, the seasons, the changing night sky and weather patterns throughout the year. Since the sun, moon and stars, as well as the winds and rains, are mentioned in the Koran, a truly Islamic cosmology (quite independent from the tradition adapted from Greek sources by Muslim scientists) developed in the vast corpus of the Koranic commentaries and in separate treatises on the glory of God as revealed by His creation. Since, in addition, the Koran encourages Man to use the stars for guidance, a basic knowledge of the heavens was considered advantageous. Folk astronomy, based on what could actually be seen in the sky throughout the year and innocent of any underlying theory or associated computus, thus became widespread in the Islamic Near East and remained so throughout medieval times. The basics of this subject are outlined in encyclopaedias and a series of special treatises compiled over many centuries, and its application to religious needs is discussed in books dealing with the sacred law of Islam.

The period from the eighth to the fourteenth or fifteenth centuries saw the flourishing in the Near East of a different kind of astronomical knowledge. Muslim astronomers, heirs to the sophisticated astronomical traditions of the Hellenistic world, and also of Iran and India, made new observations, developed new theories, compiled new tables, and invented new instruments. They produced an enormous corpus of scientific literature covering all subjects from cosmology to computational techniques, and they made progress in all branches of their discipline. But the scientists did not have a wide audience. They wrote mainly technical treatises which circulated only within the scientific community, and few of them compiled popular summaries. In particular, the solutions they proposed for problems relating to religious ritual were generally considered to be too complicated or even completely irrelevant.

We now consider three facets of Islamic religious practice involving astronomy. As we shall see, the simple techniques of folk astronomy were applied to these practical problems by the legal scholars, and the complicated techniques of mathematical astronomy were applied to the same problems by the astronomers. The former, generally disinclined to listen to the opinions of scientists, had far greater control over the practice of the people than had astronomers. On the other hand, the solutions developed by Muslim scientists, invariable too complicated for widespread application in the medieval milieu, are impressive indeed from a scientific point of view.

The regulation of the lunar calendar

The Islamic calendar is strictly lunar. The beginnings and ends of the lunar months, in particular of the holy month of Ramadan, and various festivals throughout the twelve-month 'year', are regulated by the first appearance of the lunar crescent.

Since twelve lunar months add up to about 354 days, the twelve-month-cycles of the Islamic calendar occur some eleven days earlier each year, and the individual months move forward through the seasons. To keep the lunar months in line with the seasons of the solar year it was the custom in pre-Islamic times to insert an additional 'intercalary' month in the lunar calendar every few years. This practice the Prophet Muhammad abandoned. The Koran expressly forbids such intercalation, and the exegetes explain that the proscription was necessary because intercalation caused months that God had intended to be holy to be confused with other months.

For scholars of the sacred law, the month began with the first sighting of the crescent moon. This observation is a relatively simple affair, provided that one knows roughly where and when to look and the western sky is clear. Witnesses with exceptional eyesight were sent to locations that offered a clear view of the western horizon, and their sighting of the crescent determined the beginning of the month; otherwise they would repeat the process the next day. If the sky was cloudy, the calendar would be regulated by assuming a fixed number of days for the month just completed. Also, the crescent might be seen in one locality and not in another. Unfortunately the historical sources contain very little information on the actual practice of regulating the calendar.

Astronomers, on the other hand, knew that the determination of the possibility of sighting on a given day was a complicated mathematical problem, involving knowledge of the positions of the sun and moon and the mathematical investigation of the positions of the both celestial bodies relative to each other and to the local horizon (see figure 1). In short, the lunar crescent will be seen after sunset on a given evening at the beginning of a lunar month if it is far enough away from the sun, and if it is high enough above the horizon not to be overpowered by the background sky glow. Conditions required to assure crescent visibility on most occasions can be determined by observation, but the formulation of a definitive set of conditions has defied even modern astronomers. The positions of the sun and moon must be investigated to see whether the assumed visibility conditions are satisfied, but, even if they are, the most ardent astronomer can be denied the excitement of sighting the crescent at the predicted time if clouds or haze on the western horizon restrict his view.

The earliest Muslim astronomers adopted a lunar visibility condition which they found in Indian sources. It was necessary to calculate the positions of the sun and moon

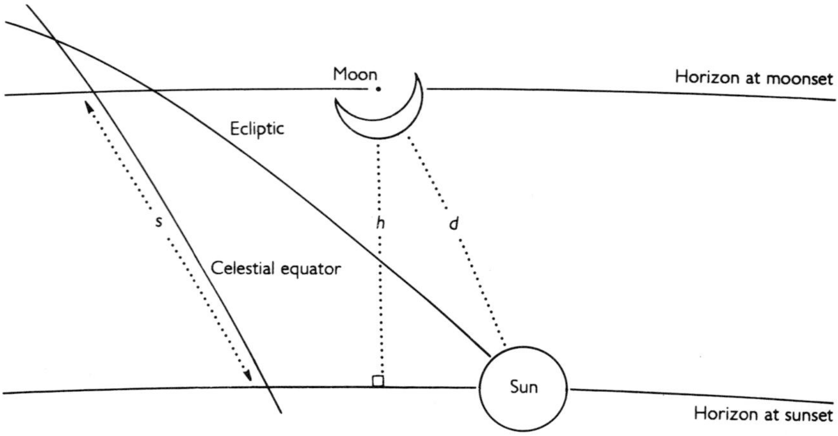

Figure 1. The western horizon at sunset on the evening of first visibility of the crescent moon. For predicting visibility Muslim astronomers devised sets of conditions on such quantities as the apparent distance between the sun and moon (*d*), the altitude of the moon above the horizon at sunset (*h*), and the difference in setting times of the sun and moon (*s*).

from tables and then to calculate the difference in setting times over the local horizon. If the latter was 48 minutes or more, the crescent would be seen, if it was less the crescent would not be seen. Using this condition and computing specifically for the latitude of Baghdad, the astronomer al-Khwarizmi in the early ninth century compiled a table showing the minimum distances between the sun and moon (measured on the ecliptic) to ensure crescent visibility throughout the year.

During the following centuries Muslim astronomers not only derived far more complicated conditions for visibility determination but also compiled highly sophisticated tables to facilitate their computations. Some of the leading Muslim astronomers proposed conditions involving three different quantities, such as the apparent angular separation of the sun and moon, the difference in their setting times over the local horizon, and the apparent lunar velocity. Annual ephemerides or almanacs gave information about the possibility of sighting at the beginning of each month (see figure 2). In brief, the achievements of the Muslim astronomers in this area were impressive.

In modern times the regulation of the calendar has led to controversy between religious authorities and scientists. The main problems are the difficulty of making sure predictions for a multiplicity of locations and the unwillingness of the religious authorities to listen to the scientists. For example, Ramadan has sometimes been announced one, or even two, days early in some Islamic countries. (see, for example, *Al-Ahram*, Cairo, 26 and 27 September 1973). This occurrence, unthinkable in medieval times, resulted not only from the enthusiasm to begin the fast, but also from the ineptitude of the responsible authorities in matters scientific. Modern communications and divergent political interests also played a role. An international commission has recently been formed to handle problems associated with the Islamic calendar, happily under the enlightened leadership of an astronomer, Dr. Mohammed Ilyas of Malaysia.

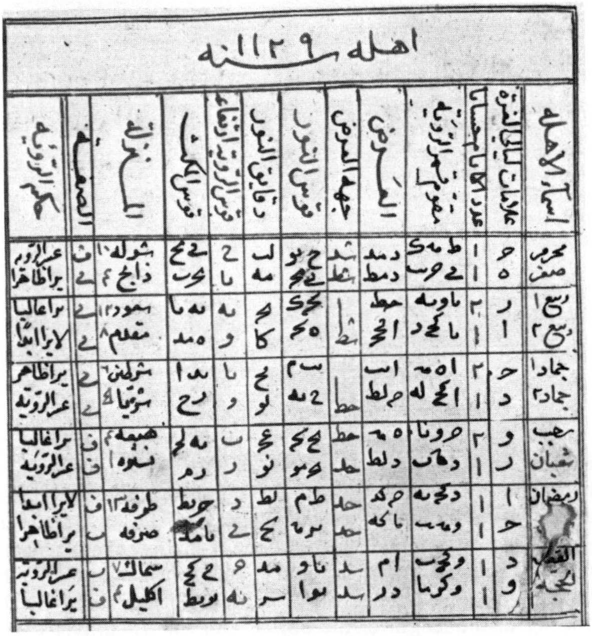

Figure 2. A table showing visibility predictions for the first evening in each month of the *civil* calendar during the year 1129 Hijra (=1716/17). Calculations of the positions of the sun and moon relative to each other and relative to the local horizon lead to pronouncements such as the crescent 'will be seen clearly', 'will be seen with difficulty', or—in the case of Ramadan that year—'will not be seen at all'. In the last case the religious authorities would have announced the beginning of Ramadan the following evening. Numbers in these and other medieval tables were written in the standard alphanumerical system (called *abjad* in Arabic) and are expressed sexagesimally, that is to base 60—as in modern notation for angles in degrees and minutes or time in hours and minutes. (Courtesy of the Egyptian National Library.)

The regulation of the five daily prayers

The times of the five daily prayers in Islam are defined in terms of astronomical phenomena dependent upon the position of the sun in the sky. More specifically, the times of daylight prayers are defined in terms of shadows, and those of night prayers in terms of twilight phenomena. They therefore vary with terrestrial latitude, and unless measured with respect to a local meridian, also with terrestrial longitude.

Because the months begin when the new moon is seen for the first time shortly after sunset the Islamic day is considered to begin at sunset. Each of the five prayers in the Islamic day (see figure 3) may be performed during a specified interval of time, and the earlier during the interval that the prayer is performed, the better.

Figure 3.

The five standard prayers of Islam and their timings. The three prayers at night are regulated by horizon and twilight phenomena and the two daylight prayers by means of shadow lengths.

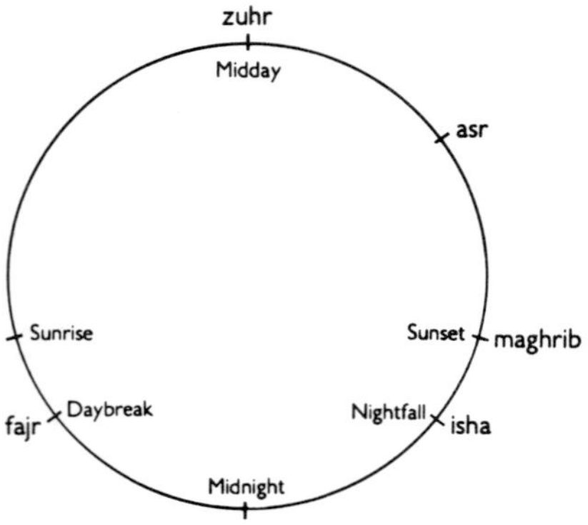

The day begins with the *maghrib* or sunset prayer. The second prayer is the *isha* or evening prayer, which begins at nightfall. The third is the *fajr* or dawn prayer, which begins at daybreak. The fourth is the *zuhr* or noon prayer, which begins shortly after astronomical midday when the sun has crossed the meridian. The fifth is the *asr* or afternoon prayer, which begins when the shadow of any object has increased beyond its midday minimum by an amount equal to the length of the object casting the shadow. In some medieval circles, the *zuhr* prayer began when the shadow increase was one quarter of the length of the object, and the *asr* prayer continued until the shadow increase was twice the length of the object—see figure 4. In other communities, a prayer at midmorning, called the *duha*, began at the same time before midday as the *asr* began after midday. This prayer is mentioned in the hadith, and there are varying accounts about it. In some the Prophet is said to have performed it himself. In others, he is said to have declared it a heretical innovation. This is a clear indication that later authorities were undecided about whether or not to include it in the daily ritual.

The five prayers adopted by the Muslim community are not specifically mentioned in the Koran. In the hadith literature, more than five prayers are mentioned but no specific definitions are given in their times. The *duha* prayer at midmorning was clearly practised by some in the early Muslim community, but later it was generally, although not completely, abandoned. Also there is reference to the night-vigil called *tahajjud*; this was later made optional. The standard definitions of the times for the daytime prayers in terms of shadow *increases* rather than shadow *lengths* (as mentioned in the Prophetic hadith quoted above) first appear in the eighth century.

The reason why five prayers were adopted by the early Muslim community is clear from the definitions of their times. The definitions of the *duha*, *zuhr*, and *asr* prayers in terms of shadow increases provide simple and practical means for regulating them at the ends of the third, sixth and ninth hours of daylight, the hours being seasonal hours, that is, one-twelfth divisions of the length of daylight. Seasonal hours, which vary in duration throughout the year, were in standard use in the Near East in Antiquity. The relationship between these and the shadow increases is provided by a simple, approximate formula for timekeeping of Indian origin known to the Muslims in the eighth century. Even the names of the prayers in Islam are the same as those of the corresponding seasonal hours recorded by some of the Arab lexicographers. Their

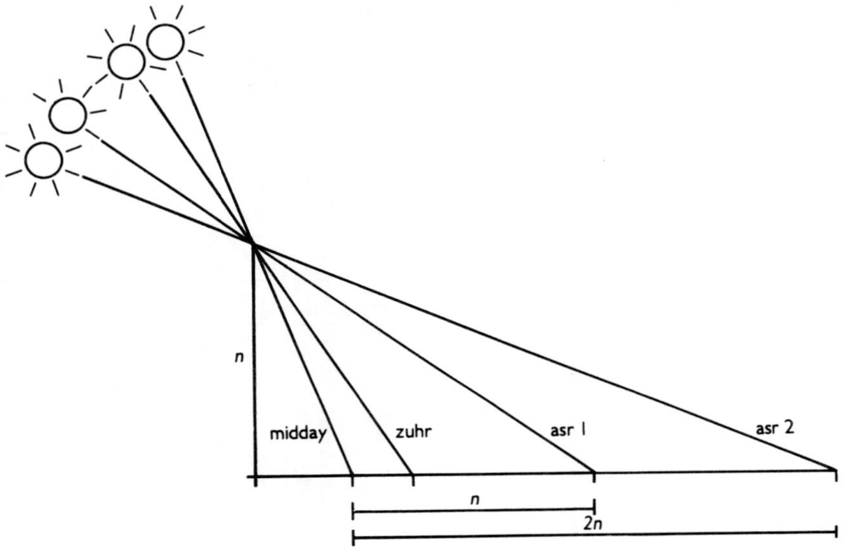

Figure 4. The times of the *zuhr* and the *asr* prayers are defined in terms of the increase of the shadow of a vertical object over its midday minimum. Thanks to medieval texts on folk astronomy these curious definitions have now been explained.

times correspond to the times of the seven prayers of early Syrian Christianity, but with the omission of the prayer at sunrise because it was expressly forbidden by the Prophet and the dropping of the prayer at midmorning in all but a few communities.

In the first few decades of Islam, the times of prayer were regulated by observation of shadow lengths by day and of twilight phenomena in the evening and early morning. Precisely how either the daylight or the night-time prayers were regulated is unfortunately not clear from the available historical sources. Muezzins who performed the call to prayer from the minarets of mosques were chosen for their piety and the excellence of their voices, but their technical knowledge was limited to the basics of folk astronomy.

On the other hand, the determination of the precise moments (expressed in hours and minutes, local time) when the prayers should begin, according to the standard definitions, required complicated mathematical procedures in spherical astronomy, that is the study of problems associated with the apparent daily rotation of the celestial sphere. Accurate as well as approximate formulae for reckoning time of day or night from solar or stellar altitudes were available to Muslim scholars from Indian sources and these were improved and simplified by Muslim astronomers. Certain individual astronomers from the ninth century onwards applied themselves to the calculation of tables for facilitating the determination of the prayer-times. The earliest prayer-tables were prepared by al-Khwarizmi for the latitude of Baghdad. The first tables for finding the time of day from the solar altitude or the time of night from the altitudes of certain prominent fixed stars appeared in Baghdad in the ninth and tenth centuries. The extent

Figure 5. An extract from the main corpus of astronomical tables for regulating the times of the prayers in medieval Cairo. These tables, discovered only in 1970, cast considerable light on contemporary religious practice. This extract displays for each degree of solar longitude, corresponding roughly to each day of the year, the time from sunset until the moment when the muezzin should extinguish the candles on the minarets during the month of Ramadan. (Courtesy of the Egyptian National Library.)

to which these tables deriving from mathematical procedures were used before the thirteenth century is unknown; the earliest examples are contained in technical works which must have had fairly limited circulation. The muezzins certainly had no need of them. One had to be an astronomer, for they had to be used together with some kind of observational instrument for measuring the sun's altitude and reckoning the passage of time.

It was not until the thirteenth century that the institution of the *muwaqqit* appeared in mosques and madrasas. These professional astronomers associated with a religious institution not only regulated the prayer-times, but constructed instruments, wrote treatises on spherical astronomy, and gave instruction to students. In thirteenth-century Cairo, new tables were compiled which set the tone for astronomical time-keeping all over the Islamic world in the centuries that followed. In medieval Cairo there was a corpus of some 200 pages of tables available for time-keeping by the sun and for regulating the times of prayer (see figure 5).

Impressive innovations in astronomical time-keeping were made in other medieval cities, especially Damascus, Tunis and Taiz, although by the sixteenth century Istanbul had become the main centre of this activity. We may mention, for example, highly sophisticated tables of special trigonometric functions especially designed to solve problems of spherical astronomy for any latitude. Tables for finding the time of day from the solar altitude at any time of year were compiled for Cairo, as we have mentioned, and also for Damascus, Tunis, Taiz, Jerusalem, Maragha, Mecca, Edirne and Istanbul. Medieval tables for regulating the times of prayer have been found for a series of localities between Fez in Morocco and Yarkand in China. Such tables have a history spanning the millennium from the ninth century to the nineteenth.

As noted above astronomical tables for regulating the prayer-times had to be used together with instruments. Only in this way could one ascertain that the time advocated in the table had actually arrived. The most popular of these instruments were the astrolabe and the quadrant. Hundreds of Islamic astrolabes and several dozen quadrants are preserved in the museums of the world, a small fraction of the instruments actually made by Muslim astronomers. An alternative means of regulating the daytime prayers was available to the Muslims in the form of the sundial. Many mosque sundials from the later period of Islamic astronomy survive to this day, though most are non-functional.

The call of the muezzin is to be heard every day in towns and villages all over the Islamic commonwealth, and the call to prayer is also broadcast on radio and television. But muezzins and technicians now read the prayer-times from tables found in pocket diaries, wall calendars, and daily newspapers. The times are usually computed by local survey departments or other agencies acceptable to the religious authorities, who apply modern methods to definitions which have been standard for over a millennium. Recently clocks and watches have appeared on the market which beep at the prayer-times and reproduce a recorded prayer-call—a far cry indeed from observing shadow lengths or reckoning prayer-times with an astrolabe or sophisticated astronomical tables.

The determination of the sacred direction

The *Kaaba* is a shrine of uncertain historical origin which served as a sanctuary and centre of pilgrimage for the Arabs for centuries before the advent of Islam. It was adopted by the Prophet Muhammad as the focal point of the new religion, and the Koran advocates prayer towards it. For Muslims it is a physical pointer to the presence of God. Thus since the early seventh century Muslims have faced the Sacred Kaaba in Mecca during their prayers. Mosques are built with the prayer-wall facing the Kaaba, the direction being indicated by a *mihrab* or prayer-niche. In addition, certain ritual acts such as reciting the Koran, announcing the call to prayer, and slaughtering animals for food, are to be performed facing the Kaaba. Also Muslim graves and tombs were laid out so that the body would lie on its side and face the Kaaba. (Modern burial practice is slightly different but still Kaaba-oriented.) Thus the direction of the Kaaba—called *qibla* in Arabic and all other languages of the Islamic commonwealth—is of prime importance in the life of every Muslim.

In the first two centuries of Islam, when mosques were being built from Andalusia to Central Asia, the Muslims had no truly scientific means of finding the qibla. Clearly they knew roughly the direction they had taken to reach wherever they were, and the direction of the road on which pilgrims left for Mecca could be, and in some cases actually was, used as a qibla. But they also followed two basic procedures, observing tradition and developing a simple expedient.

In the first case, some authorities observed that the Prophet Muhammad had prayed due south when he was in Medina (north of Mecca) and they advocated the general adoption of this direction for the qibla. This explains why many early mosques from Andalusia to Central Asia face south.

Other authorities held that the Koran required one to stand precisely so that one faced the Kaaba. Now the Muslims of Meccan origin knew that when they were standing in front of the walls or corners of the Kaaba they were facing directions specifically associated with the risings and settings of the sun and certain fixed stars. The major axis of the rectangular base of the edifice is said to point towards the rising point of Canopus, and the minor axis is said to point towards summer sunrise and winter sunset (see figure 6). These assertions about the Kaaba's astronomical alignments, found in newly-discovered medieval sources, have been confirmed by modern measurements. In addition, Arabic folklore associates the sides of the Kaaba with the winds and rain. These features and associations cast new light on the origin of the edifice, and in a sense confirm the Muslim legend that the Kaaba was built in the style of a celestial counterpart called *al-bayt al-ma'mur*: indeed it seems to have been an architectural model of a pre-Islamic Arab cosmology in which astronomical and meteorological phenomena are represented. The religious association was achieved first by a number of statues of the gods of the pagan Arabs which were housed inside it. With the advent of Islam these were removed, and the edifice has for close on 1400 years served for Muslims as a physical pointer to the presence of God.

The corners of the Kaaba were associated even in pre-Islamic times with the four main regions of the surrounding world, Syria, Iraq, the Yemen, and 'the West'. Some Muslim authorities said that to face the Kaaba from Iraq, for example, one should stand in the same direction as if one were standing right in front of the north-eastern wall of the Kaaba. Thus the first Muslims in Iraq built their mosques with the prayer-walls towards winter sunset because they wanted the mosques to face the north-eastern

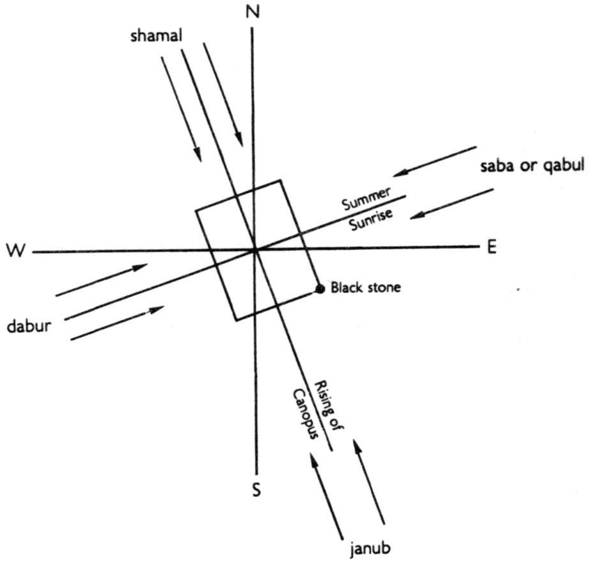

Figure 6.

The orientation of the rectangular base of the Kaaba towards the rising of Canopus and summer solstice, as recorded in various medieval sources. The 'cardinal' winds are shown, each one striking a wall of the Kaaba head-on.

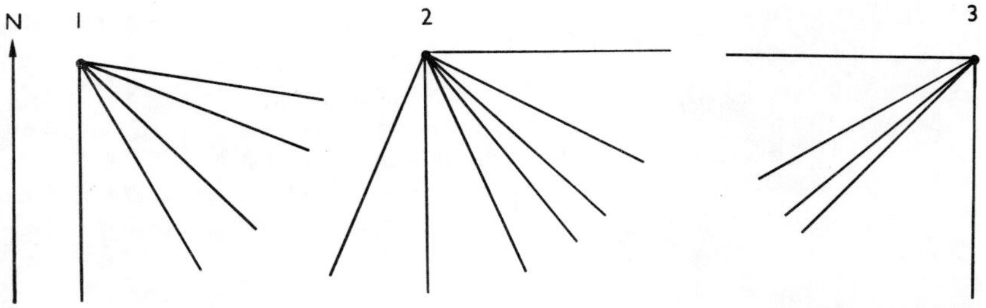

Figure 7. The various qiblas used for mosque orientations in (1) Cordova, (2) Cairo, and (3) Samarqand, as reported in medieval sources. Cardinal directions were used, as well as solar and stellar risings and settings, and—last but not least—mathematically-computed directions based on complicated accurate and/or simple approximate formulae. (For details see King, 'The Sacred Direction in Islam...', p. 325.)

wall of the Kaaba. Likewise the first mosques in Egypt were built with their prayer-walls facing winter sunrise so that the prayer-wall was 'parallel' to the north-western wall of the Kaaba. Inevitably there were differences of opinion, and different directions were favoured by particular groups. Indeed, in each major region of the Islamic world, there was a whole spectrum of directions used for the qibla (see figure 7). Only rarely do the orientations of medieval mosques correspond to the qiblas derived by computation (see below). Recently some medieval texts have been identified which deal with the problem of the qibla in Andalusia, the Maghrib, Egypt, Iraq and Iran, and Central Asia. Their study has done much to clarify the orientation of mosques in these areas. In order that prayer in any reasonable direction be considered valid, some legal texts assert that while facing the *actual* direction of the Kaaba (*ayn*) is optimal, facing the *general* direction of the Kaaba (*jiha*) is also legally acceptable.

In various texts on folk astronomy, popular encyclopaedias and legal treatises, we find the notion of the world divided into sectors about the Kaaba, with the qibla in each sector having an astronomically-defined direction. Some twenty different schemes have been discovered recently in the manuscript sources, attesting to a tradition of sacred geography in Islam far more sophisticated than the corresponding Jewish and Christian notions of the world centred around Jerusalem.

The earliest schemes of Islamic sacred geography date from the ninth century, but the main contributor to the development of Islamic sacred geography was a Yemeni legal scholar named Ibn Suraqa, who studied in Basra about the year 1000. Ibn Suraqa devised three different schemes of sacred geography, with the world arranged in 8, 11 and 12 sectors around the Kaaba. Each sector of the world faces a particular section of the perimeter of the Kaaba. Simpler versions of his 12-sector scheme occur in the popular geographical works of Yaqut al-Rumi (*ca.* 1200) and al-Qazwini (*ca.* 1250) and the encyclopaedia of al-Qalqashandi (*ca.* 1400). From the fifteenth century to the nineteenth, we find a proliferation of schemes with different numbers of divisions between eight and 72 divisions of the world around the Kaaba; one rather spectacular example for a sixteenth-century Tunisian nautical atlas is reproduced as figure 8.

Muslim astronomers from the eighth century onwards concerned themselves with the determination of the qibla as a problem of mathematical geography. This activity

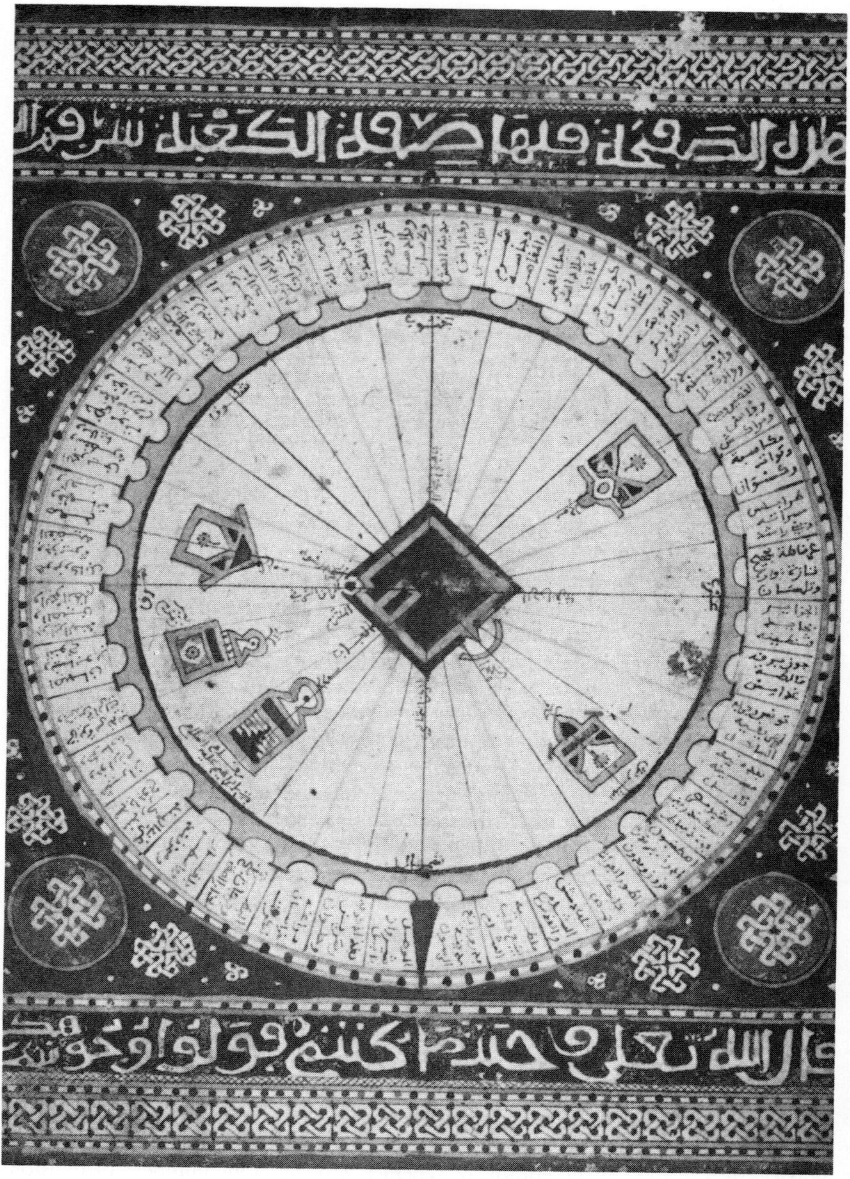

Figure 8. A late scheme of sacred geography in which localities all over the Islamic world are arranged around the Kaaba. Their positions are derived by tradition not by calculation and in many cases are not in accord with geographical reality. (Reproduced from MS Paris B.N.ar.2278 with kind permission of the Bibliothèque Nationale, Paris).

involves the measurement of geographical coordinates and the computation of the direction of one locality from another by procedures of geometry or trigonometry. The qibla at any locality was defined as the direction of Mecca along the great-circle on the terrestrial sphere. The basic problem, illustrated in figure 9, is to determine the direction of Mecca M from any locality X, given the latitudes of both localities measured by MB ($=b$) and XA ($=a$), and the longitudinal difference AB ($=c$). The qibla is measured by the angle AXM ($=q$).

Muslims inherited the Greek tradition of mathematical geography, together with Ptolemy's lists of localities and their latitudes and longitudes. Already in the early ninth century observations were conducted in order to measure the coordinates of Mecca and Baghdad as accurately as was possible, with the express intention of computing the qibla at Baghdad. Indeed, the need to determine the qibla in different localities inspired much of the activity of the Muslim geographers. The most important Muslim contribution to mathematical geography was a treatise by the eleventh-century scientist al-Biruni. His purpose was to determine for his patron the qibla at Ghazna (in what is now Afghanistan), a goal which he achieved most admirably.

Once the geographical data are available, a mathematical procedure is necessary to determine the qibla. The earliest Muslim astronomers who considered this problem developed a series of approximate solutions, all adequate for most practical purposes, but in the early ninth century, if not before, an accurate solution by solid trigonometry was formulated. The modern formula is rather complicated, namely:

$$q=\cot^{-1}\left(\frac{\sin a\cos c-\cos a\tan b}{\sin c}\right)$$

but the formulae derived by the Muslim astronomers from the ninth century onwards were mathematically equivalent to this. Muslim astronomers also compiled a series of tables displaying the qibla for each degree of latitude and longitude difference from Mecca based on both approximate and exact formulae, the first of these being prepared in Baghdad in the ninth century.

Over the centuries, numerous Muslim scientists discussed the qibla problem, presenting solutions by spherical trigonometry, or reducing the three-dimensional situation to two dimensions and solving by geometry or plane trigonometry. They also formulated solutions using calculating devices. But one of the finest medieval mathematical solutions to the qibla problem was reached in fourteenth-century

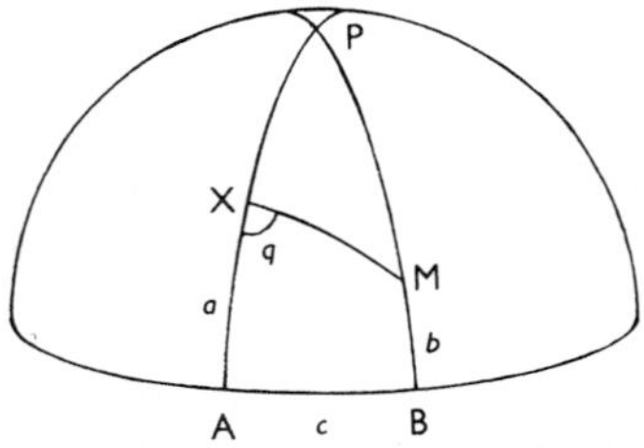

Figure 9. In this diagram, AB represents the equator and P the North Pole. It is required to find the direction of Mecca from any locality X. The latitudes of both localities are represented by MB and XA and their longitude difference by AB.

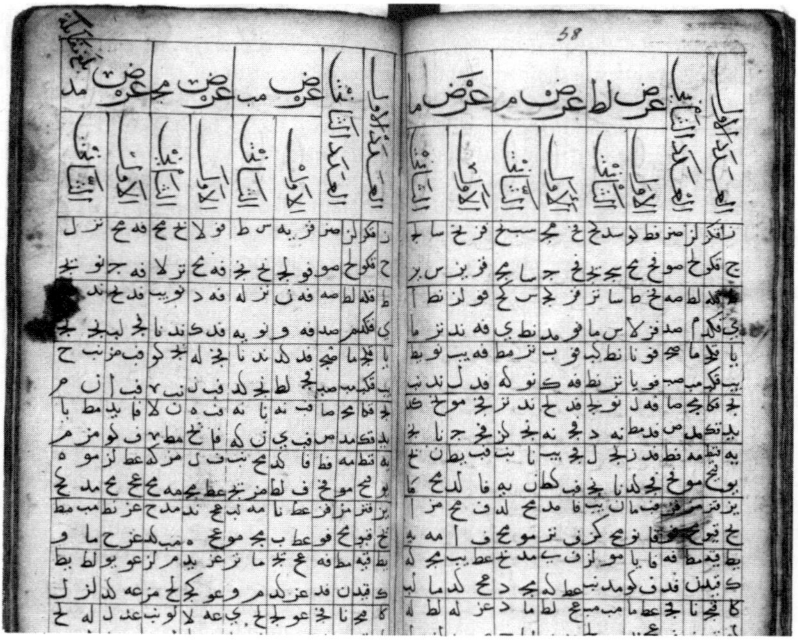

Figure 10. An extract from al-Khalili's table for finding the qibla. For each degree of latitude (here 39°, 40°,..., 44°) and each degree of longitude difference the qibla is given in degrees and minutes. The values are invariably accurately computed. (Reproduced from MS Paris B.N.ar.2558 with kind permission of the Bibliothèque Nationale, Paris).

Damascus: a table by al-Khalili displays the qibla for each degree of latitude from 10° to 56° and each degree of longitude from 1° to 60° east or west of Mecca, with entries correctly computed according to the accurate formula (see figure 10). This splendid table (rediscovered only in the early 1970s) was not widely known in later Muslim scientific circles. Muwaqqits, or professional mosque astronomers of later centuries, wrote treatises about the determination of the qibla but did not mention this Syrian table. By the fourteenth century the correct values of the qibla of each major city had long been established (correct, that is, for the medieval coordinates used in the calculations). Simple qibla-indicators fitted with a magnetic compass and a gazetteer of localities and qiblas became common, and the modern variety (see below) represents a continuation of this tradition.

Nevertheless, al-Khalili's table did not mark the end of serious Muslim activity in this field. In 1989 there was sold at Sotheby's in London a qibla-indicator, probably made in Isfahan about 1700, which bears a cartographic grid so devised that one can read qiblas directly off the map (see figure 11). Mecca is the centre of the grid and one has only to lay the diametrical rule over any city marked on the map (between Spain and China, Europe and the Yemen) to read off the qibla on a circular scale around the

Figure 11. A spectacular cartographic grid for finding the qibla of any locality in the Islamic world. Mecca is situated at the centre, and the names of numerous localities are written alongside points which represent their geographical coordinates. The projection is so devised that the qibla can be read directly from the scale around the grid. (Courtesy of the owner.)

grid. So much for the achievements of the Muslim scientists in this single small area of their activities.

The alignment of medieval mosques reflects the fact that the astronomers were not always consulted on their orientation. But now we know from textual sources which directions were used as a qibla in each major locality, we can better understand not only the mosque orientations but also recognize numerous cities in the Islamic world that can be said to be qibla-oriented. In some, such as Taza in Morocco and Khiva in Central Asia, the orientation of the main mosque dominates the orientation of the entire city. In the case of Cairo (see figure 12) various parts of the city and its suburbs are oriented in three different qiblas. The new Fatimid city of al-Qahira, founded in the tenth century, faces winter sunrise, which was the qibla of the Companions of the

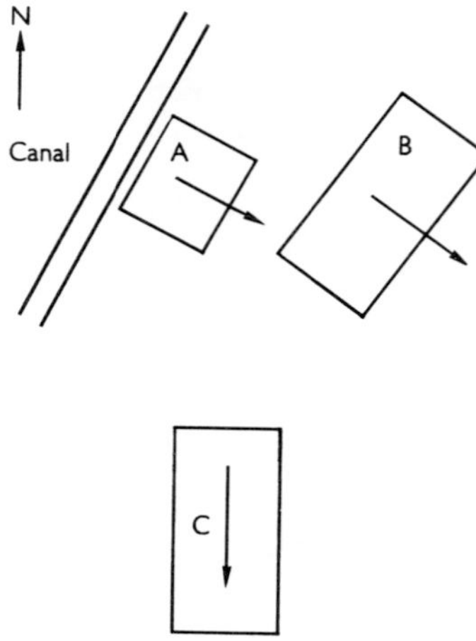

Figure 12. The main orientations of three parts of medieval Cairo. Each part is qibla-oriented, one (A) in the qibla of the Companions of the Prophet (winter sunrise at 27° south of east), another (B) in the qibla of the astronomers (at 37° south of east), and the third (C) in the qibla of the Prophet when he was in Medina (due south).

Prophet who erected the first mosque in nearby Fustat some three centuries previously. The later Mamluk 'City of the Dead' faces the qibla of the astronomers. The predominant orientation of architecture in the suburb of al-Qarafa is towards the south, another popular qibla. The splendid Mamluk mosques and madrasas built along the main thoroughfare of the old Fatimid city are aligned externally with the street plan, and internally with the qibla of the astronomers: one can observe the varying thickness of the walls when standing in front of the windows inside the mosque overlooking the street outside.

This is an area of the history of urban development in the Islamic world which has only recently been studied for the first time, not least because, prior to the discovery of the textual evidence, it was by no means clear which directions were used as qiblas: even if a qibla at variance from the true qibla was clearly popular, it was not known why.

The first accurate longitude values of localities in the Islamic world become available only with the systematic scientific cartographic surveys of the eighteenth and nineteenth centuries. Thus most of the accurately-computed qiblas of the medieval astronomers could be judged as being in error by a few degrees anyway. But who are we to judge the 'accuracy' of any particular qibla?

Nowadays urban Muslims are content to use the qiblas found by calculation from modern geographical coordinates. In rural areas where there is no mosque nearby, astronomical horizon phenomena are still used for the sacred direction. In recent years, various devices for finding the qibla have appeared on the market, usually in the form of a magnetic compass with a list of directions for the qiblas of the major cities of the world. Some computerized prayer-clocks will purr when properly aligned with the qibla. Few who use such devices realize that they are heirs to a tradition going back over a millennium.

Other applications of science to daily life

There are two other topics which ideally merit inclusion in an article bearing this title, but which—not just for reasons of space—must be mentioned briefly. The first is the algebra of inheritance. The rules for the distribution of estates as formulated in the Koran are complicated, and their application involves some skill in arithmetic. Both the legal scholars and certain mathematicians wrote on this subject, but only two or three simple works by legalists have been studied and until recently no research of consequence had been conducted on the numerous available sources. There is also a vast corpus of literature on weights, measures and arithmetical techniques.

Secondly, the Muslims developed geometric designs for the decoration of religious architecture and also secular artefacts. The acceptability of such ornamentation is discussed by various legal scholars, but their writings have yet to be properly studied. Only two Muslim mathematicians are known to have included remarks on geometric design in their writings, a fact which confirms the suspicion that this was an art passed down amongst the practitioners. Some years ago a manuscript of an artisan's manual with guidelines for generating numerous patterns was discovered but it has yet to be published.

Concluding remarks

As we have shown, the legal scholars of medieval Islam used methods for regulating the calendar and prayer-times and for finding the the sacred direction which were simple and adequate for practical purposes. Their ingenuity in coping with differences of opinion never lost sight of the basic purpose of Koranic and Prophetic injunctions.

Some of the greatest of the Muslim scientists dealt with the calendar, prayer-times, and the qibla, and in these areas, as in others, their mathematical creativity and their quest for greater accuracy was impressive. In later centuries (after the thirteenth), competent astronomers were appointed to the staffs of major mosques in order to advise on these specific subjects. But the solutions developed by Muslim scientists were invariably too complicated for widespread application in the medieval milieu.

Although the scholars of the sacred law and the scientists proposed different solutions for the same individual problem, there are few records of serious discord between the two groups in the medieval sources. The legal scholars criticized mathematical astronomy mainly insofar as it was used by some as the handmaiden of astrology, which was anathema to them. The scientists seldom spoke out against the simple procedures adopted by the legal scholars.

Only through recent studies of the vast manuscript sources available for the further documentation of Islamic civilization have we come to appreciate the dual nature of science in Islam, and the ways in which it was applied in daily life. We now have a much better idea of the role of folk science in Islamic society, as well as a clearer understanding of the outstanding achievements of the Muslim scientists, which were remarkable in many respects and by no means restricted to the three areas which we have discussed here. The Islamic Near East, far more than any other religio-cultural area, witnessed truly remarkable developments in scientific research directed towards the requirements of religion.

To delve more deeply

The Encyclopaedia of Islam, 2nd. ed., Leiden: E. J. Brill, 1960 onwards: articles: *Ilm al-hay'a* (=astronomy); *Kibla* (=sacred direction) and '*Makka*: As Centre of the World' (=sacred geography); '*Ilm al-mikat*' (=astronomical time-keeping). [Survey articles.]

A. HEINEN, *Islamic Cosmology*, Beirut (in commission for Franz Steiner Verlag, Wiesbaden), 1982. [Rediscovers a tradition of sacred cosmology in medieval Islam.]

M. ILYAS, *A Modern Guide to Astronomical Calculations of Islamic Calendar, Times and Qibla*, Kuala Lumpur: Berita Publishing, 1984. [A modern scientific approach providing little historical information.]

M. ILYAS, *Astronomy of Islamic Times for the Twenty-first Century*, London and New York: Mansell, 1988. [Contains universal tables for the prayers but little historical information.]

E. S. KENNEDY, *A Commentary upon Biruni's Kitab Tahdid al-Amakin, An 11th Century Treatise on Mathematical Geography*, Beirut: American University of Beirut Press, 1973. [A key to the most important medieval treatise on mathematical geography.]

E. S. KENNEDY, Colleagues and Former Students, *Studies in the Islamic Exact Sciences*, Beirut: American University of Beirut Press, 1983. [Reprints of about 70 articles by the leading scholar in the field.]

Kennedy Festschrift: D. A. KING and G. SALIBA, eds., *From Deferent to Equant: Studies in the History of Science in the Ancient and Medieval Near East in Honor of E. S. Kennedy, Annals of the New York Academy of Science* 500 (1986). [Some 35 articles by the world's leading experts.]

D. A. KING, *Islamic Mathematical Astronomy* and *Islamic Astronomical Instruments*, London: Variorum Reprints, 1986 and 1987. [Reprints of about 40 articles, including considerable material on the regulation of the prayer-times and the determination of the qibla.]

D. A. KING, 'Ibn Yūnus on Lunar Crescent Visibility', *Journal for the Hist*[illegible]*stronomy* 19 (1988), pp. 155–168 [on a sophisticated set of conditions from tenth-[illegible]Cairo]; 'Ibn Yūnus' *Very Useful Tables* for Reckoning Time by the Sun', *Archive f*[illegible]*y of Exact Science* 10 (1973), pp. 342–394 [on the corpus of tables used for time-k[illegible] medieval Cairo]; 'A Survey of Medieval Islamic Shadow Schemes for Simple [illegible]eckoning', *Oriens* 32 (1990), pp. 191–249 [on time-keeping in the folk tradition]; 'O[illegible]entation of the Kaaba' (with G. Hawkins), *Journal of the History of Astronomy* 13 (1982), pp. 102–109 [discussion of a medieval text on the subject and a comparison with the actual alignment of the edifice]; 'al-Khalīlī's Qibla Table', *Journal of Near Eastern Studies* 34 (1975), pp. 81–122 [detailed analysis of a highly sophisticated medieval table]; 'Astronomical Alignments in Medieval Islamic Religious Architecture', *Annals of the New York Academy of Sciences* 385 (1982), pp. 303–312 [the first study of the subject]; and 'The Sacred Direction in Islam . . .', *Interdisciplinary Science Reviews* 10:4 (1985), pp. 315–328 [a survey article].

D. A. KING, *Astronomy in the Service of Islam*, Aldershot (U.K.): Variorum Reprints, in press. [Contains reprints of various articles (including some of those mentioned above), notably analyses of the earliest mathematical methods for finding the qibla and the first tables for determining lunar crescent visibility.]

P. KUNITZSCH, *The Arabs and the Stars*, Northampton (U.K.): Variorum Reprints, 1989. [Reprints of articles dealing with Arabic star-names and their transmission to the West.]

S. H. NASR, *Islamic Science: An Illustrated Study*, London, 1976. [A beautifully illustrated book which, however, virtually ignores the three topics treated in this article!]

II

Some Early Islamic Tables for Determining Lunar Crescent Visibility

INTRODUCTION

IN THIS PAPER I discuss various medieval Islamic astronomical tables for predicting the visibility of the lunar crescent at the beginnings of the lunar months. Each of these appears to be based on a theory related to a simple visibility criterion adapted from Indian astronomy. Only two have been published and discussed in the modern literature— appropriately enough, both are treated in monographs by E.S. Kennedy and former students;[1] the remainder are preserved in the vast manuscript sources available for the further study of the history of Islamic astronomy.

The theory underlying the tables is that visibility can occur if the difference in setting times between the sun and moon is 12 equatorial degrees (or 48 minutes of time). The difference in setting times depends on three factors: the longitudes of the sun and moon and their difference, the lunar latitude, and the local terrestrial latitude. The tables discussed in this paper all display as a function of solar or lunar longitude the difference in longitude between the sun and moon for which the difference in setting times is 12°. The tables are computed either for a fixed latitude or for a range of latitudes, and, with one exception, are independent of lunar latitude.

I begin with a brief description of the simple Indian theory of crescent visibility as used by certain Muslim astronomers, in order to provide the necessary mathematical background, and then discuss various Islamic tables based on this theory. I conclude the study with a brief discussion of some less sophisticated Islamic tables for crescent visibility that have come to my attention.

The reader should bear in mind that several Muslim astronomers used considerably more sophisticated visibility theories and tables than those which I now describe, although there is still very little published material available on the topic as treated by these astronomers.[2] I wish to stress at the outset that the tables discussed in this particular study have been selected only because they are all related by their underlying theory; by no means do these tables represent a random sample of Islamic crescent visibility tables. Also, these tables are surely only a selection of Islamic crescent visibility tables based on Indian theory, although I have included all such examples currently known to me. Most of these tables, but not all, are contained in the astronomical handbooks known as *zījes*, of which some two hundred were compiled in the millenium beginning with the Muslim encounter with mathematical astronomy in the 8th century.[3] All of the entries in the tables are expressed in standard medieval Arabic alphanumerical notation.[4]

INDIAN LUNAR CRESCENT VISIBILITY THEORY

The criterion for crescent visibility outlined in the *Sūryasiddhānta* (*ca.* 600) and the *Khaṇḍakhādyaka* (665) is that the difference in setting time between the sun and moon be 12 equatorial degrees (1 degree = 4 minutes of time).[5]

FIGURE 1 shows in plane section the western horizon at sunset at a locality with latitude φ on the evening when visibility is in question. The celestial equator is shown as VW where V is the equinox and W is the west point.

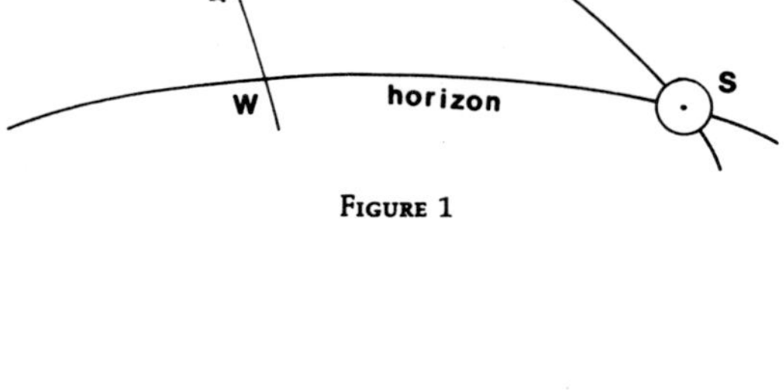

FIGURE 1

Its inclination to the horizon is $\bar{\varphi}$. The sun is at S on the horizon, and the moon is at M. The ecliptic longitude of the sun is λ_s (= VS), and the latitude of the moon is β (= MN), measured perpendicular to the ecliptic. The longitude of the moon is λ_m (= VN), and the difference in longitudes is $\Delta\lambda$ (= SN).

We denote by P and Q the points of the ecliptic and the equator which set simultaneously with the moon. The difference in the setting times of the sun and moon s is measured by the arc WQ on the equator: it can be expressed as the difference in setting times of the arcs VS and VP, whose longitudes are λ_s and λ_m'. If we denote by $\sigma(\lambda)$ the descensions or the time taken to set by an arc λ of the ecliptic, measured from the vernal equinox in the direction of signs, then clearly:

$$s = QW = \vec{VQ} - \vec{VW} = \sigma(\lambda_m') - \sigma(\lambda_s).$$

Finally, the Indian visibility condition is that if $s \geqslant 12°$ the crescent will be seen.

The problem for the medieval astronomer using this theory was first to determine λ_s, λ_m, and β, then to determine λ_m' from λ_m and β, and finally to determine the descensions $\sigma(\lambda_m')$ and $\sigma(\lambda_s)$ and investigate their difference. The determination of λ_s, λ_m and β is a standard procedure in ancient and medieval astronomy. To determine the correction μ (λ_m, β) such that

$$\lambda_m' = \lambda_m + \mu$$

is a nontrivial problem of spherical trigonometry. Approximations of the form

$$\mu = \beta \cot \bar{\varphi} \quad \text{or} \quad \mu = \tfrac{2}{3}\beta \quad \text{or} \quad \mu = \beta$$

were popular amongst various early Muslim astronomers. (Note that the first two approximations are virtually identical for the latitude of Baghdad since when $\varphi \approx 33\tfrac{1}{2}°$, $\cot \bar{\varphi} \approx \tfrac{2}{3}$.) To compute the descensions of λ_m' and λ_s the Muslim astronomers by the beginning of the 9th century had at their disposal tables of oblique ascensions $\rho(\lambda)$ for the latitudes of the seven climates of Greek geography. These tables they found in the *Almagest* and *Handy tables* of Ptolemy of Alexandria (*fl. ca.* A.D. 125).[6] In the *Almagest* $\rho(\lambda)$ is tabulated for the seven climates and three subclimates, for each 10° of λ. In the *Handy tables* it is tabulated for the seven climates, for each 1° of λ. Most Islamic astronomical handbooks (*zījes*) contain tables of $\rho(\lambda)$ for specific latitudes or for a range of latitudes.[7] A simple consideration of the celestial sphere shows that $\sigma(\lambda) = \rho(\lambda + 180°) - 180°$.

The above outline provides the necessary background to an investigation of the visibility tables presented below. In some of the tables the minimum longitude difference for visibility $\Delta\lambda$ is a function of solar longitude; in others,

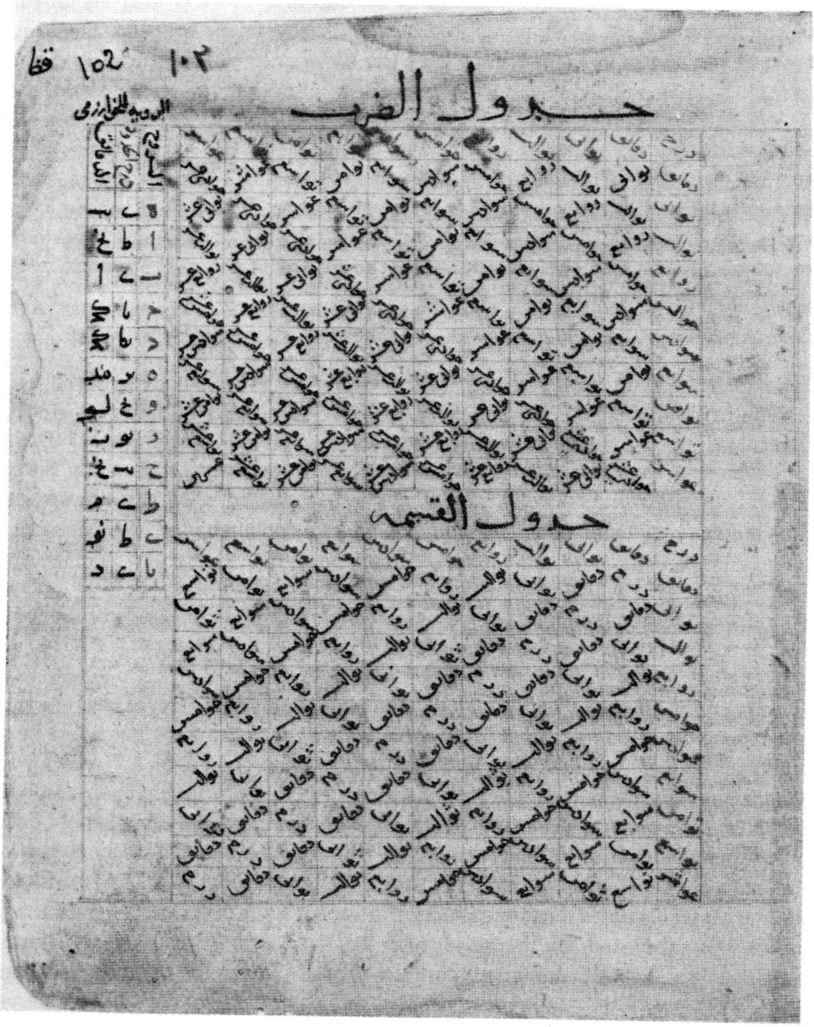

Plate 1. Al-Khwārizmī's table as it appears on fol. 102r of MS Paris B.N. ar. 6913, of the *Rīqānī Zīj*. (Reproduced with the kind permission of the Bibliothèque Nationale, Paris.)

a function of lunar longitude. Clearly the formulae for computing these two quantities are:

$$\rho\ (\lambda_s + \Delta\lambda + 180°) - \rho\ (\lambda_s + 180°) = 12°$$

when $\Delta\lambda = \Delta\lambda\ (\lambda_s)$, and

$$\rho\ (\lambda_m + 180°) - \rho\ (\lambda_m - \Delta\lambda + 180°) = 12°$$

when $\Delta\lambda = \Delta\lambda\ (\lambda_m)$. My recomputations are based on these formulae, using an 'Apple 2 Plus' computer programmed in BASIC.[8]

Clearly it makes more sense to tabulate $\Delta\lambda$ as a fraction of λ_s, but, as we shall see, good sense did not always prevail amongst those who prepared visibility tables. Also we might mention that the astronomers were rarely consulted anyway on the matter of crescent visibility. In the medieval period, as nowadays, it was the religious authorities who were responsible for regulating the lunar calendar. But that is another story, and does not concern us here.

AL-KHWĀRIZMĪ'S CRESCENT VISIBILITY TABLE

The earliest known table for determining crescent visibility was compiled by the celebrated early-9th-century astronomer - mathematician of Baghdad, al-Khwārizmī.[9] It is preserved in three different sources.

MS Paris Bibliothèque Nationale ar. 6913 is the only known copy of an anonymous *zīj* called *al-Zīj al-Rīqānī*, said to be a shorter version of *al-Zīj al-Kirmānī*,[10] and apparently based on the parameters of the *Zīj* of al-Battānī.[11] The *Rīqānī Zīj* was compiled about the year 438 Hijra (= A.D. 1047), probably in Rayy (near modern Teheran), and the manuscript is dated 489 Hijra (= A.D. 1095). The lunar crescent visibility theory outlined in the text (fols. 36r–37v) is based on the Indian condition $s \geqslant 12°$. The accompanying tables (fols. 86v–87r) are for finding the ecliptic longitude which sets with the moon, and are briefly discussed below (pages 215–216). Of considerably greater interest is a smaller table copied in the margin of the manuscript by the side of a table displaying the units of products and quotients in sexagesimal arithmetic (fol. 102r)—see PLATE 1. This table is entitled *al-ru'ya li-'l-Khwārizmī*, which means "(crescent) visibility according to al-Khwārizmī," and displays a function to two sexagesimal digits for each zodiacal sign. The argument is called *al-burūj*, "signs," and the labels corresponding to the digits of the tabulated function are *daraj al-ḥudūd*, "degrees of the limits," and *al-daqā'iq*, "minutes." The entries are displayed in TABLE 1. There are no instructions accompaning the table, but it is reasonable to suppose that the tabu-

TABLE 1
al-Khwārizmī's Crescent Visibility Table
A: MS Paris B.N. ar. 6913, fol. 102r; B: MS Cairo ṬFF 11, fol. 61r;
C: MS London B.L. Add. 27,261, fol. 445; D: MS Escorial ar. 927, fol. 6r.

Zodiacal Sign	*Visibility Function*	*Reconstructed Value*	*Recomputation*	*Error (in minutes)*
0	10;12° A–D	10;12°	10;11°	+1
1	9;18 CD 9;48 B 9;58 A	9;58	9;58	0
2	10; 1 A–D	10; 1	10; 1	0
3	11;23 A–D	11;23	11;22	+1
4	14;29 D 15;23 B 19;23 AC	14;29	14;28	+1
5	17;44 A–D	17;44 (?)	17;35	+9 (?)
6	18;36 A–D	18;36 (?) 18;16 (?)	18;13 18;13	+23 (?) +3 (?)
7	16; 2 D 16; 7 A 16;12 B 17; 2 C	16; 7	16; 6	+1
8	12;18 CD 12;58 AB	12;58	12;58	0
9	10;40 A–D	10;40	10;41	−1
10	9;16 D 9;56 A–C	9;56	9;55	+1
11	10; 4 A–D	10; 4	10; 4	0

lated function is such that when the solar and lunar longitudes at sunset differ by $\Delta\lambda$, visibility is assured if $\Delta\lambda \geqslant f(\lambda)$, where λ is either λ_m or λ_s.

Another copy of this table, this time without the attribution to al-Khwārizmī, occurs in MS Escorial ar. 927 (fol. 6r) of the anonymous Mosul recension of the *Mumtaḥan Zīj* of Yaḥyā ibn Abī Manṣūr, originally compiled *ca.* A.D. 830 for the Abbasid Caliph al-Ma'mūn.[12] The table is entitled simply *jadwal ru'yat al-ahilla* and is presented without any explanation — see PLATE 2; the entries are shown in TABLE 1 and differ slightly from those in the *Rīqānī Zīj*. There follows an even simpler procedure for determining crescent visibility, which is discussed below (pages 213–214).

PLATE 2. Al-Khwārizmī's table as it appears on fol. 6r of MS Escorial ar. 927 of the *Mumtaḥan Zīj*. Note the second set of entries in the text - see (a) on pages 213-214. (Reproduced with the kind permission of the Biblioteca de El Escorial).

The same table, also with minor variants and again explicitly associated with al-Khwārizmī, is found in the Persian astrological handbook entitled *Rawḍat al-munajjimīn*, compiled by the 11th-century astronomer Shāhmardān ibn Abi 'l-Khayr Rāzī, a student of the scholar Abu 'l-Ḥasan ʿAlī ibn Aḥmad al-Nasawī.[13] I have consulted the tables in two copies of this work, namely, MS Cairo Ṭalʿat *falak fārisī* 11, fol. 61r, (copied *ca.* A.D. 1400) – see PLATE 3 – and MS London B.L. Add. 27,261, fol. 445, (copied 814 Hijra = A.D. 1411). The accompanying text, which is in Persian, indicates that al-Khwārizmī stated that visibility was possible if

$$\rho\,(\lambda_m + 0;42\beta + 180°) - \rho\,(\lambda_s + 180°) \geqslant 12°,$$

and there is no explanation of the table. We note that the closest integral value of φ corresponding to cot φ = 0;42 is 35°: it seems probable that this modification to the lunar latitude is a later addition to the original *Zīj*. Indeed it may be that this condition was intended to serve localities with latitude 36° (or 35;34,40°– see especially fol. 78v of the Cairo manuscript), perhaps Rayy, where al-Nasawī and al-Rāzī were active.

It is known that al-Khwārizmī compiled two separate *zījes*, both called *al-Sindhind*. Now as far as we can ascertain from the commentary of the 10th-century Andalusian astronomer Ibn al-Muthannā on one of al-Khwārizmī's *zījes*[14] and the commentary of the contemporary (?) ʿAbd Allāh ibn Masrūr to the same work,[15] there was no such table in the original *zīj*. However, analysis of the table confirms the attribution to al-Khwārizmī. It can be shown that:

(1) the table is based on Indian visibility theory;

(2) the underlying latitude is 33;0° (Baghdad), a parameter used by al-Khwārizmī in various other works;[16]

(3) the underlying obliquity is 23;51°, a parameter also securely associated with al-Khwārizmī.[17]

Recomputation of the table – see TABLE 1 – reveals that the original entries are remarkably accurately computed. I suspect that al-Khwārizmī had a table of $\rho(\lambda)$ for the latitude of Baghdad at hand, with values for each degree of λ. With such a table, values of the tabulated function can be generated with reasonable facility.

THE CRESCENT VISIBILITY TABLE OF AL-QALLĀS

Maslama ibn Aḥmad al-Majrīṭī was a scholar of considerable renown who worked in Cordova in the late 10th century.[18] One of his works was a recension of the *Zīj* of al-Khwārizmī, which is extant only in a Latin translation of a modified version: this Latin translation has been published by H.

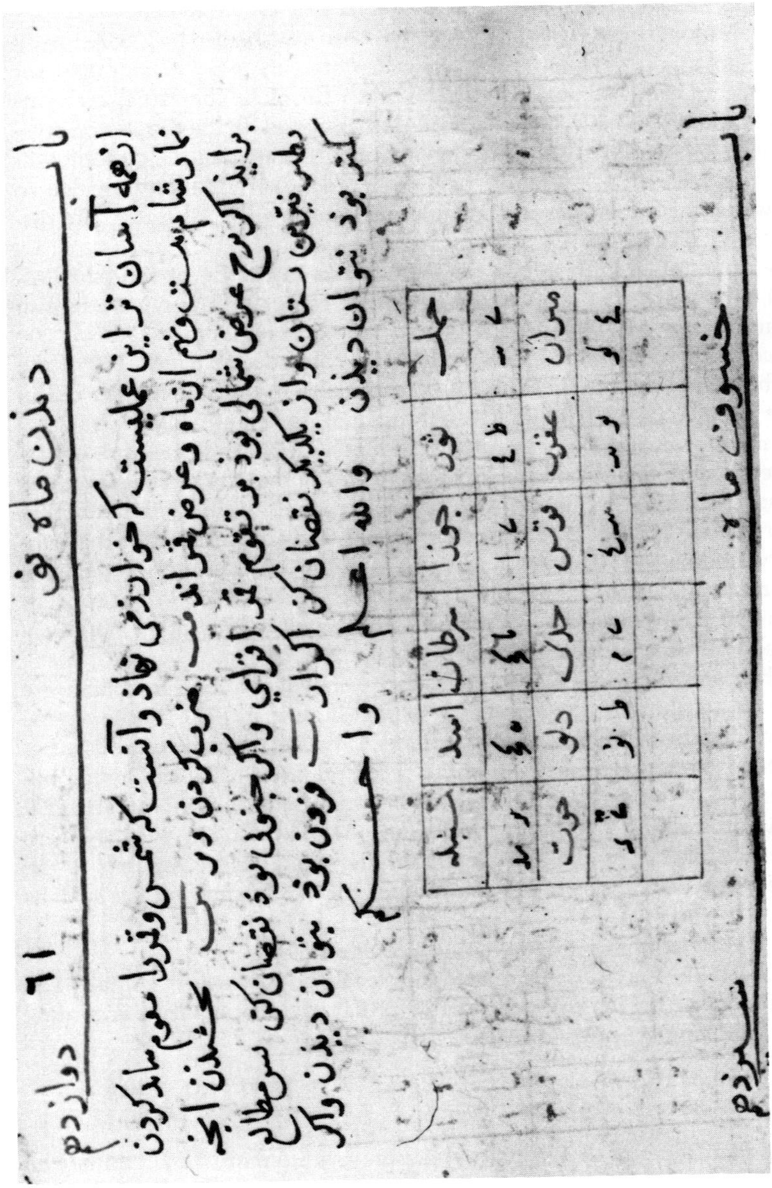

PLATE 3. Al-Khwārizmī's table as it appears on fol. 61r of MS Cairo Ṭalʿat *falak fārisī* 11 of the *Rawḍat al-munajjimīn* of Shāhmardān Rāzī. (Courtesy of the Egyptian National Library.)

Suter, and translated into English and analyzed by O. Neugebauer.[19] It contains a table for crescent visibility which has been investigated by E.S. Kennedy and M. Janjanian.[20] The table displays values of a function $f(\lambda)$ computed for each "face" of each zodiacal sign, that is, each 10° of λ. The values are symmetrical about $\lambda = 180°$ and no latitude is mentioned. In the chapter on crescent visibility in al-Majrīṭī's work the instructions state that the crescent will be seen if $\Delta\lambda > f(\lambda_m)$. Clearly neither instructions nor table are due to al-Khwārizmī. The entries in the table, as published by H. Suter, are displayed in TABLE 2.

The same table, with minor variants, occurs in the *Zīj* of Ibn Isḥāq al-Tūnisī, who worked in Tunis in the early 13th century.[21] His *Zīj* is extant in the unique copy MS Hyderabad Andra Pradesh State Library 298 (*ca.* 200 fols., *ca.* 800 Hijra ≈ A.D. 1400),[22] where the table occurs as no. 160 of the 360 tables (see PLATE 4). The table is here and also in the 31st chapter of Ibn Isḥāq's introduction attributed to an individual named al-Qallās, on whom I have no information whatsoever beyond that his name indicates that he was a maker of skull-caps. Ibn Isḥāq does not mention that the table is compiled for a specific latitude. (Two other tables for crescent visibility in his *Zīj* are respectively specifically for Tunis and universal, and are attributed to the 11th-century Andalusian scholar Ibn Muʿādh, and the 12th-century Andalusian scholar and student of Maimonides, al-Sabtī.[23] These tables do not concern the present study but do merit detailed investigation.) For other tables associated with Ibn Isḥāq, see I and J on page 203.

In their analysis of al-Majrīṭī's table, E.S. Kennedy and M. Janjanian were unable to account for the symmetry of $f(\lambda)$ in terms of the mathematical methods used by early Muslim astronomers. However, they did establish that the entries correspond fairly closely to recomputation using the Indian parameter of 12° for the time between sunset and moonset (with an artificial condition to produce symmetry) and a latitude of about 42°. But they were unable to explain why the latitude underlying the table corresponds to the Pyrenees and Cantabrica rather than some center of Islamic culture such as Toledo.

My investigations of the table reveal the following:

(1) The table was clearly computed for the latitude of the fifth climate. The precise value of the latitude φ is related to that of the obliquity ε according to the following scheme

ε = 23;33°	φ = 41;17°	ε = 23;51	φ = 40;52
23;35	41;14	24; 0	40;41

Recomputed values of the visibility function vary no more than a minute or two with any of these pairs of parameters.

(2) The fact that entries are given for each "face" of each sign, corresponding to each 10° of ecliptic longitude, leads me to suspect that tables of $\rho(\lambda)$ for

Table 2

Lunar Crescent Visibility Table in Adelard's Translation of al-Majrīṭī's Recession of al-Khwārizmī's *Zīj* (A) and Attributed to al-Qallās in the *Zīj* of Ibn Isḥāq al-Tūnisī (B)

A: Suter, 1914, 168, edited from three Latin manuscripts
B: MS Hyderabad State Central Library 298, table no. 160.

Signs/Faces		*Visibility Function* (A)	(B)	*Faces/Signs*	
I	1	9;26°	9;26°	3	XII
	2	9;25	9;25	2	
	3	9;21	9;21	1	
II	1	9;19[a]	9;19	3	XI
	2	9;18[b]	9;18	2	
	3	9;21	9;21	1	
III	1	9;33	9;33	3	X
	2	9;57	9;57	2	
	3	10;37[c]	10;37[j]	1	
IV	1	11;29	11;29	3	IX
	2	12;48	12;48[i]	2	
	3	14;15	14;15	1	
V	1	15;58[d]	15;58[h]	3	VIII
	2	17;31[e]	17;31[e]	2	
	3	19;11	19;11	1	
VI	1	20;20	20;20	3	VII
	2	21; 4	21; 4[g]	2	
	3	21;17[f]	21;17[f]	1	

[a] A has one reading 29. [b] B: 19. [c] A has one reading 9;37; B: 10;57. [d] B: 12;38. [e] B: 14;31. [f] Suter has 21;57, which Neugebauer (1962, 103) emends to the other attested value 21;17; B: 21;57. [g] A: 20;4. [h] B: 15;28. [i] B: 14;43. [j] B: 10;57.

Recomputation

$\varphi = 40;52°$ $\varepsilon = 23;51°$

Argument is Lunar Longitude

10°	9;29°	130°	15; 2°	250°	14;14°
20	9;27	140	16;29	260	12;45
30	9;24	150	17;53	270	11;31
40	9;20	160	19;10	280	10;36
50	9;20	170	20;12	290	9;58
60	9;24	180	20;53	300	9;35
70	9;36	190	21; 9	310	9;23
80	9;58	200	20;56	320	9;20
90	10;34	210	20;13	330	9;21
100	11;23	220	19; 3	340	9;24
110	12;25	230	17;33	350	9;27
120	13;40	240	15;53	360	9;29

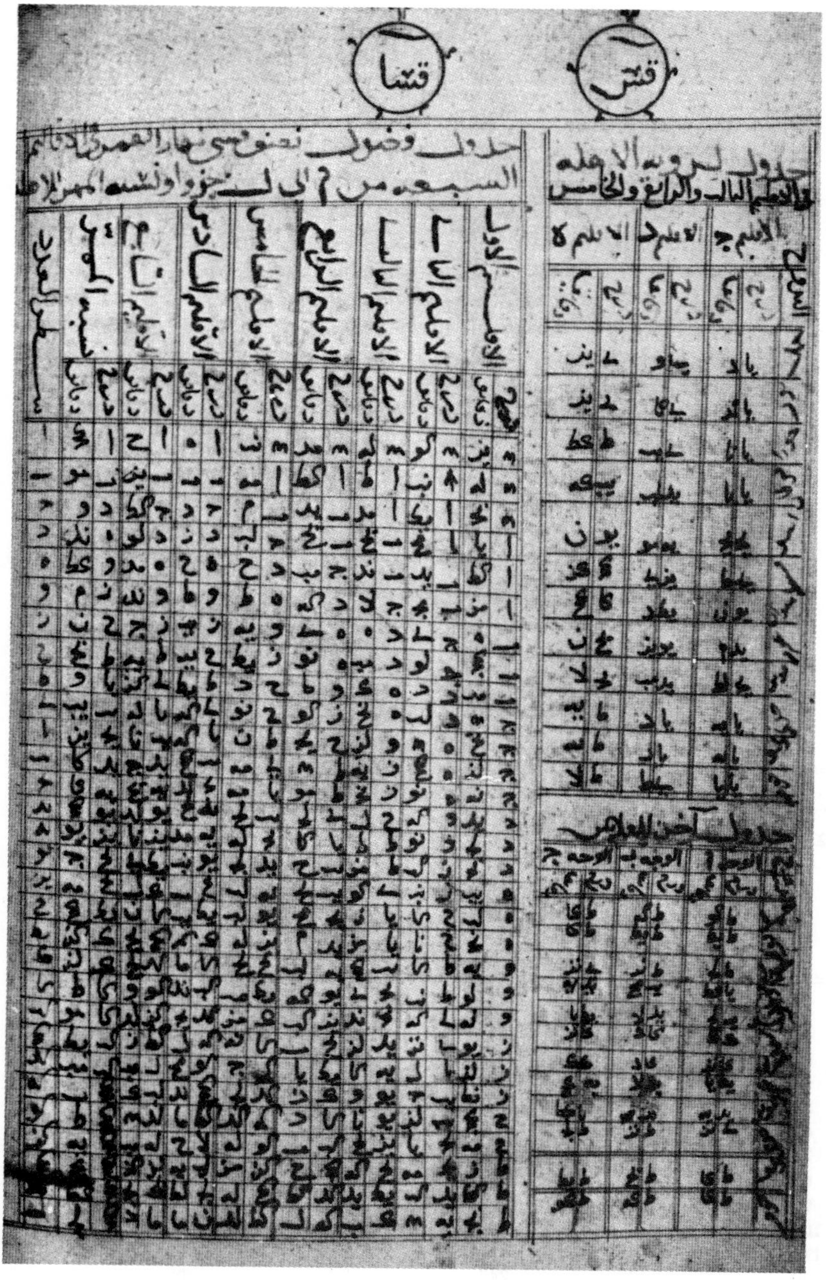

قشا   قشر

جدول فضول نصف قوسي نهار القمر في الاقاليم السبعة من ... الى ...

جدول لرؤية الاهلة في الاقاليم الثالث والرابع والخامس

الاقليم ه	الاقليم د	الاقليم ج	البروج

الاقليم الاول	الاقليم الثاني	الاقليم الثالث	الاقليم الرابع	الاقليم الخامس	الاقليم السادس	الاقليم السابع	نسبة الجوزهر	سطر المعدل

جدول آخر للعرض

الوجه ا	الوجه ب	الوجه ج

the fifth climate with entries for each 10° of λ were used to compute this table. Such tables (with values for arguments corresponding to the end limits for each "face") are contained, for example, in the *Zīj* of al-Battānī.[24] However, even using linear interpolation in such tables produces values of $\Delta\lambda(\lambda)$ within a few minutes of the proper value. So the method used was even cruder than that. We should also mention that this table is unrelated to the other one for the fifth climate described in the next section.

(3) Unlike Kennedy and Janjanian, I do not see the need to seek a mathematical method that reproduces a symmetrical function. The compiler, I would suggest, merely calculated values for $\lambda < 180°$ and then simply assumed the symmetry of the function.

(4) Like Kennedy and Janjanian I am unable to explain how the table was generated. It is not clear whether the arguments are the beginnings, midpoints, or endpoints of each 10° interval of lunar longitude. An accurate procedure does not appear to have been used throughout. Whilst agreement with recomputation is reasonably good for the first three signs, the values in the table are as much as *ca.* 1° too high in the next three signs.

(5) There is every reason to suppose that this table was one of a set serving several, if not all, of the climates. None of the others from this set is known to have survived.

[**Note added in proof:** My colleague Dr. J.-P. Hogendijk of the Mathematical Institute, Utrecht, has shown (in a paper to appear) that the entries in the first half of the table correspond *very well* to recomputation with a slightly larger φ than I had assumed. The problem which vexed Neugebauer, Toomer, Kennedy and *al-ʿabd al-miskīn* can now be regarded as solved. But the motivation for using such a latitude remains obscure.]

AN EARLY ANDALUSIAN CRESCENT VISIBILITY TABLE

A table for crescent visibility for each of the seven climates occurs in several sources mentioned below. I suspect that it is of early (9th or 10th century?) Andalusian origin.

The function tabulated is a double argument function $f(\varphi_i, \lambda_j)$ where φ_i ($i = 1, 2, \ldots, 7$) are the latitudes of the seven climates and λ_j ($j = 1, 2, \ldots, 12$) are the longitudes of the twelve zodiacal signs. To use the table one simply determines the lunar longitude λ and the solar longitude $\lambda - \Delta\lambda$ at sunset on the evening when visibility is anticipated: then if, and only if, $\Delta\lambda > f(\lambda_i, \lambda)$ the crescent will be seen in the i-th climate.

The entries in each of the copies of the table are extremely corrupt. Values are displayed in TABLE 3. In the manuscripts they are written in standard Arabic

←

PLATE 4. A table for three of the seven climates and the table of al-Qallās as they appear in MS Hyderabad 298. (Courtesy of the Andra Pradesh State Library.)

Table 3

An Anonymous Andalusian Table of Lunar Crescent Visibility for the Climates

Sources A–N are identified on pp. 200–205. (B) *etc.* indicates that an alternative value is given in B. [A] *etc.* indicates that the value given in A can be restored to a more sensible value. The last entries in each frame are recomputed. Values which correspond fairly closely to recomputation are marked with an asterisk.

	1	2	3	4	5	6	7
I	11;10 N 11;14 (B) 11;24 BCG 11;25 AH 12;21 F 12;24 DEM 11;35 recomputation	11; 3 F * 11; 4 ABCDEGHI[M]N * 51; 4 M 10;58	10;19 N 11;19 BCDEFG[M] 11;49 AH 51;19 M 10;25	10; 4 M 10; 6 ABCDEFGHIJK 10;40 LN 9;55	10;14 N 10;17 BDEFIM 10;40 H 11;17 CG 11;40 A 9;29	9; 9 BCDEFGMN * 9;19 AH 9; 6	9;18 ADEFH 9;28 BCG 9;38 MN 8;45
II	11;10 N 11;11 ABCDEFGHM 11;23	5;24 M 11;22 AHN 11;24 BCDEFGI[M] 10;48	10;13 ABEHMN * 10;33 CG 10;53 DF 10;17	10;21 ABCDEFGHIJKLMN 9;49	10;12 ABCGHM 10;17 I 10;52 DEF 12;12 N 9;24	8;13 H 8;18 DEFN 8;28 M 9;18 BCG 18;13 A 9; 1	9; 0 N 9; 8 [A]DEF[H]M 9;28 BCG 19; 8 AH 8;40
III	11; 2 BCDEFH 11; 9 N 11;10 G 11;12 AM 10;50	11;11 BCDEFGM 11;21 A 11;41 N 11;51 HI 10;26	10; 8 ABFHMN * 10;10 CG * 11; 8 DE 10; 4	10;12 ADEFHIJKM 10;32 BCG 10;42 LN 9;43	9;24 B * 9;29 CDEFGIMN * 9;39 AH 9;24	9;14 N 9;24 BCDEFG 9;25 M 19;24 AH 9; 5	9; 0 N 9; 3 BCDEFGM 20; 8 A 21; 8 H 8;47
IV	11; 4 N 11;10 ABCDEF 11;50 H 13;10 G 10;51	11;11 I 11;15 ABCDEFG[H]M 11;45 M 51;15 H 10;47	11;18 AHMN 11;33 B 11;38 CDEFG 10;43	10;32 CG * 12;12 L 12;32 BN 14;12 K 14;32 AFHIJM 14;37 DE 10;38	12;20 N 12;25 ABCDEFGHIM 10;34	12;17 N 12;36 AH 12;46 BCDEFGM 10;29	12; 9 ABCDFGHM 12;21 E 14; 9 N 10;24
V	13;10 AHM 13;13 N 13;14 BCEF 13;24 DG 13;54 (D) 12; 1	13;13 DEFIM 13;18 ABCGH 13;28 N 12;23	13; 2 AH 13; 4 BCG[DEF] 13; 5 N 13;30 M 18; 4 DEF 12;47	14; 5 AH 14; 9 J 15; 0 BCDEFGKLM 15; 1 J 15; 5 N 15;46 I 13;12	16; 2 M 16; 7 (B)C 16;17 B 16;50 AGHI 17;30 N 17;50 DEF 13;40	16;17 MN 16;37 BCDEFG 17;17 AH 14; 9	13; 2 H 13;15 CDEFGM 18; 2 A 18;15 BN 14;42

VI	14;17 N 14;20 F 14;24 M 14;26 AH 14;27 BCG 14;29 D(?)E 13;50	13; 9 I 16; 9 F 16;12 AH 16;19 BCG 17; 9 M 17;15 N 17;19 DE 14;41	16; 2 AH 17; 2 BCDEFGMN 15;38	14;10 K 15; 1 J 17; 0 DEF 17;10 ABCGIMN 17;13 H 18;20 J --;10 L 16;42	23;27 ABCDEFGHI 28;24 MN 17;53	23;21 ABCDEFGHM 27;24 N 19;14	24; 2 M 24;12 N 24;50 ABCD[E]FGH 34;50 E 20; 6
VII	5; 3 N 14; 2 AH 14;12 DEFG 15; 2 BCM * 15; 1	14; 7 F 16; 7 BCDEGN * 16;50 AHI 17; 2 M 16;11	8;19 M 18;19 BCDEFGH[M]N 18;49 A 17;32	18;20 J 19; 4 ABCDEFGIMN * 19;10 K * 19;14 H 19;20 L 19; 6	21;28 ABCDEFGHIMN 20;55	22;14 AH 22;24 BCDEFGM 22;35 N 22;59	22; 1 AH 24; 1 BCFGMN 24;24 (B) 24;30 D[E] 34;30 E 25;26
VIII	12;31 AH 12;42 N 14; 0 DE 14;12 BCGM 15;52 F 14;34	12;30 M 14; 1 I 14;32 ABCDEFGHN 15;40	14;19 N 16; 9 E 16;19 ABCDFGHM 16;57	14;50 J 16;16 H 16;17 I 17;16 ADEFKLN 17;17 BCGM 18;27	13; 2 CGH 18; 2 AB[CGH]N 18; 7 DEFM 18;50 I 20;13	19;22 AH 19;42 BCFGMN 19;47 DE 22;20	21; 2 BCDEFG 21;12 N 21;30 M 21;31 AH 24;54
IX	11; 0 D 11;10 AH 12; 0 BCEFGM 13; 2 N 12;47	13;19 MN 13;39 A[BC]DEF[G]HI 18;39 BCG 13;24	13; 8 H 13;18 ABCFGN 13;21 D[E] 13;33 M 18;21 E 14; 7	13;39 J 14;12 A 14;21 N 14;42 BCD[E]FGIM 14;52 H * 15;21 L 16;12 K 31;42 E 14;56	13; 0 M 13;31 ADFHIN 14;31 BCG 20;31 E 15;53	12;12 N 14; 2 ABCDEFGH 14; 7 M 17; 1	11;45 AH 12;31 DEFM 14;13 B 14;31 (B)CGN 18;22
X	11;10 BCDEFGMN * 11;13 AH * 11; 9	11; 4 A 11; 5 H 11;12 N * 11;14 (B) * 11;15 DEFIM * 11;25 B 11;45 CG 11;14	11;21 ABCDEFGHMN **! 11;20	11; 4 I 11;15 J 11;17 K 11;24 L * 11;26 BCDEFG * 11;27 AHM * 11;34 N 11;25	9;15 I 11; 0 ABCDEFGMN 11;45 H 11;31	9; 9 E 11; 9 ABCDFGHMN 11;38	9;51 AH 11;44 N * 11;45 BCDEFGM * 11;45
XI	11; 0 M 11; 3 ABCDEFGH 11;24 N 10;43	10;50 AH 11; 2 M 11; 7 B 11;12 DEF 11;15 I 11;30 N 11;47 CG 10;26	10; 2 AM * 10; 3 F * 10;42 N 10;50 H 11; 2 BCG 11; 3 DE 10; 8	9; 4 A 11; 4 CDEFGIM 11;10 J 11;20 LN 11;22 K 11;27 H 14; 4 B 9;52	9;15 CGIN 9;19 BDEF 9;35 AH * 11; 0 M 9;35	9; 7 BCDEFGM 9;10 N 9;50 AH 9;19	9;15 BCDEFGM 9;18 N 9;54 AH 9; 3
XII	11;24 ABCDEFGHM 11;29 N 11;10	9;51 H 11;11 BCGIMN 11;40 E 11;41 DF 11;51 A 10;39	9;18 AH 10; 9 N * 11; 9 BCDEFGM 10;11	9; 9 M 9;18 A 9;28 [H] 10; 9 BCDEFGIK 10;29 L 11;11 J 35;28 H 49; 9 N 9;44	9;11 BCDEFG 9;18 M * 9;28 N 9;29 A 9;31 I 9;51 H 9;21	9; 0 BCDEFG * 9; 5 A[H] * 9; 8 MN 19; 5 H 8;59	8;14 BCDEFGHMN 8;54 A 8;38

alphanumerical notation, and in this system it is easy to account, for example, for a 58 changing to a 13, or a 21 changing into a 9. However, in addition to such errors, each of the copies of the table contains numerous obvious additional errors which defy any interpretation of this kind. Undaunted by this situation, I shall present the tables one by one and attempt a recomputation, bearing in mind Berman's corollary to Roberts' Axiom: "One man's error is another man's data."

The first source for the table (hereafter, A) is MS Paris Bibliothèque Nationale ar. 2513 of a recension of the anonymous 13th-century Egyptian astronomical handbook called *al-Zīj al-Muṣṭalaḥ*.[25] This was the most popular *zīj* in Egypt between the 13th and 15th centuries, and was based on earlier *zīj*es such as the *Ḥākimī Zīj* of the 10th-century Egyptian astronomer Ibn Yūnus[26] and the *Shāhī Zīj* of the 13th-century Persian astronomer Naṣīr al-Dīn al-Ṭūsī.[27] In this particular copy of the *Muṣṭalaḥ Zīj* there are numerous tables and astronomical notices extracted from yet other works: one such is the table for lunar crescent visibility which occurs on fol. 71v (see PLATE 5). There is no accompanying text and the table bears the inaccurate title *jadwal ru'yat al-hilāl bi-'l-buʿd bayna 'l-nayyirayn ʿinda 'l-ghurūb*, "table of crescent visibility, (enter) with the distance between the sun and moon at sunset." The table is not contained in MS Paris Bibliothèque Nationale ar. 2520, which is the only other known copy of the *Muṣṭalaḥ Zīj* (see (j) on page 217).

A second copy of this table (B) is found on fol. 58v of an unnumbered Arabic manuscript preserved in the Museo Naval de Madrid. Here it occurs amongst other tables (fols. 58v–59v) said to be taken from *al-Zīj al-Qawīm* of the early-14th-century Tunisian astronomer Ibn al-Raqqām.[28] The visibility table is entitled *jadwal ru'yat al-ahilla bi-'l-ʿashīyāt*, "table for the visibility of lunar crescents in the evenings." However this set of tables is preceded by a short astronomical poem by the late-16th-century Moroccan astronomer Abū Zayd ʿAbd al-Raḥmān al-Fāsī,[29] and the crescent visibility table is not mentioned elsewhere in the same manuscript (fols. 31r, 39v, 41r, 41v) where the different visibility theory of the *Qawīm Zīj* and two accompanying tables occur. This manuscript was copied (see fol. 48v) in 1232 Hijra (= A.D. 1817).

The third source (C) is MS Escorial ar. 909,2 of an astronomical poem with tables compiled in 14th-century Fez by Abu 'l-Ḥasan ʿAlī ibn Abī ʿAlī al-Qusanṭīnī for the Merinid Sultan Ibrāhīm al-Mustaʿīn.[30] Al-Qusanṭīnī was more competent as a poet than an astronomer, and his tables are all lifted from earlier sources. Nevertheless his work is of considerable interest and deserves detailed study. It contains the only known set of planetary equation tables based on Indian rather than Ptolemaic planetary theory which survives in Arabic, and other interesting tables for trepidation. The table for lunar

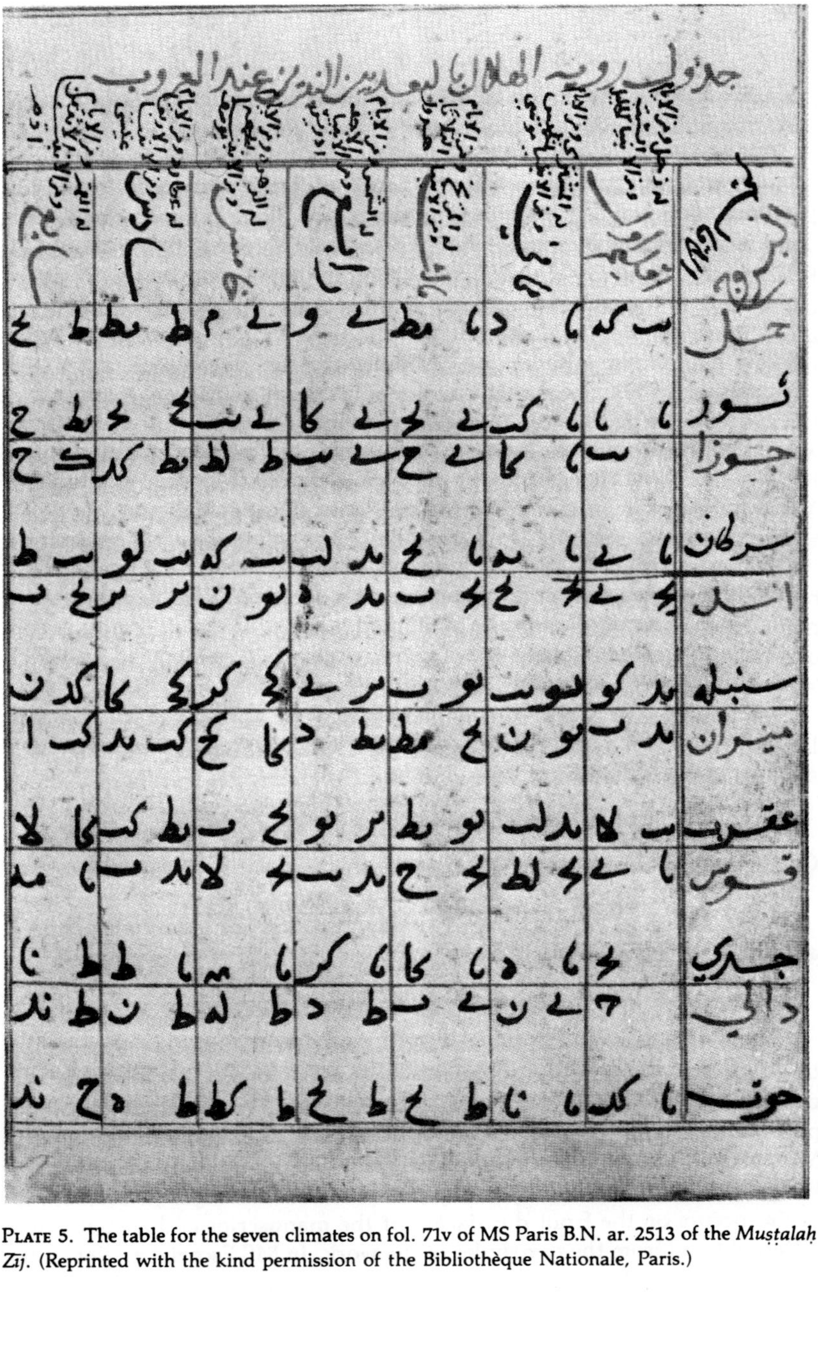

PLATE 5. The table for the seven climates on fol. 71v of MS Paris B.N. ar. 2513 of the *Muṣṭalaḥ Zīj*. (Reprinted with the kind permission of the Bibliothèque Nationale, Paris.)

crescent visibility occurs on fol. 61v and bears the same title *jadwal ru'yat al-ahilla bi-'l-ʿashīyāt* as found in the Madrid manuscript.

The fourth source (D) is MS Cairo Dār al-Kutub Muṣṭafā Fāḍil *mīqāt* 23 of a short work, actually a baby *zīj*, entitled *Tashīl al-maṭālib fī taʿdīl al-kawākib*, which means "A simple approach to planetary astronomy." The work was compiled by Aḥmad ibn Muḥammad ʿAyyād in Cairo about 1110 Hijra (= A.D. 1698/99).[31] The lunar visibility table occurs on fol. 9r after a blank page at the end of the treatise and bears the title *jadwal rāyat* [sic] *al-ahilla bi-'l-ʿashāt* [sic]. It is written in the same hand as the main work, namely, that of the author's son ʿAbd al-Raḥmān. The copy is dated 1153 Hijra (= A.D. 1740). The author claims that he compiled his work after noticing the errors in the work entitled *al-Yasāra* by Ibn al-Bannā' (*fl.* Marrakesh, *ca.* A.D. 1300).[32] This is, however, not so, since his tables are virtually identical to those in the *Tashīl al-maṭālib fī taʿdīl al-kawākib* of Ibn Qunfudh (*fl.* Tlemcen, Algeria, *ca.* A.D. 1375), which is introduced in the same way.[33] The earlier work, preserved in fols. 45r–48v of the unnumbered manuscript in the Museo Naval de Madrid (see above), does not contain the crescent visibility table. Note however that the table does occur on fol. 58v of the Madrid manuscript, amidst other tables attributed to Ibn al-Raqqām. Another copy of ʿAyyād's treatise, namely, MS Cairo Taymūr *riyāḍa* 49 (copied 1305 Hijra = A.D. 1887/88) is our fifth source (E).

The table in MS Cairo Muṣṭafā Fāḍil *mīqāt* 23 also shows values for the latitudes of the climates. These are inconsistent with reality for any value of ε. They are as follows:

16;30° 24;55 30;0 36;20 41;54 45;27 48;53

Correct values for $\varepsilon = 23;51,20°$ are:

16;27° 23;48 30;20 36;1 40;52 45;0 48;31

and for $\varepsilon = 23;35°$:

16;39° 24;5 30;40 36;22 41;14 45;22 48;53.

The sixth and seventh sources (F and G) are two manuscripts of an Arabic translation of the perpetual almanac by the Jewish scholar Abraham Zacuto, compiled in Salamanca in the late 15th century.[34] The translation was prepared in 912 Hijra (= A.D. 1506/07) by Mūsā Gālīnūs in Istanbul. In MS Milan Ambrosiana C82 of Gālīnūs' translation, the table for crescent visibility, entitled *jadwal maʿrifat ru'yat al-ahilla bi-'l-ʿashiyāt fi 'l-aqālīm al-sabʿa*, occurs on the front dust-jacket of the manuscript and is copied in a different hand from the remainder of the work. In MS Escorial ar. 966 of the same work, the table entitled *jadwal ru'yat al-ahilla fī sā'ir al-aqālīm*, occurs on fol. 192v at the end of the main tables of the *zīj* and copied in the same hand. The table is not contained in MS Cairo Dār al-Kutub *mīqāt* 1081 of a Maghribi version of Zacuto's tables.

The eighth copy (H) of this table is found on the title folio of MS Cairo Ṭalʿat *mīqāt* 119, a copy of the astrological treatise *al-Masāʾil fi ʾl-aḥkām* of ʿUmar ibn al-Farrukhān al-Ṭabarī (*fl.* Baghdad, *ca.* A.D. 800), prepared in 1185 Hijra (= A.D. 1771/72), apparently in Cairo.[35] The table bears no title and there is no indication of its purpose (see PLATE 6).

Yet another source (I) is the *Zīj* of Ibn Isḥāq al-Tūnisī (see page 194). in which tables for three of the climates, namely, the third, fourth, and fifth, are displayed separately (see PLATE 4). In the Hyderabad manuscript they are tabulated alongside the table attributed to al-Qallās (see above). In his introduction Ibn Isḥāq does not mention the name of any compiler for the tables, and states simply that one should enter in the table for the appropriate climate with the sign of the moon and compare the entry with the longitude difference between the sun and moon, which is to be computed for six hours after midday on the 29th day of the lunar month. The entries in Ibn Isḥāq's table correspond more or less to those in the Andalusian table for the second (not third!), fourth, and fifth climates.

A single table for the fourth climate (J), which differs from those described above, is attributed to Ibn Isḥāq in MS Paris B.N. ar. 2520,2, fol. 128r in the margin of a passage on crescent visibility theory in a recension of the *Zīj* of Ibn al-Shāṭir by the Damascus astronomer Ibn Zurayq.[36] The table is entitled *Jadwal li-ruʾyat al-hilāl fi ʾl-iqlīm al-rābiʿ li-Abī Isḥāq al-Tūnisī*, "table for crescent visibility in the fourth climate by Abū [*sic*] Isḥāq al-Tūnisī." The entries are displayed separately in TABLE 4. The first four entries correspond to variant readings in the Andalusian table for the fourth climate, and the last four entries are close to but not the same as the last four values in the Andalusian table for the third climate!

Another copy of the table for the seven climates is contained on fol. 165v of this same manuscript. The hand is different from that of the rest of the manuscript. This is the only copy in which the zodiacal sign is the horizontal argument and the climate is the vertical argument. The values from this copy of the table are *not* included in TABLE 3.

Our next source (K and L) is a treatise on folk astronomy entitled *al-Bayān or al-Tibyān bi-wāḍiḥ (?) al-burhān* written by a Moroccan astronomer named Abū ʿAbd Allāh Muḥammad ibn ʿAbd Allāh ibn ʿAmr al-Muʾaddib ("the teacher").[37] This work is extant in the unique MS Oxford Bodleian Pococke 249 (63 fols., copied *ca.* 800 Hijra ≈ A.D. 1400?). On fol. 43v of the treatise the author presents a simple table for visibility—see (h), page 217—together with another pair which concerns us here—see PLATE 7.

وما لكها ابو الاتفاق مد
لخالقه علي النعماح مد

طلعت
١١٩

هذا كتاب المسايل في الاحكام تاليف عمر بن
الفرخان الطبري عفا الله عنه وعن كاتبه والناظر
فيه ولمن دعا لهم بالغفران امين يارب العالمين

العدد	الاول	الثاني	الثالث	الرابع	الخامس	السادس	السابع
الحمل	سه كه	يا د	يا مط	ي و	ي م	ط نط	ط كح
الثور	يا يا	يا كه	ي كح	ي كا	ي س	ح كح	نط ح
الجوزا	يا ب	يا نا	ي ح	ي يب	ط لط	نط كد	كا ح
السرطان	يا ن	نا ه	يا كح	ند لس	س كه	س لو	س ط
الاسد	كح ي	كح ع	كح ب	ند ه	نو ن	ر ر	كح ب
السنبلة	ند كو	نو يب	نو ب	ر كح	كح كر	كح كا	كد ن
الميزان	ند ب	نو ن	كح نط	نط ند	كا كح	كه ند	كه ا
العقرب	سه لا	ند له	نو نط	نو نو	كح ب	نط كه	كا لا
القوس	يا يح	كح لط	كح ح	ند نه	كح لا	ند ب	يا مه
الجدي	يا كح	يا ه	يا كا	يا كر	يا مه	يا ط	ط نا
الدالي	يا ح	ي ن	ي ن	يا كر	ط له	ط ن	ط ند
الحوت	يا كد	ط نا	ط كح	له كح	ط نا	نط ه	ح ند

PLATE 6. The table for the seven climates as it occurs on fol. 1r of MS Cairo Ṭalʿat *mīqāt* 119 of an astrological treatise by Ibn Farrukhān al-Ṭabarī. (Courtesy of the Egyptian National Library.)

TABLE 4
Crescent Visibility Table for the Fourth Climate by Ibn Isḥāq al-Tūnisī
MS Paris Bibliothèque Nationale ar. 2520, fol. 128r.

0^s	10; 6°
1	10;21
2	10;12
3	14;32
4	14; 9
5	15; 1
6	18;20
7	14;50
8	13;39
9	11;15
10	11;10
11	11;11

The tables are labeled *ru'yat al-ahilla bi-'l-ʿashiyāt / bi-'l-ghadāt li-'l-iqlīm al-rābiʿ*, "lunar crescent visibility in the evenings / mornings for the fourth climate." The city of Baghdad is specifically mentioned, although it lies roughly 3° south of the middle of the fourth climate. This is the only one of our sources containing a table for last visibility in the mornings—the entries for sign *n* in this table correspond roughly to those for sign $n + 6$ in the table for evening visibility. The entries in the two tables, often mutually inconsistent, are displayed (labeled K for evening and L for morning) in TABLE 3.

The last source (M and N), which came to my attention only in April, 1984, is a collection of astronomical and astrological treatises of *ca.* 150 folios in a manuscript in the private collection of Rabbi Kafah of Jerusalem.[38] The manuscript is clearly of Yemeni provenance and was copied *ca.* 1800 (?). There are two tables for lunar crescent visibility, one for the evenings (M) and the other for the mornings (N), each serving the seven climates. The entries are displayed in TABLE 3, and those in M and N are often mutually inconsistent. Clearly the entries for the fourth climate in sources L and N are related. In an introductory note it is implied that the tabulated function is a function of the lunar longitude (*wa-khudh mā tajid qibālat burj al-qamar . . .*).

My reconstruction of the table is only tentative. I have not been able to determine precisely how it was compiled, nor am I able to account for the inordinate number of copyists' errors. TABLE 3 shows the accurately computed values for $\varepsilon = 23;51,20°$ when the argument is the lunar longitude. In very few cases, values given in or reconstructed from the text agree closely (to less than $\pm 6'$) with recomputation: these are indicated with asterisks in TABLE 3.

II

جدول رؤية الأهلة في البروج

درج	الحمل	الثور	الجوزا	السرطان	الأسد	السنبلة	الميزان	العقرب	القوس	الجدي	الدلو	الحوت
ا	لا يرا	لا يرا	لا يرا	لا يرا	لا يرا	لا يرا	لا يرا	لا يرا	لا يرا	لا يرا	لا يرا	لا يرا
ب	لا يرا	لا يرا	لا يرا	لا يرا	لا يرا	لا يرا	لا يرا	لا يرا	لا يرا	لا يرا	لا يرا	لا يرا
ج	لا يرا	لا يرا	لا يرا	لا يرا	لا يرا	لا يرا	لا يرا	لا يرا	لا يرا	لا يرا	لا يرا	لا يرا
د	لا يرا	لا يرا	لا يرا	لا يرا	لا يرا	لا يرا	لا يرا	لا يرا	لا يرا	لا يرا	لا يرا	لا يرا
ه	لا يرا	لا يرا	لا يرا	لا يرا	لا يرا	لا يرا	لا يرا	لا يرا	لا يرا	لا يرا	لا يرا	لا يرا
و	لا يرا	لا يرا	لا يرا	لا يرا	لا يرا	لا يرا	لا يرا	لا يرا	لا يرا	لا يرا	لا يرا	لا يرا
ز	لا يرا	لا يرا	لا يرا	لا يرا	لا يرا	لا يرا	لا يرا	لا يرا	لا يرا	لا يرا	لا يرا	لا يرا
ح	لا يرا	لا يرا	لا يرا	لا يرا	لا يرا	لا يرا	لا يرا	لا يرا	لا يرا	لا يرا	لا يرا	لا يرا
ط	لا يرا	لا يرا	لا يرا	لا يرا	لا يرا	لا يرا	لا يرا	لا يرا	لا يرا	لا يرا	لا يرا	لا يرا
ي	يرى	يرى	لا يرا	لا يرا	لا يرا	لا يرا	لا يرا	لا يرا	لا يرا	لا يرا	يرى	[illegible]
يا	يرى	يرى	يرى	لا يرا	لا يرا	لا يرا	لا يرا	لا يرا	لا يرا	يرى	يرى	[illegible]
يب	يرى	يرى	يرى	لا يرا	لا يرا	لا يرا	لا يرا	لا يرا	لا يرا	يرى	يرى	[illegible]
يج	يرى	يرى	يرى	يرى	لا يرا	لا يرا	لا يرا	لا يرا	يرى	يرى	يرى	[illegible]
يد	يرى	يرى	يرى	يرى	لا يرا	لا يرا	لا يرا	لا يرا	يرى	يرى	يرى	[illegible]
يه	يرى	يرى	يرى	يرى	لا يرا	لا يرا	لا يرا	لا يرا	يرى	يرى	يرى	[illegible]
يو	يرى	يرى	يرى	يرى	يرى	لا يرا	لا يرا	يرى	يرى	يرى	يرى	[illegible]
يز	يرى	يرى	يرى	يرى	يرى	لا يرا	لا يرا	يرى	يرى	يرى	يرى	[illegible]
يح	يرى	يرى	يرى	يرى	يرى	لا يرا	لا يرا	يرى	يرى	يرى	يرى	[illegible]
يط	يرى	يرى	يرى	يرى	يرى	يرى	يرى	يرى	يرى	يرى	يرى	[illegible]
ك	يرى	يرى	يرى	يرى	يرى	يرى	يرى	يرى	يرى	يرى	يرى	[illegible]

بغداد

رؤية الأهلة بالعلامات [illegible]

بروج	درج	دقائق
الحمل	ط	ك
الثور	[illegible]	يو
الجوزا	[illegible]	[illegible]
السرطان	يا	ند
الأسد	يا	ك
السنبلة	ي	لط
الميزان	ي	م
العقرب	ي	ك
القوس	ي	[illegible]
الجدي	[illegible]	[illegible]
الدلو	[illegible]	[illegible]
الحوت		[illegible]

رؤية الأهلة بالعيان [illegible]

بروج	درج	دقائق
الحمل	ي	و
الثور	يا	كا
الجوزا	ي	[illegible]
السرطان	[illegible]	[illegible]
الأسد	[illegible]	[illegible]
السنبلة	[illegible]	[illegible]
الميزان	يط	[illegible]
العقرب	[illegible]	يو
القوس	[illegible]	[illegible]
الجدي	يا	[illegible]
الدلو	يا	[illegible]
الحوت	[illegible]	ط

This table takes the prize as the most corrupt table in the known medieval Islamic sources.

MISCELLANEOUS CRESCENT VISIBILITY TABLES I

The Table of Abū Ja'far al-Khāzin

Abū Ja'far al-Khāzin was an astronomer of some renown who worked in the early 10th century.[39] A crescent visibility table attributed to him is contained on fols. 157v–158v of MS Paris B.N. ar. 5968 (copied *ca.* 650 Hijra ≈ A.D. 1250) of the *Dastūr al-munajjimīn*, an Ismā'īlī *Zīj* consisting of material culled mainly from the works of earlier scholars such as al-Battānī, Kūshyār ibn Labbān, al-Bīrūnī, and al-Khāzin.[40] The table is entitled *ru'yat al-ahilla*, "crescent visibility," and is divided into two parts labeled *wa-'arḍ al-qamar fi 'l-shamāl* and ... ... *fi 'l-janūb*, "for lunar latitudes in the north" and ". . . in the south." The function displayed is a double argument function of lunar latitude and lunar longitude, which I label $f(\lambda, \beta)$. Values are given to two sexagesimal digits for each 10° of λ and each 0;30° of β from $-5°$ to $+5°$, except that no values are given for $\beta = 0$. A text attributed to al-Khāzin on fol. 157v of the Paris manuscript explains how to use the table, and states that visibility will occur if $\Delta\lambda > f(\lambda, \beta)$.

The same table, without any attribution to al-Khāzin, occurs in the *Athīrī Zīj* of the early-13th-century scholar Athīr al-Dīn al-Abharī.[41] This *zīj* is extant in the unique copy MS Dublin Chester Beatty 4076 (61 fols., *ca.* 800 Hijra ≈ A.D. 1400). It is based on the earlier *Zīj* of Abu 'l-Wafā' al-Būzajānī, who worked in Baghdad in the late 10th century,[42] and contains several tables computed specifically for the latitude of Mardin, taken as 37;25°. In the text of the *Athīrī Zīj* there are two sections on crescent visibility, one of which deals with the use of this particular table (fols. 16v–17r) and immediately precedes it (fol. 16r). Further information on the other section is given under (k) on pages 217–218.

A similar (and related) table, with entries for arguments for the same range of λ and integral values of β is contained in MS Cambridge Browne O.1 (copied *ca.* 700 Hijra ≈ A.D. 1300), fols. 129r–129v, of the *Mufrad Zīj* of Muḥammad ibn Ayyūb al-Ṭabarī, a Persian astronomer of the late 10th or early 11th cen-

←

Plate 7. Two small tables of a visibility function for mornings and evenings in the fourth climate, specifically associated with Baghdad, as they occur on fol. 43v of MS Oxford Bodleian Pococke 249 of a medieval Moroccan treatise on folk astronomy—see K and L on pages 203 and 205. The large table is also for lunar crescent visibility: Entries "*lā yurā*" (= will not be seen) or "*yurā*" (= will be seen) are shown for each degree of modified longitude difference for each zodiacal sign—see (h) on page 217. (Reproduced with the kind permission of the Bodleian Library, Oxford.)

TABLE 5
Extracts from Visibility Table in the *Mufrad Zīj*
MS Cambridge University Library Browne O.1, fols. 129r–129v.

Lunar Longitude	*Lunar Latitude* −5°	0°	+5°
0°	10;30°	9;46°	8;44°
30	11;58	10; 2	8;28
60	14;50	11; 6	8;19
90	19;58	13;39	9;13
120	25;53	16;57	10;58
150	29;30	18;59	12;14
180	28;17	18;38	11;49
210	24;30	16;28	10;37
240	19;32	13;42	9;19
270	15;12	11;29	8;32
300	12;20	10;12	8;22
330	11; 9	9;50	8;36

tury. [43] Extracts from this table, in which values for $\beta = 0$ *are* given, are presented in TABLE 5. I have not investigated the values further than noting that those for $\beta = 0$ are not based on an accurate method and that the values for $\beta \neq 0$ are not linearly related to β.

[**Note added in proof:** Dr. J.-P. Hogendijk has cracked these tables too.]

A TABLE ATTRIBUTED TO AL-BĪRŪNĪ

In MS Cambridge Browne O.1, fol. 108r, of the *Mufrad Zīj* of al-Ṭabarī — see preceding paragraph — there is a table for lunar crescent visibility attributed to Abu'l-Rayḥān, that is, the celebrated early-11th-century scientist al-Bīrūnī.[44] The instructions indicate that the argument is lunar longitude, and that visibility occurs if

$$\Delta\lambda + \beta \geqslant f(\lambda_m),$$

where f is the tabulated function. In view of the crudity of the table and the fact that it is unrelated to al-Bīrūnī's crescent visibility theory as outlined in his major work *al-Qānūn al-Masʿūdī*, the attribution to al-Bīrūnī is probably incorrect.

The entries are arranged in a circle rather than in columns — see PLATE 8 — and are displayed in TABLE 6. The tabulated function is symmetrical about

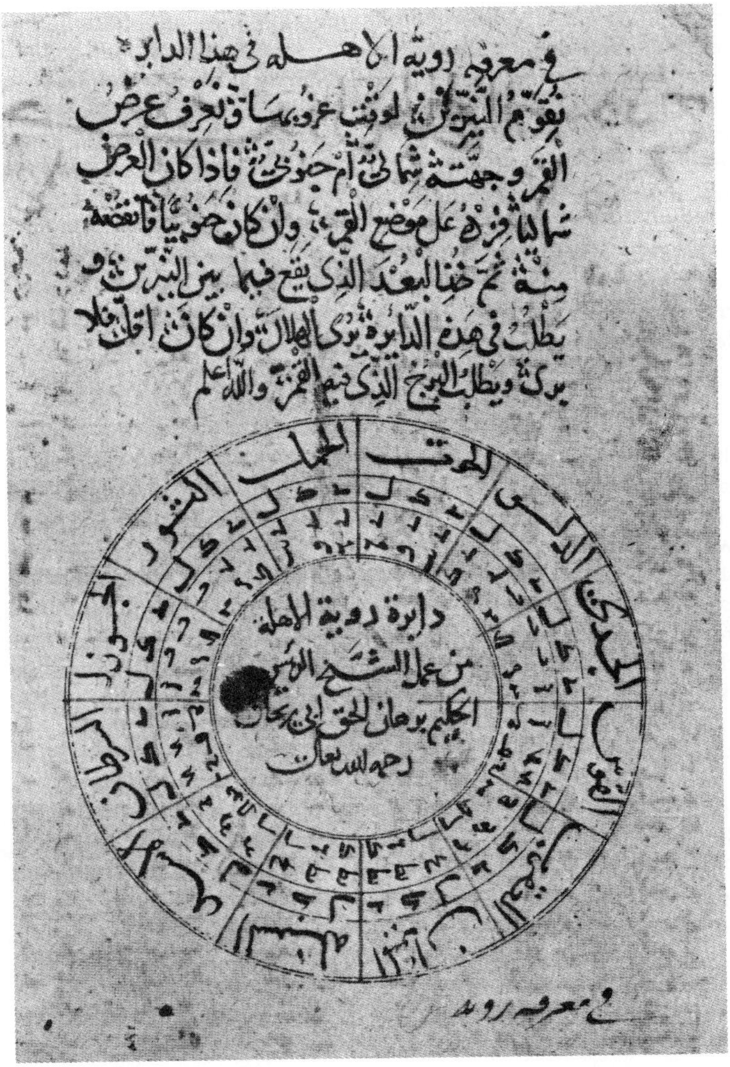

PLATE 8. A table attributed to al-Bīrūnī on fol. 108r of MS Cambridge Browne O.1 of the *Mufrad Zīj* of Muḥammad ibn Ayyūb al-Ṭabarī. (Reproduced with the kind permission of the Syndics of Cambridge University Library.)

$\lambda = 180°$ and displays no secondary minima. Linear interpolation has been used to generate some of the values.

A slightly different and still cruder version of the same table, without the attribution to al-Bīrūnī, occurs in MS Istanbul Nurosmaniye 2933, fol. 2v (copied *ca.* 700 Hijra ≈ A.D. 1300). This manuscript consists of a copy of the *Īlkhānī Zīj* of Naṣīr al-Dīn al-Ṭūsī (*fl.* Maragha, *ca.* A.D. 1250),[45] made up of fragments of various earlier copies. There is no such table in more respectable copies of the *Īlkhānī Zīj*.

It is idle to speculate about the latitude for which the table was intended, save perhaps to note that the limits 10;0° and 19;0° used in the second copy correspond roughly to those for the fourth climate, which would be reasonable for Tabaristan (the area at the south-eastern corner of the Caspian Sea).

The Table of al-Zarqāllu

It is worth noting the existence in the Islamic source material of another crescent visibility table in which the arguments are the zodiacal signs and the climates. This occurs in the unique MS Munich ar. 853 (copied 655 Hijra = A.D. 1257) of the *Canon* of Ammonius in the recension of the early-11th-century Andalusian astronomer al-Zarqāllu.[46] The use of the table (fol. 45v) is explained in the text (fols. 5v–6r). I cannot claim to fully understand al-Zarqāllu's crescent visibility theory, which is strange indeed and unrelated to the present discussion. However, the function displayed in the table is apparently the setting time corresponding to an ecliptic elongation of 12°. Values are given in degrees and minutes for each 15° of solar longitude and for each of the seven climates, and are displayed in Table 7. Tables of $\rho(\lambda)$ for each 10° of λ are also contained in al-Zarqāllu's work (fols. 42v–43v). The text of al-Zarqāllu's introduction and the tables have been published by J. Millás Vallicrosa, who used the Munich manuscript and also a Latin version preserved in Paris. However, the crescent visibility table published by Millás contains over thirty unrecorded variants from the table in the Munich manuscript. I have not used the Paris manuscript, but Table 7 accurately represents the table in the Munich manuscript.

The Table of al-Lādhiqī

A table which is at first sight similar to that of al-Khwārizmī (see Plate 8) is contained in MS Cairo Ṭalʿat *mīqāt* 92 (copied *ca.* 1150 Hijra ≈ A.D. 1750), fol. 22v, of a small *zīj* entitled *Nihāyat al-tashīl li'-l-ʿibāra wa-'l-ikhtiṣār li'-l-ghāya li-taqwīm al-kawākib al-sayyāra*, compiled by Muḥammad ibn ʿAbd al-Maḥmūd al-Ḥakīm al-Lādhiqī in 1110 Hijra (= A.D. 1698/99).[47]

TABLE 6

Lunar Crescent Visibility Table Attributed to al-Bīrūnī MS Cambridge Browne O.1, fol. 108r.				Anonymous Lunar Crescent Visibility Table MS Istanbul Nurosmaniye 2933, fol. 2v.			
10°	10; 0°	10; 0°	30°	10°	10; 0°	10;40°	30°
20	10; 6	10; 6	20	20	10; –	10;30	20
30	10;12	10;12	10	30	10;20	10;20	10
10	10;20	10;20	30	10	10;30	10;30	30
20	10;40	10;40	20	20	10;40	10;40	20
30	11; 0	11; 0	10	30	10; 0	11; 0	10
10	11;20	11;20	30	10	11;20	11;20	30
20	11;40	11;40	20	20	11;40	11;40	20
30	12; 0	12; 0	10	30	12;10	12;10	10
10	12;35	12;35	30	10	13;10	12;30	30
20	13; 9	13; 9	20	20	13;50	13;10	20
30	13;55	13;55	10	30	14;50	13;50	10
10	15; 0	15; 0	30	10	15;50	14;50	30
20	16;20	16;20	20	20	16;20	16;20	20
30	17;30	17;30	10	30	17; 0	17; 0	10
10	18;30	18;30	30	10	18; 0	18; 0	30
20	19; 0	19; 0	20	20	19; 0	19; 0	20
30	19;20	19;20	10	30	19; 0	19; 0	10

In his introduction al-Lādhiqī states that he had seen the tables in a work *Tashīl al-ʿibāra fī taqwīm al-kawākib al-sayyāra* by the Egyptian astronomer Ibn Yūnus,[48] which only served dates up to 900 Hijra, and that he had extended the range of the tables in this work also using the tables of Riḍwān Efendī of Cairo based on the *Zīj* of Ulugh Beg (*fl.* Samarqand, *ca.* A.D. 1430).[49] It is of interest that there is a manuscript of a small Maghribi *zīj* entitled *Tashīl al-ʿibāra fī taqwīm al-kawākib al-sayyāra* preserved in the Aḥmadīya Library in Aleppo which is falsely attributed to Ibn Yūnus on the title folio and actually copied by al-Lādhiqī![50]

Our present concern is with the lunar visibility table, which al-Lādhiqī claims to have computed himself for latitude 34;30°, probably Lattakia. The instructions state that one should find $\Delta\lambda$ two thirds of an hour after sunset. If $\Delta\lambda$ (there is no mention of any modifications for $\beta \neq 0$) is greater or equal to the tabulated value for the zodiacal sign corresponding to the lunar longitude the crescent will be seen, otherwise not. The values are displayed in TABLE 8. They are not computed according to the proper method, either for lunar

TABLE 7
Crescent Visibility Table of al-Zarqāllu
MS Munich ar. 853, fol. 45v. (Variants given in M = Millás, 228 are noted.)

	1	2	3	4	5	6	7
15°	9;45°	8;42	8;42	7;12	6;59[a]	6;25	5;49
30	9;56[a]	9;18	8;36	8;15	7;58[b]	6;40	6; 5
15	10;35	9;58[a]	9;25	8;49	8;32[c]	7;48	7;15
30	11;17	10;46	10;15	9;50	9;19	8;35[a]	8;39
15	12; 3	11;44	11[a];28	11; 9[a]	10; 9	10;49	10;21
30	12;56	12;48	12;44	12;38	12;33	12;28	12;24
15	13;10	13;17	13;24	13;31	13;39	13;44	13;49
30	13;26	13;48	14; 7	14;28	14;51	15; 8	15;23
15	13;11[b]	13;43[b]	14; 8	14;36	15; 7	15;31	15;54[a]
30	12;56	13;39[c]	14; 8	14;45	15;23	15;54[b]	16;26
15	12;41	13;25	13;59[b]	14[b];38	15;19	15;32	16;27
30	12;27	13;12	13;50	14;32	15;16	15;51[c]	16;28
15	12;33	13;17	13;45	14;36	15;20	15;53[d]	16;29
30	12;47	13;23	13;40	14;41	15;23	15;56[e]	16;31
15	12[c];58	13;35[d]	13;37[c]	14;42	15;19	15;47	16;23
30	13;17	13;48	13;55[d]	14;44	15;16	15;39	16;15
15	13;11	13;35	13;50	14; 7	14;25[d]	14;39	14;57[b]
30	13;16	13;22	13;26	13;52	13;37[e]	13;42	13;46
15	12;37	12;28	12;55[e]	11;53[c]	11;29	11;52	11;43
30	11;19	11;37	11;58[f]	10;56	10;34	10;57	9;51
15	11; 9	10;35	10;51[g]	9;41	8;25	8;48	8;57
30	10;22	9;39	9;10	8;33	7;15[f]	7;24	6;12
15	9;48	9;14	8;40	8; 0	7;10[g]	6;46	6;11
30	9;35	8;50	8;12	7;30	6;56[h]	6;11	5;54[c]

Variants (by climate)

1: [a] MS: 36, M: 56 [b] MS: 51, M: 51 [c] MS: 13, M: 12
2: [a] MS: 18, M: 18 [b] M: 13;39(!) [c] M: 13;29(!) [d] M: 34
3: [a] MS: 10, M: 10 [b] M: 49 [c] M: 57 [d] M: 15 [e] M: 15 [f] M: 18 [g] M: 11
4: [a] MS: 38, M: 9 [b] MS: 15 [c] M: 33
5: [a] M: 49 [b] M: 53 [c] M: 31 [d] M: 23 [e] M: 27 [f] M: 35 [g] M: 0 [h] M: 56
6: [a] M:55 [b] M: 44 [c] M: 41 [d] M: 43 [e] M: 46
7: [a] M: 44 [b] M: 17 [c] MS: 14 or 54, M: 14

TABLE 8
al-Lādhiqī's Crescent Visibility Table
MS Cairo Ṭalʿat *mīqāt* 92, fol. 22v.

Zodiacal Sign	*Visibility Function*	*Recomputation* $\varphi = 34;30°$ $\varepsilon = 23;35°$ *Longitude*	*Lunar*	*Solar*
1	10; 8°	0°	10; 4°	10; 4°
2	9;14	30	9;57	9;53
3	9; 9	60	9;51	9;58
4	10;31	90	10;41	11;25
5	12;26	120	13; 4	14;39
6	15; 3	150	16;18	17;52
7	16;30	180	17;30	18;30
8	15;27	210	17;52	16;18
9	13;30	240	14;39	13; 4
10	10;22	270	11;25	10;41
11	9;11	300	9;58	9;51
12	10;25	330	9;53	9;57

longitude as argument or for solar longitude. (Immediately below these values is a second unrelated set copied in the same hand – see PLATE 9: for this second set, see (n) on page 218 below.)

MISCELLANEOUS CRESCENT VISIBILITY TABLES II

In this section I discuss a number of very simple Islamic visibility tables that display a set of values f_n ($n = 1, 2, \ldots, 12$) for the twelve signs such that if $\Delta\lambda > f_n$ for the lunar longitudes in the nth sign then the crescent will be seen. Clearly the values f_n should correspond to a specific latitude, but since the values are invariably given in integral degrees it is impossible to derive the underlying latitudes with any certitude. For each of the sources labeled (a) to (p), the values are displayed in TABLE 9. The most remarkable feature of the seventeen different sets of values discussed in this section is that they are all different!

(a) In MS Escorial ar. 927, fol. 6r, of the introduction to the anonymous Mosul recension of the *Mumtaḥan Zīj* (see above, page 190), immediately preceding the visibility table attributed in other sources to al-Khwārizmī – see PLATE 2 – there is a section that translates as follows:

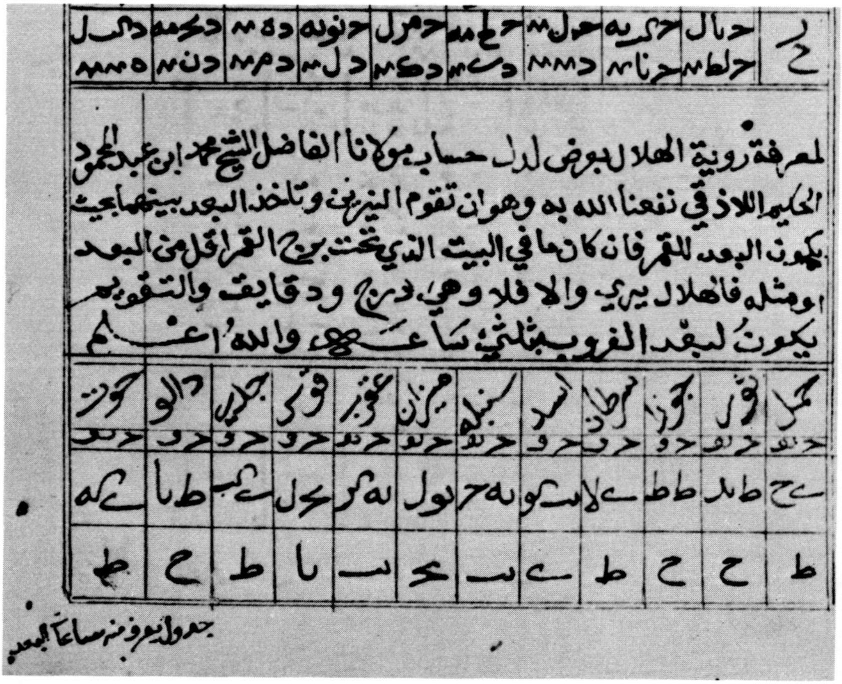

Plate 9. al-Lādhiqī's table for Lattakia as it appears on fol. 22v of MS Cairo Ṭal'at *mīqāt* 92 of his *Zīj*. Note that there are two sets of values tabulated – see pages 210–211, 213 and (n) on page 218. (Courtesy of the Egyptian National Library.)

> In (signs) 0 (and) 11 (the crescent) will be seen from 11 degrees. (In signs) 1 (and) 10 it will be seen from 12 (degrees). (In signs) 2 (and) 9 it will be seen from 12 degrees, 30 minutes. (In signs) 3 (and) 8 it will be seen from 14 degrees, 30 minutes. (In sign) 4 it will be seen from 15 degrees. (In signs) 5, 6, (and) 7, it will be seen from 18 degrees. Know this, if God Almighty wills.

The values given in the text are clearly independent of the values in the main table (*cf.* Table 1).

(b) The treatise on theoretical astronomy and astrology entitled *al-Mudkhal* compiled in 365 Hijra = A.D. 975/76 by Abū Naṣr al-Qummī[51] contains in *Maqāla* 2, *Faṣl* 12 a set of values of f_n for an unspecified latitude. I have examined MSS Cairo Ṭal'at *mīqāt* 222,2 (fols. 60r–177r, 617 Hijra = A.D. 1220/21) of this work, in which the values occur in the text of fols. 99v–100r, and also the later copy MS Cairo Dār al-Kutub Muṣṭafā Fāḍil *mīqāt* 208 (91 fols., *ca.* 1150 Hijra ≈ A.D. 1750), where the same values are arranged in tabular form on fol. 28v. al-Qummī's treatise contains a table of oblique ascensions for

TABLE 9

Crescent Visibility Tables for Various Latitudes[a]

Signs		(a)	(b)	(c)	(d)	(e)	(f)	(g)	(h)	(i)	(j)	(k) (1)	(k) (2)	(l)	(m)	(n)	(o)	(p)
0	1	11°	10	8	8	11	10	9/10	10	12	10	12	12	10	9	12	9	13
1	2	12	11	10	8	11	10	10	10	13	10	13	13	10	10	13	8	12
2	3	12½	12	12	9	13	13	13	11	14	13	14	14	13	11	14	8	14
3	4	14½	14	14	11	14	15	15	13	15	14	15	15	14	14	15	9	14
4	5	15	16	16	13	15	15	18	16	16	15	17	16	15	16	16	10	16
5	6	18	18	17	16	17	19	18	19	17	19	18	18	19	19	18	12	16
6	7	18	18	19	16	17	19	18	19	17	19	18	18	19	18	18	13	15
7	8	18	16		15	15	19	15	16	16	15	16	16	15	16	16	12	16
8	9	14½	14		12	15	15	15	13	15	14	15	15	15	14	15	11	14
9	10	12½	12		11	14	15	12	11	14	12	14	14	13/14	12	14	9	14
10	11	12	11		8	11	10	9	10	13	10	13	12	10	11	13	8	12
11	12	11	10		8	11	10	9	10	12	10	12	12	10	10	12	9	11

[a] See Miscellaneous Crescent Visibility Tables II.

$\varphi = 36;0°$ (fol. 78v of the Ṭalʿat manuscript), apparently intended to serve the middle of the fourth climate (see fol. 76v).

(c) In the margin of the main crescent visibility tables on fol. 87r of MS Paris Bibliothèque Nationale ar. 6913 of the 11th-century *Rīqānī Zīj* (see above pages 190–191) there is a simple table labeled *taqrīb* "approximation," giving values of f_n labeled *darajāt al-sawāʾ* "equatorial degrees," for zodiacal signs 0 to 6. The values are different from those attributed to al-Khwārizmī in the same manuscript (see above).

In passing I note that the main lunar visibility tables in the *Rīqānī Zīj* (fols. 86v–87r), which are of a kind not attested in any other Islamic source known to me, display two functions $f(\beta,\lambda)$ and $g(\beta,\varphi)$ for modifying the lunar longitude in visibility determinations. The first function is tabulated to degrees and minutes for each degree of $|\beta|$ from 1° to 5° and each degree of lunar elongation from the nearer equinox, λ', from 1° to 90°. The second function is tabulated likewise for each degree of $|\beta|$ and $\varphi = 12°, 15°, 18°, 19°, \ldots, 42°, 45°, 48°, 51°$. The tabulated functions are defined by

$$f(\beta,\lambda') = \beta \cdot \frac{\text{Sin}\ \varepsilon}{60} \cdot \frac{\text{Cos}\ \lambda'}{60} \qquad (\text{Sin}\ \varepsilon \approx 24)$$

and

$$g(\beta,\varphi) = \beta \tan \varphi.$$

The instructions prescribe forming $\lambda \mp f(\beta,\lambda)$ according as β is in the same or different direction as $\delta(\lambda)$ (*idhā kāna mawḍiʿuhu fī jihat al-ʿarḍ*). Then form

$$\lambda \mp f(\beta,\lambda) \pm g(\beta,\varphi)$$

according as $\beta \gtrless 0$. Finally form the difference in descensions between this modified lunar longitude and the solar longitude. Visibility will depend on whether this difference is greater or less than 12°. The reasoning behind this procedure appears to be inspired by the fact that the angle between the ecliptic and the horizon at sunset varies between $\bar{\varphi} - \varepsilon$ at the summer solstice and $\bar{\varphi} + \varepsilon$ at the winter solstice.

A similar set of instructions, without the associated tables, occurs in MS Utrecht Or. 23, fols. 2r–2v, of a fragment (47 fols.) of an anonymous *zīj*, related to the anonymous *Shāmil Zīj* (see (k) below) but perhaps closer to the original *Zīj* of the 10th-century Baghdad astronomer Abu 'l-Wafā' al-Būzajānī.[52]

(d) In MS Cambridge Browne O.1, fol. 172r of the 10th- or 11th-century *Mufrad Zīj* (see page 208 ff. above) there are two other methods presented for determining crescent visibility. The first is that the condition

$$\sigma(\lambda_m + 0;42\beta) - \sigma(\lambda_s) \geqslant 12°$$

implies visibility. The factor 0;42 (see page 187 above) may have been intended to serve a specific latitude. We note that cot $\bar{\varphi} = 0;42$ for $\varphi \approx 35°$ but that the spherical astronomical tables in the *Mufrad Zīj* are computed for latitudes 29;30° (Shiraz), 32° (Isfahan), and 36° (Raqqa). The second method involves the use of a table of values of $\Delta\lambda$ for each zodiacal sign, for which there are no instructions.

(e) MS Florence Laurenziana Or. 106 (fols. 1v - 72r, copied in Mosul in 1330/31) is a unique copy of a *zīj* based on the observations of Abu 'l-Wafā' al-Būzajānī (see above). In a marginal note on fol. 72r there is a table of values of $\Delta\lambda$ for each sign. See also (f) below.

(f) MS Florence Laurenziana Or. 106 (fols. 72v - 170r, copied in Mosul in 1330/31) is a unique copy of a *zīj* called *al-Zafīr* (?) based on the ʿAlāʾī observations, that is, on the *ʿAlāʾī Zīj* of the 12th-century astronomer al-Fahhād.[53] This work contains a table (fol. 163r) in which one enters the quantity $\Delta\lambda + \beta$ ($\beta \gtrless 0$) vertically and the zodiacal sign horizontally and reads either zero (in the *abjad* notation) for "not seen" or the word *yurā*, "it will be seen." The entries are based on the values shown in Table 9 (f). In the text of Chapter 11 of this work the condition $\Delta\lambda + \beta \geqslant 12°$ is stated for visibility, and the latitude used for the tables of spherical astronomical functions is 36°. (The geographical tables on fol. 141v give 36;30° for the latitude of Mosul.)

(g) In MS Istanbul Topkapi Seray A3337 of a Persian astronomical treatise entitled *Kitāb Faṣṣ al-khātam fī hay'at al-ʿālam* attributed to Mu[a'yyad ibn] Muḥammad al-Hajmāmī (???) and copied in 680 Hijra (= A.D. 1280),[54] there is an additional folio at the end (fol. 58r) containing a lunar crescent visibility table. The instructions are to enter the quantity $\Delta\lambda + \beta$ ($\beta \gtrless 0$) in the table vertically (and the lunar longitude horizonally): if the entry is zero the crescent will not be seen, otherwise the word *yurā* "it will be seen" indicates visibility. Several of the entries in the table have been altered, as indicated in TABLE 9 (g). Clearly the table is based on the assumption that visibility occurs when $\Delta\lambda + \beta > f(\lambda_m)$.

(h) In MS Oxford Bodleian Pococke 249, fol. 43v, of the treatise on folk astronomy by the Moroccan astronomer Abū ʿAbd Allāh al-Mu'addib (see K and L on pages 203 and 205), there is a table similar to those described in (f) and (g) above—see PLATE 7. The entries are either *lā yurā* or *yurā*, "it will not be seen" or "it will be seen," and are based on the symmetrical values of $\Delta\lambda$ shown in TABLE 9 (h).

(i) In MS Cairo Dār al-Kutub *mīqāt* 1106,[55] at the end of a fragment of the introduction to the *Muṣṭalaḥ Zīj* (see A on page 200) copied about A.D. 1750, there is scribbled a small crescent visibility table (fol. 32v). The instructions state that one should compute $\Delta\lambda + \beta$ ($\beta \gtrless 0$) and that visibility will occur if this quantity is larger than the set of values of f_n extracted in TABLE 9(i). See also (j) below.

(j) In MS Paris Bibliothèque Nationale ar. 2520 of the *Muṣṭalaḥ Zīj* (see A on page 200) there is a crescent visibility table scribbled in a later hand in the margin of the portion of the text dealing with the same subject (fol. 128r). Text and table are unrelated, and the entries in the table vary linearly between 12° for Aries and 17° for Libra. I have not seen this table in any other manuscripts of Egyptian provenance. See also (i) above.

(k) MS Paris Bibliothèque Nationale ar. 2528 is a copy of the *Shāmil Zīj*, compiled by an anonymous astronomer who relied on the *Zīj* of the late-10th-century Baghdad astronomer Abu 'l-Wafā' al-Būzajānī (see (c) above). In the introduction to this *Zīj* (fol. 8v) two methods are prescribed for determining crescent visibility. The same methods are presented in the *Athīrī Zīj* of the 13th-century astronomer Athīr al-Dīn al-Abharī, compiled in Mardin and preserved in MS Dublin Chester Beatty 4076 (see page 207). These two methods presented by al-Abharī (fols. 10r–10v) are unrelated to the visibility tables that occur elsewhere in the same manuscript.

The first method in the *Shāmil / Athīrī Zīj*es prescribes forming the quantity $\Delta\lambda + \beta$ ($\beta \gtrless 0$) and comparing this with a set of values of f_n presented in TABLE 9 (k). The second method involves computing the descensional differences of the sun and the moon using the modified lunar longitude $\lambda + 0;44\beta$, and then investigating whether or not this is greater than 12°. Note that cot

$\bar{\varphi}$ = 0;44 for $\varphi \approx 36°$, which corresponds to the latitude of the middle of the fourth climate. Since al-Abharī takes the latitude of Mardin as 38°, a factor of 0;47 would be more appropriate.

(l) In MS Paris Bibliothèque Nationale ar. 2515 (68 fols., *ca.* 800 Hijra ≈ A.D. 1400) of a treatise on theoretical astronomy by Athīr al-Dīn al-Abharī (see (k) above) the instructions for determining crescent visibility (fol. 62v) are to compute $\Delta\lambda \pm |\beta|$ and then investigate whether this is greater than or equal to a set of values of f_n different from those presented in al-Abharī's *Zīj*. The only other copy of al-Abharī's treatise that I have inspected, namely, MS Cairo Ṭalʿat *hayʾa* 48,2 (fols. 1v–62r, *ca.* 700 Hijra ≈ A.D. 1300) contains the same table (fol. 58r). In the Paris manuscript the fourth entry from the end has been altered to 14 in a later hand, no doubt in order to make the values symmetrical.

(m) In MS Cairo Dār al-Kutub Muṣṭafā Fāḍil *mīqāt* 213,4, fol. 29r, there is a table of values of f_n. It occurs amidst various anonymous notes included at the end of an 18th-century copy of part of the introduction to the *Jāmiʿ Zīj* of Kūshyār ibn Labbān, who worked in Iran at the beginning of the 11th century.[56] The instructions, which are quite unrelated to the visibility theory of Kūshyār, prescribe forming the quantity $\Delta\lambda + \frac{1}{2}\beta$ ($\beta \gtrless 0$) and comparing this with the entry in the table corresponding to the zodiacal sign of the conjunction.

(n) In MS Cairo Dār al-Kutub Ṭalʿat *mīqāt* 92, fol. 22v, of the small *Zīj* of the Syrian astronomer al-Lādhiqī (see pages 210–213), there are two distinct sets of values in the crescent visibility table - see PLATE 9. The second set of values is displayed in TABLE 9(n).

(o) In MS Cairo Dār al-Kutub K8526 of the planetary tables of ʿAbd al-Raḥmān ibn Banafshā al-Ṣāliḥī, a *muwaqqit* at the Umayyad Mosque in Damascus who flourished about A.D. 1500,[57] there is a table of values f_n (fol. 112r) apparently related to the values in the *Shāmil Zīj* (see (k) above). The former set is displayed in TABLE 9(o). al-Ṣāliḥī's planetary tables, entitled *al-Durr al-naẓīm fī tashīl al-taqwīm*, are based on the parameters of Ulugh Beg and are compiled for the longitude of Damascus. The lunar crescent visibility table in the Cairo manuscript is not contained in any of the other copies of al-Ṣāliḥī's tables that I have examined.

(p) Finally, we note that a similar set of twelve integers can be generated by an entirely unrelated function. In MS Cairo Ṭalʿat *mīqāt* 242,2 (fol. 32r), at the end of a commentary by the 16th-century writer Ibn Ghānim on his poem entitled *al-Nasama al-nafḥīya*, in turn a commentary on the very popular treatise on the sine quadrant by the 15th-century Egyptian astronomer Sibṭ al-Māridīnī,[58] we find the following note:

The sun enters (*nazala*) the first point of Aries on the thirteenth of Ādhār (March = III), and enters Taurus on the twelfth of Nīsān (April = IV), and so on, as you see in the following table:
(with modern notation)

Syrian month:	II	III	IV	V	VI	VII	VIII	IX	X	XI	XII	I
Date on which sun enters	11	13	12	14	14	16	16	15	16	14	14	12
Zodiacal sign	XII	I	II	III	IV	V	VI	VII	VIII	IX	X	XI

The date numbers corresponding to each zodiacal sign are displayed in TABLE 9 (p).

ACKNOWLEDGMENTS

My research on Islamic lunar crescent visibility theory was funded by a grant from the National Science Foundation, Washington, D.C. This support is gratefully acknowledged.

It is a pleasure to record my gratitude to the various libraries where the manuscripts used for this study are preserved, particularly the Egyptian National Library in Cairo and the Bibliothèque Nationale in Paris.

NOTES

1. See Kennedy and Janjanian, and Kennedy and King, 19–20. See also note 2 below.
2. The following is a list of modern works dealing with lunar crescent visibility in the Islamic sources: Kennedy, 1968 (Ya'qūb ibn Ṭāriq); Nallino (al-Battānī); Kennedy, 1960, and Morelon (Thābit ibn Qurra); Millás (al-Zarqāllu); Neugebauer, 1962, Kennedy and Janjanian, and Toomer, 1964, 208 (al-Majrīṭī); Neugebauer, 1956 (Maimonides); Vernet, 1950 (Ibn al-Bannā'); and Sédillot-*fils* (Ulugh Beg). On these last two sources, see note 33 below.
3. For a survey of Islamic *zīj*es, see Kennedy, 1956. On the different categories of Islamic astronomical tables not contained in *zīj*es, see King, 1975.
4. On this notation, see Irani.
5. See Goldstein, 222, note 5, for references to the Indian sources.
6. See *Almagest*, II.8 (Toomer, 1984, 99–103) and Stahlman, 30–46 and 206–242.
7. See Kennedy, 1956 1, 140, and my article "*Maṭāli'*" (= ascensions) in EI_2.
8. On the recomputation of medieval Islamic tables, see Kennedy, 1967 and King, 1975, 40.
9. On al-Khwārizmī, see the article by G. Toomer in *DSB*. On this table and various other newly discovered minor works by al-Khwārizmī, see King, 1983b.

10. Neither of these *zījes* is listed in Kennedy, 1956. It may be that the appellation al-Kirmānī refers to an individual named al-Ḥasan ibn Aḥmad ibn ʿAbd Allāh al-Ṣūfī al-Kirmānī, author of a treatise on astrology extant in MSS Cairo Ṭalʿat *mīqāt* 188 (314 fols., copied 1301 Hijra = A.D. 1883/84) (see Sezgin, VII, 67) and Princeton Mach 5067 = Yahuda 2501 (30 fols., 1302 Hijra = A.D. 1884/85). An individual named Ibn al-Ṣūfī al-Kirmānī is mentioned on fol. 214v of MS Paris B.N. pers. supp. 1488 of the early-14th-century *Ashrafī Zīj* (Kennedy, 1956, no. 4) and another treatise on astrology by Abu 'l-Qāsim al-Kirmānī is contained in MS Oxford Bodleian Hunt. 663 (fols. 127r–132r, *ca.* 650 Hijra ≈ A.D. 1250).
11. On al-Battānī, see the article by W. Hartner in *DSB* and Sezgin, VI, 182–87. On his *Zīj*, published in Nallino, see Kennedy, 1956, no. 55.
12. On the *Mumtaḥan Zīj*, see Kennedy, 1956, no. 51, and 145–47, where this table is mentioned on 146b. The Escorial manuscript is listed in Renaud, 1941, 135–37, and a brief survey of its contents is presented in Vernet, 1950, and Sezgin, VI, 136–37.
13. On Shāhmardān Rāzī, see Storey, 45, and *Cairo Survey*, no. B91. On the Cairo manuscript, see *Cairo Catalogue*, I, *sub* ṬFF 11, and for a table of contents see II, Section 5.2.30. Several other manuscripts of his handbook which I have not consulted are listed by Storey. A detailed study of this work has been prepared by Prof. E. S. Kennedy.
On al-Nasawī, see Kennedy, 1956, no. 44, and Sezgin V, 345–48 and VI, pp. 245–46, and the article by A. S. Saidan in *DSB*.
14. See Goldstein, 216–25.
15. On Ibn Masrūr, see Sezgin VII, 166–67. I have consulted the unique copy MS Cairo Taymūr *riyāḍa* 99 (164 pages, *ca.* 600 Hijra ≈ A.D. 1200), on which, see *Cairo Catalogue*, I, *sub* TR 99; II, Section 2.1.3.
16. See King, 1983c, 2.
17. *Ibid.*
18. On al-Majrīṭī, see the article by J. Vernet in *DSB*, and Sezgin, V, 334–36 and VI, 226–27.
19. See Suter, 1914, esp. 168, and Neugebauer, 1962, esp. 42–4 and 103, and also Toomer, 1964, 208.
20. See Kennedy and Janjanian.
21. On Ibn Isḥāq, see Suter 1900, no. 356, and note 22 below.
22. This manuscript is an important new source for the history of Islamic astronomy. On the importance of his *Zīj* in the medieval Maghrib see the account of the 15th-century historian Ibn Khaldūn translated in Rosenthal, III 136–37. For a survey of the history of astronomy in the Maghrib, see Kennedy and King, 5–9.
23. On Ibn Muʿādh, see the article "al-Jayyānī" by Y. Dold-Samplonius and H. Hermelink in *DSB*, and also Villuendas, xxii–xxxiv. (See also the paper by G. Saliba in this volume.) On al-Sabtī, see Suter, 1900, no. 342. It would be interesting to investigate whether or not al-Sabtī's tables are related to the crescent theory of his teacher Maimonides (on which, see Neugebauer, 1956).
24. *Cf.* Nallino, II, 65–7.
25. The *Muṣṭalaḥ Zīj* is listed as no. 47 in Kennedy, 1956. See also King, 1983b, 535–36. Two different recensions of it are preserved in MSS Paris B.N. ar. 2513 and 2520,1, both incorrectly attributed to the 10th-century Cairo astronomer Ibn Yūnus (see note 26). On the Paris manuscripts, see de Slane, 446 and 448. A detailed investigation of both of these manuscripts would be a worthwhile undertaking.
26. On Ibn Yūnus, see my article in *DSB*. The main lunar crescent visibility theory in the *Muṣṭalaḥ Zīj* is due to Ibn Yūnus.
27. The identity of the *Shāhī Zīj* (Kennedy, 1956, no. 32) and its relation to al-Ṭūsī is still obscure. MS Paris B.N. ar. 2523 (see de Slane, 450) is a Yemeni *zīj* based on the *Ḥākimī Zīj* and the *Shāhī Zīj*: see King, 1983a, II, Section 23.

28. On Ibn Raqqām, see Suter, 1900, nos. 388 and 417 (and also 221, note 85) and further Kennedy and King, 7. On the Madrid manuscript, see Vernet, 1956b.
29. On al-Fāsī, see Renaud, 1932, no. 541 and *Cairo Survey*, no. F50.
30. On al-Qusanṭīnī, see Suter, 1900, no. 371, and on the Escorial manuscript, see Renaud, 1941, 8–9. Some of his tables are analyzed in Kennedy and King – see especially 19–20.
31. On these two Cairo manuscripts, see *Cairo Catalogue*, I, *sub* MM 23 and TR 49, and II, Section 2.1.17, and *Cairo Survey*, nos. D67 and F23.
32. On Ibn al-Bannā', see the article by J. Vernet in *DSB*.
33. On Ibn Qunfudh, see Suter, 1900, no. 422, and Renaud, 1932, no. 422. The condition for crescent visibility proposed by Ibn Qunfudh in the introduction to his tables (fol. 45v of the Madrid manuscript) is that either one or the other of $\Delta\lambda$ or s should be at least 10° and the other at least 13°. This condition, which is part of the visibility theory of Ibn al-Bannā' (*cf.* Vernet, 1951, text, 62–4, and trans. 131–32), is repeated by ʿAyyād (fols. 3v–4r of the Cairo manuscript), who then presents a different visibility theory which is in fact that of Ulugh Beg of Samarqand (*cf.* Sédillot-*fils*, 190–92).
34. On Zacuto, see the article by J. Vernet in *DSB*, and on his tables, see Vernet, 1950. On the Ambrosiana, Escorial, and Cairo manuscripts, see respectively Griffini, 105–6; Renaud, 1941, 110–11; and *Cairo Catalogue*, I, *sub* DM 1081, and II, Section 2.2.9, and *Cairo Survey*, no. F31. Now that it is established that there were two different Arabic versions of Zacuto's tables, a new investigation of the various manuscripts would be worthwhile.
35. On Ibn Farrukhān al-Ṭabarī, see Sezgin, VII, 111–13. On this manuscript, see *Cairo Catalogue*, I, *sub* ṬM 119.
36. On Ibn al-Shāṭir, see my article in *DSB* and also King, 1983b, 538–39 and 547. On Ibn Zurayq, see Suter, 1900, no. 426; Renaud, 1932, no. 426; and *Cairo Survey*, no. C 116. On this Paris manuscript, see de Slane, 448–49.
37. This individual is new to the modern literature.
38. I am extremely grateful to Dr. Tzvi Langermann of the Institute of Asian and African Studies at the Hebrew University of Jerusalem for providing me with photocopies of the relevant pages of this manuscript.
39. On al-Khāzin, see the article by Y. Dold-Samplonius in *DSB*, and Sezgin, VI, 189–90. Two recent studies on his works not listed by Sezgin are Samsó and King, 1980.
40. On this work, not listed in Kennedy, 1956, see now Zimmermann, and Sezgin, VI, 63–4.
41. On Athīr al-Dīn al-Abharī, see Suter, 1900, no. 364; Krause, no. 364; and the article "al-Abharī" by C. Brockelmann in *EI₂*. This manuscript of his *Zīj* is catalogued in Arberry, no. 4076.
42. On Abu 'l-Wafā', see the article by A.P. Youschkevitch in *DSB*, and also Kennedy, 1956, nos. 73 and 28, and Sezgin, V, 321–25 and VI, 222–24.
43. On al-Ṭabarī, see Suter, 1900, no. 360; Krause, no. 360; Kennedy, 1956, no. 65; Storey, nos. 5 and 79; and Sezgin, V, 385–86 (he was inadvertently omitted from Sezgin, VI).
44. On al-Bīrūnī, see the article by E. S. Kennedy in *DSB* and also Sezgin, V, 375–83, VI, 261–76, and VII, 188–92.
45. On the *Īlkhānī Zīj*, see Kennedy 1956, no. 6, and Storey, 58–60.
46. On al-Zarqāllu, see the article by J. Vernet in *DSB*. His lunar crescent visibility table is published in Millás, 228.
47. On al-Lādhiqī and the Cairo manuscript, see *Cairo Catalogue*, I, *sub* ṬM 92, and II, Section 2.2.13; and *Cairo Survey*, no. C132.
48. On Ibn Yūnus, see note 26.
49. On the *Zīj* of Ulugh Beg, see Kennedy, 1956, no. 12. The introduction is published with

translation in Sédillot-*fils*. On Riḍwān Efendī, see Kennedy, 1956, no. X209 and *Cairo Survey*, no. D58.

50. MS Aleppo Awqāf 947 – see further *Cairo Survey*, no. D66.

51. On al-Qummī (Suter, 1900, no. 174), see Sezgin, VII, pp. 174–75. On the Cairo manuscripts, see *Cairo Catalogue*, I, *sub* ṬM 222 and MM 208, and II, Section 5.2.16; and *Cairo Survey*, no. B68.

52. On al-Būzajānī, see note 42. On the *Shāmil Zīj*, which merits detailed investigation, see Kennedy, 1956, no. 29. On the Paris manuscript see de Slane, no. 2528 on 451–52, and *Cairo Catalogue*, I, *sub* DM 902 (photos of the Paris manuscript), and II, Section 2.1.10.

53. On al-Fahhād and his various *zījes*, see Kennedy, 1956, nos. 23, 53, 54, 58, 62, 64, 84, and 99.

54. On the manuscript, see Karatay, no. 227. The catalogue has A2337 for the number, and the A3337 in my notes may be in error. Karatay gives the name as al-Jajarami and states that he worked for (?) the Sultan of Delhi, Quṭb al-Dīn Aybak al-Turkī, *ca*. 1210. The same author with the name given as Jājarmī is listed in Storey, 105, no. 14, together with two manuscripts of the *Faṣṣ al-khātam* preserved in Oxford and Hyderabad.

55. On this Cairo manuscript, see *Cairo Catalogue*, I, *sub* DM 1106, and II, Section 2.1.15.

56. On this manuscript, see *Cairo Catalogue*, I, *sub* MM 213. On Kūshyār, see the article by A.S. Saidan in *DSB*, and also Sezgin, II, 343–45, VI, 246–49, and VII, 182–83. His *zīj* is listed as Kennedy, 1956, nos. 7 and 9.

57. On al-Ṣāliḥī, see Suter, 1900, no. 454, *Cairo Survey*, no. C87, and King 1983b, no. 70. On the Cairo manuscript, see *Cairo Catalogue*, I, *sub* K8526.

58. On Sibṭ al-Māridīnī, see Suter, 1900, no. 445; *Cairo Survey*, no. C97; and King, 1983b, no. 63 and p. 551. On Ibn Ghānim, see Brockelmann, II, 404–05 and SII, 429, and *Cairo Survey*, no. C127.

LIST OF MANUSCRIPTS CONSULTED

Aleppo Aḥmadīya Awqāf 947
Cairo Egyptian National Library
  DM (= Dār al-Kutub *mīqāt*) 902, DM 1081, DM 1106, Dār al-Kutub K 8526
  MM (= Muṣṭafā Fāḍil *mīqāt*) 23, MM 208, MM 213
  ṬM (= Ṭal'at *mīqāt*) 92, ṬM 119, ṬM 188, ṬM 222, ṬM 242, ṬH (= Ṭal'at *hay'a*) 48, ṬFF (= Ṭal'at *falak fārisī*) 11
  TR (= Taymūr *riyāḍa*) 49, TR 99
Cambridge University Library Browne O.1
Dublin Chester Beatty 4076
Escorial ar. 909, 927, 966
Florence Medicea Laurenziana Or. 106
Hyderabad Andra Pradesh State Library 298
Istanbul Nurosmaniye 2933
Istanbul Topkapı Seray A3337
Jerusalem, manuscript in the private collection of Rabbi Kafah
London British Library Add. 27,261 (Rieu, II, 870a)
Madrid Museo Naval, unnumbered manuscript
Milan Ambrosiana C 82
Munich ar. 853
Oxford Bodleian Hunt. 663, Pococke 249
Paris Bibliothèque Nationale ar. 2513, 2515, 2520, 2523, 2528, 5698, 6913, supp. pers. 1488

Princeton Mach 5067 = Yahuda 2501
Utrecht Or. 23

BIBLIOGRAPHY

Aaboe, A.
1960 On the tables of planetary visibility in the Almagest and the Handy Tables. *Kgl. Danske Vidensk. Hist.-Fil. Meddelelser* 37,8: 3–20.

Arberry, A. J.
1955–66 *The Chester Beatty Library: A Handlist of the Arabic manuscripts,* 8 vols. Dublin: Emery Walker.

Brockelmann, C.
1937–49 *Geschichte der arabischen Litteratur.* 2 vols. 2nd ed. Leiden: E.J. Brill, and Supplementbände: 3 vols. Leiden: E.J. Brill, 1937–42.

Cairo Catalogue and *Survey* D. A. King, *A catalogue of the scientific manuscripts in the Egyptian National Library* (in Arabic), 2 vols. Cairo: General Egyptian Book Organization, 1981–1986, and *A survey of the scientific manuscripts in the Egyptian National Library,* Winona Lake, IN: Eisenbrauns, 1986.

de Slane, MacG.
1883–95 *Catalogue des manuscrits arabes,* Paris: Imprimerie Nationale.

DSB *Dictionary of Scientific Biography,* 14 vols. and 2 supp. vols. New York: Charles Scribner's Sons, 1970–1980.

EI$_2$ *Encyclopaedia of Islam,* 2nd ed., 5 vols. to date. Leiden: E.J. Brill, 1960–present.

Goldstein, B. R.
1967 *Ibn al-Muthannā's commentary on the astronomical tables of al-Khwārizmī.* New Haven: Yale University Press.

Griffini, E. Lista dei manuscritti arabi nuovo fondo della Biblioteca Ambrosiana di Milano. In *Rivista degli Studi Orientali,* VII.

Irani, R. A. K.
1955 Arabic numeral forms. *Centaurus* 4: 1–12; reprinted in Kennedy *et al.*, 710–721.

Karatay, F. E.
1961 *Topkapı Sarayı Müzesi Kütüphanesi: Farsça Yazmalar Kataloğu,* I. Istanbul: Topkapı Sarayı Müzesi.

Kennedy, E. S.
1956 A survey of Islamic astronomical tables. *Trans. Am. Phil. Soc.*, N.S. 46;2: 123–77.
1960 The crescent visibility theory of Thābit bin Qurra. *Proc. Math. Phys. Soc. UAR,* 24:71–4; reprinted in Kennedy *et al.*, 140–43.
1967 The digital computer and the history of the exact sciences. *Centaurus* 12:107–13; reprinted in Kennedy *et al.*, 385–91.
1968 The lunar visibility theory of Yaʿqūb ibn Ṭāriq. *J. Near East. Stud.*, 27:126–32; reprinted in Kennedy *et al.*, 157–63.

Kennedy, E. S. and M. Janjanian
1965 The crescent visibility table in al-Khwārizmī's *Zīj*. *Centaurus* 11:73–8; reprinted in Kennedy *et al.*, 151–56.

Kennedy, E. S. and
D. A. King
1982 Indian astronomy in fourteenth-century Fez: The versified *Zīj* of al-Qusunṭīnī. *J. Hist. Arabic Sci.* 6:3–45, reprinted in King, 1986.

Kennedy, E. S. and
H. Krikorian-Preisler
1972 The astrological doctrine of projecting the rays. *al-Abhath* 25: 3–15; reprinted in Kennedy *et al.*, 372–84.

Kennedy, E. S. *et al.*
1983 *Studies in the Islamic exact sciences*, Beirut: American University of Beirut.

King, D. A.
1975 On the astronomical tables of the Islamic Middle Ages. *Studia Copernicana* 13:37–56, reprinted in *idem*, 1986.
1980 New light on the *Zīj al-Ṣafā'iḥ* of Abū Ja'far al-Khāzin. *Centaurus* 23:105–17, reprinted in *idem*, 1987.
1983a *Mathematical astronomy in medieval Yemen: A biobibliographical survey*, Malibu, Ca.: Undena Publications.
1983b The astronomy of the Mamluks. *Isis* 74: 531–55, reprinted in *idem*, 1986.
1983c Al-Khwārizmī and new trends in mathematical astronomy in the ninth century. *Occasional papers on the Near East* 2. New York: New York University, Hagop Kevorkian Center for Near Eastern Studies.
1986 *Islamic Mathematical Astronomy*. London: Variorum Reprints.
1987 *Islamic Astronomical Instruments*. London: Variorum Reprints.

Krause, M.
1936 Stambuler Handschriften islamischer Mathematiker. *Quellen und Studien zur Geschichte der Mathematik, Astronomie und Physik* B3;4:437–532.

Millás Vallicrosa, J.
1943–50 *Estudios sobre Azarquiel*. Madrid-Granada: Consejo Superior de Investigaciones Científicas, Instituto Miguel Asín.

Morelon, R.
1981 Fragment arabe du premier livre du Phaseis de Ptolémée. *J. Hist. Arabic Sci.* 5:3–21.

Nallino, C. A.
1899–1907 *Al-Battani sive Albatenii opus astronomicum*, (*Pubblicazioni del Reale Osservatorio di Brera in Milano*, XL), 3 vols., Milan and Rome; (reprinted Frankfurt: Minerva G.m.b.H., 1969).

Neugebauer, O.
1956 Astronomical commentary in *The Code of Maimonides, Book 3, Treatise 8, Sanctification of the new moon*. By S. Gandz, J. Obermann, and O. Neugebauer. New Haven: Yale University Press.
1962 The astronomical tables of al-Khwārizmī. *Kgl. Danske Vidensk. hist.-fil. Skrifter* 4:2.
1975 *A history of ancient mathematical astronomy*. 3 Pts. Berlin-Heidelberg-New York: Springer Verlag.

Pingree, D.
1968 The fragments of the works of Ya'qūb ibn Ṭāriq. *J. Near East. Stud.* 26:97–125.

Renaud, H. J. P.
1932 Additions et corrections à Suter *Die Mathematiker und Astronomen der Araber*. *Isis* 18:166–83.

1941 *Les manuscrits arabes de l'Escorial*, Tôme II, Fasc. 3: *Sciences exactes et sciences occultes*. Paris: Paul Geuthner.

Rosenthal, F.
1967 Translation and commentary, *Ibn Khaldūn: The Muqaddimah*, 3 vols. 2nd ed. Princeton: Princeton University Library.

Samsó Moya, J.
1977 A homocentric solar model by Abū Ja'far al-Khāzin. *J. Hist. Arabic Sci.* 1;2:268–75.

Sédillot, L. A. (*fils*)
1847–53 *Prolegomènes des tables astronomiques d'Oulog-Beg*, 2 vols. Paris: Firmin Didot Frères.

Sezgin, F.
1974, 1978, 1979 *Geschichte des arabischen Schrifftums*, Band V: *Mathematik*; Band VI: *Astronomie*; Band VII: *Astrologie u.s.w.* Leiden: E.J. Brill.

Stahlman, W. D.
1959 The astronomical tables of Codex Vaticanus Graecus 1291. Unpublished doctoral dissertation, Brown University.

Storey, C. A.
1958 *Persian Literature: A Bio-Bibliographical Survey*, Vol. II, Part 1: *Mathematics, Weights and Measures, Astronomy and Astrology, Geography*. London: Luzac and Co.

Suter, H.
1900 Die Mathematiker und Astronomen der Araber und ihre Werke. *Abhandlungen zur Geschichte der mathematischen Wissenschaften*, 10.
1914 Die astronomische Tafeln des Muḥammed ibn Mūsā al-Khwārizmī in der Bearbeitung des Maslama ibn Aḥmed al-Madjrīṭī und der Latein. Uebersetzung des Adelard von Bath. *Kgl. Danske Vidensk. Hist.-fil. Skr.* 3:1.

Toomer, G. J.
1964 Review of Neugebauer, 1962 in *Centaurus*, 10:203–12.
1984 Translation and annotation of *Ptolemy's Almagest*. London: Duckworth.

Vernet Ginés, J.
1950 Una versión árabe resumida del Almanach perpetuum de Zacuto. *Seferad* 10:115–33.
1951 *Contribución al estudio de la labor astronómica de Ibn al-Bannā'*. Tetuán: Editora Marroquí.
1956a Las tabulae probatae. In *Homenaje a Millás-Vallicrosa*, II: 501–22. Barcelona: Consejo Superior de Investigaciones Científicas.
1956b Los manuscritos astronómicos de Ibn Al-Bannā. *Actes du VIIIe Congrès International d'Histoire des Sciences*. 297–98. Florence.

Villuendas, M. V.
1979 *La trigonometría europea en el siglo XI: estudio de la obra de Ibn Mu'āḏ El Kitāb maŷhūlāt*, Barcelona: Instituto de Historia de la Ciencia.

Zimmermann, F. W.
1976 The Dustūr al-Munajjimīn of MS Paris, BN ar. no. 5968. *Proceedings of the First International Symposium for the History of Arabic Science*, II: 184–92. Aleppo.

III

IBN YŪNUS ON LUNAR CRESCENT VISIBILITY

1. *Introduction*

The determination of the criteria that would ensure visibility of the lunar crescent shortly after sunset on the evening following a conjunction of the sun and moon is a complicated problem involving several factors, such as the positions of the sun and moon relative to the horizon, the relative brightness of the moon and the background sky, and the prevailing atmospheric conditions. The problem was of particular concern to the scholars of medieval Islam because the Muslim calendar is lunar and the beginning and end of each month are determined by the sighting of the crescent. The attempts by Muslim scholars to formulate conditions for visibility are of considerable interest to the history of science, but thus far only the theories of a few early astronomers have been investigated.[1] There remain numerous writings on the subject by Muslim astronomers of which there is as yet no account in the modern literature.[2] Of these, one of the most interesting is due to the celebrated Egyptian astronomer Ibn Yūnus, and it forms the subject of this paper.

Ibn Yūnus was one of the most distinguished astronomers of medieval Islam. He lived and worked in Fustat, outside the new Fatimid city of Cairo, at the end of the tenth century, and his major compilation was a *zīj* or astronomical handbook with tables.[3] This great work, called *al-Zīj al-Ḥākimī* because it was dedicated to the Fatimid Caliph al-Ḥākim, has not survived in its entirety, and only forty-four out of the original eighty-one chapters are available to us, in two unpublished manuscripts preserved in Leiden and Oxford. However, the *Ḥākimī Zīj* was very influential in later Egyptian and also Yemeni astronomy, and some of the lost material from the *Ḥākimī Zīj* is to be recovered in later texts from these regions.[4] In this study I shall present Ibn Yūnus's theory of lunar crescent visibility,[5] not contained in the two known manuscripts of the *Ḥākimī Zīj* but recently rediscovered in various manuscripts of Egyptian and Yemeni provenance now preserved in Paris, Cairo, and London.

The first of four main sources is an anonymous recension of the thirteenth-century Egyptian *Muṣṭalaḥ Zīj*,[6] extant in MS Paris Bibliothèque Nationale ar. 2513 (copied c. 1300).[7] The *Muṣṭalaḥ Zīj*, which seems to have been the most popular work of its kind in Egypt from the thirteenth to the fifteenth century, was based partly on the *Ḥākimī Zīj*, and Ibn Yūnus's visibility theory is found in fol. 61v of this particular manuscript. The second source is MS Cairo Dār al-Kutub *mīqāt* 1106 (copied c. 1750), which contains Chapter 2 of the introduction to the *Muṣṭalaḥ Zīj* and some additional notes:[8] Ibn Yūnus's crescent visibility theory is found amidst these notes on fols. 20v–21v. A garbled, abridged version occurs in MS Cairo Muṣṭafā Fāḍil *mīqāt* 93 (fols. 2v–3v, copied c. 1700), amidst some anonymous notes on astronomy.[9] In each of these

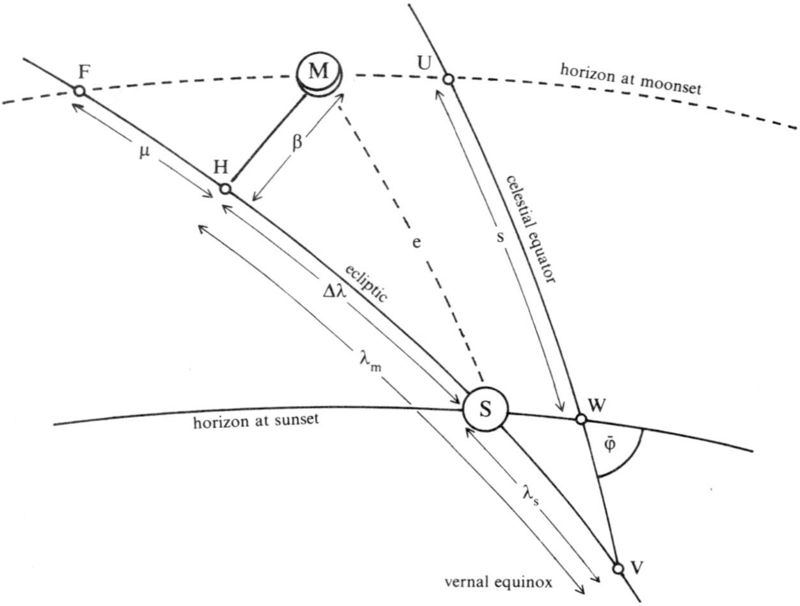

FIG. 1.

manuscripts the same crescent visibility theory is attributed to Ibn Yūnus; in the third it is stated that the theory is from his *Zīj*.

In the fourth source, the *Mukhtār Zīj* of the Yemeni astronomer Abu l-'Uqūl,[10] a modified version of the same visibility theory is introduced without mention of Ibn Yūnus. Abu l-'Uqūl worked for the Rasulid Sultan al-Mu'ayyad in Taiz about the year 1300. His *Zīj* survives in the unique MS London British Library 768 (Or. 3624) (copied 1599, probably in the Yemen)[11] and is based mainly on a *zīj* of Ibn Yūnus, apparently not the *Ḥākimī Zīj* but an earlier one. The lunar crescent visibility presented in Chapter 41 of the *Mukhtār Zīj* is contained on fols. 65r–65v of the London manuscript. The Paris-Cairo and London versions are hereafter referred to as P–C and L.

In what follows I introduce the problem as treated by Ibn Yūnus (Section 2) and present the text and translation of both versions of Ibn Yūnus's theory (Section 3) with a commentary (Section 4). In a sequel to this paper I propose to discuss the different crescent visibility theories in another recension of the *Muṣṭalaḥ Zīj*, in some anonymous commentaries purporting to relate to *zīj*es of Ibn Yūnus, and in an anonymous Egyptian *zīj* based on his.

2. *The Astronomical Problem*

Figure 1 shows the western horizon of a locality with latitude φ at sunset on the evening when visibility is in question. I denote the solar and lunar longitudes by λ_s and λ_m and the lunar latitude by β. Then with the notation in the figure:

$$SH = VH - VS = \lambda_s - \lambda_m = \Delta\lambda \text{ and } HM = \beta.$$

The problem of determining crescent visibility as treated by Ibn Yūnus involves first finding three quantities:
(1) The apparent distance between the sun and moon, $SM = e$. This determines the width of the crescent, and Ibn Yūnus uses the quantity $L = e/15$ as a measure of the amount of light on the crescent, calling it (in P only) by the standard medieval Arabic expression *qaws al-nūr*, "arc of light".
(2) The difference between the times of sunset and moonset, measured on the celestial equator by the arc $UW = s$. This influences the brightness of the sky against which the crescent may or may not be seen. Ibn Yūnus used the standard expression *qaws al-makth*, literally "arc of tarrying", and I shall simply use "difference in setting times".
(3) The lunar velocity v_m, which varies between just less than 12°/d (degrees per day) and slightly more than $14\frac{1}{2}$°/d. This is related to the distance of the moon from the earth, and hence to the brightness of the moon.

When the three quantities L, s, and v_m have been determined, visibility or non-visibility can then, according to Ibn Yūnus, be predicted by investigating whether they satisfy a given set of conditions.

3. *Ibn Yūnus's Visibility Theory*

The Arabic texts of Ibn Yūnus's visibility theory in P–C and L are presented in the Appendix. I here present free translations of the two versions. Words or phrases in the translation which are given in parentheses have no counterpart in the original and are introduced either to clarify or to question the meaning of a passage. The numbers of the conditions in the translations have no counterpart in the original texts.

TRANSLATION OF THE TEXT IN P–C

A table for (determining) lunar crescent visibility according to the opinion of the Shaykh Abu l-Ḥasan 'Alī ibn Yūnus.

Argument	(Arc of) Light
1	; 4
2	; 8
3	;12
–	–
–	–
15	1; 0
–	–
–	–
60	4; 0

You (first) compute the longitudes of the sun and moon at sunset. Then you find the distance between them in degrees [*daraj al-sawā'*], which you multiply by itself and keep in mind. Next you multiply the latitude of the moon by itself, add it to the square of the distance, and take the root of the sum: the result will be the arc of light. Reckon each 15° of this arc as one digit, and enter any remaining (degrees and) minutes in this table to find the corresponding fraction of a digit. Next find the descensions of the sun and moon and subtract the lesser from the greater: the remainder will be the 'arc of separation' [hereafter: 'difference in setting times'], which you should keep in mind. Now consider (the following cases):

(1) If the difference in setting times is 12° and there is one digit of light and the lunar velocity is greater than 13°/d, the crescent will be seen. If the difference in setting times is 12° and there is less than one digit of light and the lunar velocity is less than 12°/d, the crescent will not be seen.
(2) If there is 11° between the two (setting times) and more than one digit of light and the lunar velocity is 14°/d, the crescent will be seen. If it [i.e., the difference in setting times] is less than that, the crescent will not be seen.
(3) If there is 13° between the two (setting times) and less than one digit of light and the lunar velocity is greater than 14°/d, then the crescent will be seen. If (the difference in setting times) is less than that, the crescent will not be seen.
(4) If there is less than one digit of light and the lunar velocity is less than 13°/d and the difference in setting times is greater than 13°, the crescent will be seen. If (the difference in setting times) is less than this, the crescent will not be seen. If there is only two-thirds of a digit of light, the crescent will not be seen.
(5) If there is more than two-thirds of a digit of light and the lunar velocity is greater than 14°/d and the difference in setting times is 14°, the crescent will be seen. If (the difference in setting times) is less than this, the crescent will not be seen.
(6) If the lunar velocity is 12°/d and there is two-thirds of a digit of light and the difference in setting times is 15°, then the crescent will be seen. If (the difference in setting times) is less than this, the crescent will not be seen.

These are the boundary conditions when the air is pure, the atmosphere is clear, and one's sight is sharp. Understand this so that you may be rightly guided.

TRANSLATION OF THE TEXT IN L

Chapter 41 on lunar crescent visibility. If we want (to determine) the visibility of the lunar crescent, we first compute the longitudes of the sun and moon at midday of the 29th day of the Muslim month, and find the velocity of each, and also the hours of half daylight for that day. Then we multiply the hours of half daylight by the velocities of each of the sun and moon and divide each product by 24. We add the quotient for the sun to the longitude of the sun at midday and we add the quotient for the moon to the longitude of the moon at midday: this gives the positions of the sun and moon at sunset. Next we find the distance between the sun and moon in degrees [*daraj al-sawā'*], multiply it by itself, and keep (the result) in mind. We also work out the latitude of the moon and its direction, and keep this in mind too. Then we multiply the lunar latitude by itself and add the product to the product of the distance with itself, take the root of the sum, and divide the root we obtain by 15: the quotient will be a measure of the amount of light on the lunar crescent, expressed in digits. Distances less than 15° can be multiplied by 4 to yield minutes of a digit. We keep the result in mind: it is the amount of the light on (the lunar crescent). Next we take the sine of the lunar latitude and multiply it by the tangent of the local latitude, and divide the product by the cosine of the [lunar (text: local)] latitude. Then we take the arc sine of the quotient: this arc will be the modification to the longitude of the crescent when setting. If the lunar latitude is northerly we add this modification to the longitude of the moon, and if the lunar latitude is southerly, we subtract it; the result in either case will be the longitude of the moon adjusted for (determining) visibility.

We now find the oblique ascension of the point of the ecliptic directly opposite the longitude of the moon adjusted for computing visibility, and keep it in mind. We also find the oblique ascension of the point of the ecliptic opposite the longitude of the sun, and we subtract it from the result for the moon. The remainder will be the difference in setting times, and we keep it in mind. (Now) we consider (the following cases):

(1) If the difference in setting times is 12° and there is one digit of light and the moon is moving quickly, the crescent will be seen. If the difference in setting times is 12° and there is less than one digit of light and the moon is moving slowly, the crescent will not be seen.
(2) If the difference in setting times is 11° and there is more than one digit of light and the moon is moving at its maximum velocity, the crescent will be seen. If this (first?) condition is not satisfied, the crescent will not be seen.
(3) If the difference in setting times is 13° and there is less than one digit of light and the moon is moving quickly, the crescent will be seen. If this (first?) condition is not met, the crescent will not be seen.
(4) If there is less than one digit of light and the moon is moving slowly and the difference in setting times is greater than 13°, the crescent will be seen. If it [i.e., the difference in setting times] is less than this, the crescent will not be seen. If there is (only) two-thirds of a digit of light, visibility cannot be expected (even though the other two conditions may be satisfied).
(5) If there is more than two-thirds of a digit of light and the lunar velocity is high and the difference in setting times is 14°, the crescent will be seen. If (the difference in setting times) is less than this, the crescent will not be seen.
(6) If there is two-thirds of a digit of light and the moon is moving slowly and the difference in setting times is 15°, the crescent will be seen. If (the difference in setting times) is less than this, the crescent will not be seen.

To sum up, a high lunar velocity and a large amount of light favour visibility even if the difference in setting times is small. A low lunar velocity and a small amount of light prevent visibility even if the difference in setting times is large, unless it reaches 15°; in this case visibility will occur, and the smallness of the amount of light and the slowness of the moon's motion will not detract from (the possibility of) visibility. Mistakes always occur in predicting lunar crescent visibility because people disregard these factors. All of the conditions which we have mentioned are valid only when the air is pure and clear, because unsettled atmospheric conditions impede visibility. Success is with God.

4. *Commentary*

(a) Determination of λ_s and λ_m

The planetary tables in *zījes* usually give an epoch position, the mean motions in multiple years and single years since epoch, as well as in months, days and hours, and it is simply a question of adding these in an appropriate way to determine the mean longitudes of the sun and moon at midday (or sunset) for a given day. These must then be adjusted by the appropriate equations, also tabulated in *zījes*. Ephemerides usually give the actual positions of the sun, moon and five naked-eye planets at midday on each day of a given year.[12] The longitudes of the sun and moon at sunset λ can be found (as in L) from their longitudes at midday λ_0 using:

(1) $$\lambda = \lambda_0 + v \times d^h / 24.$$

where v is the velocity of the sun (v_s) or moon (v_m) in degrees per day and d^h is the length of half-daylight in hours for the day in question. The velocities can either be computed directly or read from tables.[13] The quantity d is a function of the local latitude φ and the solar longitude λ_s (and also of the obliquity of the ecliptic ε), and tables of $d(\lambda_s)$ in equatorial degrees or in hours and minutes for particular latitudes with values for each degree of solar longitude were standard in Islamic *zījes*.[14] Ibn Yūnus tabulated $d^h(\lambda_s)$ for latitude 30;0°, that is, Cairo, and Abu l-ʻUqūl (MS L.166v–184r) for the latitudes of four localities in the Yemen. The lunar latitude β can likewise be determined from tables in *zījes*.[15]

(b) Determination of e and L

The distance $SM = e$ between the sun and moon is derived as if $\triangle\ SMH$ were plane. In this case, by the Theorem of Pythagoras, we have:

(2) $$e^2 = SM^2 = SB^2 + BM^2 = \Delta\lambda^2 + \beta^2,$$

whereas, in fact, from the spherical $\triangle\ SMH$ we should have:

(3) $$\cos e = \cos \Delta\lambda \cos \beta.$$

However, the above approximation is quite adequate for practical purposes. The so-called ‘arc of light’, L, is then found using $L = e/15$, either by calculation or by using a simple table (only in P).[16]

(c) Determination of s

The texts now describe how to determine the differences in setting times of the sun and moon, a more complicated problem of spherical trigonometry. In P–C it is simply stated that one should take the difference between the descensions of the sun and moon, but in L there are more detailed instructions explaining precisely what this involves.

The text in L begins by introducing an arc called *ikhtilāf mughīb al-hilāl*, literally “difference in setting of the crescent” and meaning the modification to the lunar longitude needed to give the point of the ecliptic that sets with the moon, $HF = \mu$. The rule for finding μ is equivalent to the modern formula:

(4) $$\sin \mu = \sin \beta \tan \varphi / \cos \beta.$$

(In medieval Islamic astronomical works the tangent and cotangent functions (Arabic *ẓill*, "shadow") are generally used only for arguments that can be considered as altitudes:[17] note that $\bar{\varphi}$ is the altitude of the celestial pole.) Ibn Yūnus himself would have used sines to base 60 rather than base 1,[18] so if the rule stems from him it has been simplified; this confirms my impression that the text in P–C is closer to the original than that in L. The formula for μ can be derived from the spherical triangle *MFH* by assuming that $\angle$ *MFH* equals the angle between the celestial equator and the horizon, namely $\bar{\varphi}$. In fact, $\angle$ *MFH* varies between $\bar{\varphi} - \varepsilon$ and $\bar{\varphi} + \varepsilon$ (where ε is the obliquity of the ecliptic), its actual value depending on the longitude of *F*. In view of the approximation used for *e* it would not have been unreasonable to treat $\triangle$ *MNH* as plane and use:

(5) $$\mu = \beta \tan \varphi.$$

The modified lunar longitude $\lambda'_m = \lambda_m \pm \mu\ (\beta \gtrless 0)$ is then used to determine the 'arc of separation', that is, the time between sunset and moonset, measured by the arc $UW = s$, thus:

(6) $$s = UW - VW = \sigma(\lambda'_m) - \sigma(\lambda_s),$$

where the function $\sigma(\lambda)$ is the oblique descensions corresponding to ecliptic longitude λ (as in L). The quantity s may be alternatively defined (as in P) in terms of the oblique ascensions ρ where $\rho(\lambda) = \sigma(\lambda + 180°) + 180°$, using:

(7) $$s = \rho(\lambda'_m + 180°) - \rho(\lambda_s + 180°).$$

Tables of $\rho(\lambda)$ for different latitudes were standard in Islamic *zījes*.[19] In the *Ḥākimī Zīj* Ibn Yūnus tabulated the oblique ascensions for each degree of terrestrial latitude from 1° to 48°; Abu l-'Uqūl tabulated them for each of four latitudes serving localities in the Yemen (MS L.166v–184r); in both cases values are given for each degree of ecliptic longitude.[20]

(d) Conditions on s, L and v_m

We are now in a position to investigate Ibn Yūnus's conditions for visibility or non-visibility, represented in tabular form in Tables 1 and 2. Arabic expressions used in the second half of each of six conditions in both versions are linguistically ambiguous, and I am not thoroughly convinced that my rendering is what was originally intended. I use the notation $x+$ and $x-$ to represent Ibn Yūnus's imprecise expressions "greater than" and "less than" a given number x. In the second Cairo manuscript mentioned in Section 1, only the conditions for positive predictions are given. Clearly the two sets in Tables 1 and 2 are virtually equivalent since v_m varies between about 12 and about $14\frac{1}{2}$°/d, but the correspondence between the velocities in the two sources is not completely satisfactory: notice, for example, condition 1a, in which 13+°/d in P–C is probably to be interpreted as "mean velocity", whereas L has "fast" (see further below). It seems probable that Table 1 represents the theory originally formulated by Ibn Yūnus.

At first sight it seems that we have here a highly sophisticated set of boundary conditions for visibility; a three-dimensional graphical representation resembles

TABLE 1. Visibility conditions in Paris-Cairo version.

Condition	s	L	v_m	Visibility
1a	12°	1^{dgt}	$13+^{\circ/d}$	yes
b	12	1−	12−	no
2a	11	1+	14	yes
b	11−	1+	14	no
c	13+(?)	$\frac{2}{3}$	13−	no
3a	13	1−	14+	yes
b	13−	1−	14+	no
4a	13+	1−	13−	yes
b	13−(?)	1−	13−	no
5a	14	$\frac{2}{3}$+	14+	yes
b	14−	$\frac{2}{3}$+	14+	no
6a	15	$\frac{2}{3}$	12	yes
b	15−	$\frac{2}{3}$	12	no

TABLE 2. Visibility conditions in London version.

Condition	s	L	v_m	Visibility
1a	12°	1^{dgt}	fast	yes
b	12	1−	slow	no
2a	11	1+	maximum	yes
b	11−	1+	maximum	no
3a	13	1−	fast	yes
b	13−	1−	fast	no
4a	13+	1−	slow	yes
b	13−(?)	1−	slow	no
c	13+(?)	$\frac{2}{3}$	slow	no
5a	14	$\frac{2}{3}$+	fast	yes
b	14−	$\frac{2}{3}$+	fast	no
6a	15	$\frac{2}{3}$	slow	yes
b	15−	$\frac{2}{3}$	slow	no
7	15	small	slow	yes

the outline of an urban complex in Jersey City or in the Haḍramawt. However, this impression is at least partially dispelled upon closer inspection. One problem seems to be that given the limits of accuracy imposed on the quantity L, the lunar velocity v_m seems hardly relevant. I can present in this paper none of the insights that might be offered by a modern astronomer on the efficacy of Ibn Yūnus's conditions, and so I leave any further comment on this aspect to those more qualified than myself.[21] Some historical remarks are, however, perhaps more pertinent.

First we recall that already in the eighth century Ya'qūb ibn Ṭāriq had proposed two joint conditions on s and L for visibility,[22] namely:

$$s \geqslant 12^{\circ} \text{ and } L \geqslant 0;45^{dgt} \text{ or } s \geqslant 10^{\circ} \text{ and } L \geqslant 1^{dgt}.$$

Note that the limits for L are more crudely expressed than those on s. Also, another of Ibn Yūnus's predecessors, namely al-Battānī,[23] also considered the effect of the lunar distance from earth – which in turn influences the apparent velocity of the moon. al-Battānī, it should be stressed at the outset, started from the simple (Indian) premise that for visibility:

$$s \geqslant 12^{\circ}.$$

Since this lower limit is roughly equal to the mean daily elongation of the sun and moon, 12;11°, he used this figure instead (with rather dubious justification).

Also, because the condition should be valid at moonset, he then modified the 12;11° to account for the change of elongation in 4/5 of an hour, thus reducing the limit to 11;45°. This limit he then assumed to be valid for the moon at mean distance, devising a scheme to modify it by 1/12 downwards for the moon at apogee and the same upwards for perigee, since, he maintained, this was the change in the apparent lunar diameter. Another method which he advocated is less obviously related to the present discussion.

It seems that Ibn Yūnus started from the premise that visibility is possible at mean distance/velocity for a pair of values s and L, namely,

$$s \geqslant 12^\circ \text{ and } L = 1^{\text{dgt}} \rightarrow e = 15^\circ \text{ (so that } s + e = 27^\circ);$$

note that this is condition 1a in P–C (in L the velocity is falsely labelled "fast"). I suspect that he applied some simple arithmetical procedure to derive the other conditions. The actual procedure cannot be determined because of the crudeness of his conditions on L. One possibility would be that he argued that the corresponding pair of limits at apogee would have sum 26° and at perigee 28°. In any case, Ibn Yūnus was naively over-confident in supposing that the slightest deficit in s or L would prevent visibility.

In passing we mention that various other later astronomers proposed conditions on the sum of a pair of s, $\Delta\lambda$, or e. For example, Maimonides,[24] working in Cairo in the late-twelfth century, proposed the condition:

(10) $$s + \Delta\lambda \geqslant 22^\circ.$$

(It is possible that this condition was inspired by ancient Babylonian practice.) Also Shams al-Dīn al-Khalīlī,[25] working in Damascus in the late-fourteenth century, proposed the condition:

(11) $$\tfrac{1}{2}(e + s') \geqslant 7^\circ, \text{ where } s' = \sigma(\lambda'_{m}) - \sigma(\lambda_{s}).$$

But neither of these scholars considered the effect of lunar distance/velocity.

Thus al-Battānī fabricated a highly complicated theory based on the simple condition $s \geqslant 12^\circ$ at mean distance, and Ibn Yūnus probably did likewise starting from $s \geqslant 12^\circ$ and $L \geqslant 1^{\text{dgt}}$ for the same situation. Would that either had left us a description of a series of observations to verify these conditions, or indeed, any condition! It is a moot point whether either actually conducted any systematic observations for this purpose.

APPENDIX: THE ARABIC TEXTS

Note: I have left uncorrected the numerous grammatical errors in both versions of the text not only because they do not affect the meaning but also to draw attention to the kind of 'Middle Arabic' that was commonly used in late medieval scientific texts; I have, however, provided all the necessary diacritical points where these were lacking. The double points on final *yā*"s and *alif maqṣūra*s are as found in the manuscripts. Only *hamza*s found in the original texts are reproduced: their use in each of the three sources is inconsistent.

پ = مخطوطة پاريس ؛ ق = مخطوطة القاهرة ؛ ل = مخطوطة لندن

النص في پ \ ق :

العدد	النور	جدول روية الهلال علي راي الشيخ ابي *١* الحسن علي بن
ا	د	يونس تقوم النيرين لوقت مغيب الشمس ثم تاخذ البعد
ب	ح	بين النيرين بدرج السوا *٢* فتضربه في مثله وتحفظه
ج	يب	وتضرب عرض القمر في مثله وتزيده على مضروب البعد في
.	.	مثله وتاخذ جذره فما حصل من الجذر فهو قوس النور اجعل
يه	ا	كل يه درجة منه اصبع (!) فان بقي دقايق ادخل بها الى
.	.	هذا الجدول وخذ بها كسور اصابع ثم تاخذ مغارب النيرين
س	د	وتنقص الاقل من الاكثر (!) فما بقي فهو قوس المكث

فاحفظه ثم انظر

* فان كان قوس المكث يب درجة وفيه من النور اصبع وبهته اكثر من يج درجة فانه يري وان كان قوس المكث يب ونوره اقل من اصبع وبهته اقل من يب لم يري (!)

* وان كان بينهما يا وفيه من النور اكثر من اصبع وبهته يد فانه يري وان نقص من ذلك شي لم يري

* وان كان بينهما يج درجة وفيه من النور اقل من اصبع وبهته اكثر من يد فانه يري وان كان اقل من ذلك لم يري

* وان كان فيه من النور اقل من اصبع وبهته اقل من يج وكان قوس المكث اكثر من يج فانه يري وان كان اقل من ذلك لم يري

* وان كان ثلثا اصبع لم يري وان زاد علي ثلثي اصبع وكان بهته اكثر من يد وكان قوس المكث يد فانه يري وان كان اقل من ذلك لم يري

* وان كان بهته يب وفيه من النور ثلثا اصبع وقوس المكث يه فانه يري وان نقص من ذلك لم يري

وهذه الحدود مع نقا الجو وصفا الهوي وحدة البصر فافهم ذلك ترشد *٣*

التحقيق : *١* في پ : ابو *٢* في پ : السوي *٣* في ق : وعليه يقع الحكم ان شا الله تعالي

النص في ل :

الباب الحادي والاربعون في روية الاهلة اذا اردنا روية الهلال قومنا النيرين لنصف النهار من اليوم التاسع والعشرين من الشهر العربى ونعرف بهت الشمس وبهت القمر وساعات نصف النهار لذلك اليوم ثم نضرب ساعات نصف النهار فى بهت كل واحد من النيرين ونقسم ما اجتمع من كل واحد منهما على ٢٤ فما خرج من بهت الشمس زدناه على موضع الشمس نصف النهار وما خرج من بهت القمر زدناه على موضع القمر نصف النهار فيبلغ ذلك موضع النيرين عند مغيب الشمس ثم ننظر البعد بين النيرين بدرج السوا فنضربه فى مثله ونحفظه ثم نعمل عرض القمر ونحفظه ونعرف جهته ثم نضرب عرض القمر فى مثله ونزيد ما بلغ منه على مضروب البعد فى نفسه فما بلغ منه اخذنا جذره فما حصل من الجذر قسمناه على يه فيخرج منه مقدار ما فى الهلال من النور وهو اصابع وما لم يتم (!) يه ضربناه فى ٤ فيكون دقايق من اصبع فنحفظه وهو ما فيه من النور ثم نجعل عرض القمر جيبا ونضربه في ظل عرض البلد معكوسا اعنى الظل المعكوس ونقسم ما بلغ على جيب تمام عرض القمر *١* فما خرج قوسناه فى الجيب فيكون قوسه اختلاف مغيب الهلال فان كان عرض القمر شماليا زدنا قوس اختلاف مغيب الهلال على موضع القمر وان كان عرض القمر جنوبيا نقصنا قوس اختلاف مغيب الهلال من موضع القمر فيبقى موضع القمر للروية فناخذ مطالع نظير درجة القمر للروية فى بلدنا فنحفظه وناخذ مطالع نظير درجة الشمس في البلد ايضا وننقص مطالع نظير درجة الشمس من مطالع نظير درجة القمر فما بقى فهو قوس المكث فنحفظه وننظر

* فان كان قوس المكث يب درجة نظرنا فان كان فيه من النور اصبع والقمر سريع السير فان الهلال يرى وان كان قوس المكث يب وفيه من النور اقل من اصبع وهو بطى السير لم يرى (!) الهلال

* وان كان قوس المكث يا وفيه من النور اكثر من اصبع وسير القمر مسيره الاكبر فان الهلال يرى وان خالف ذلك لم ير

* وان كان قوس المكث يج درجة وفيه من النور اقل من اصبع *٢* وهو سريع السير فانه يرى وان خالف ذلك لم ير

* وان كان فيه من النور اقل من اصبع *٢* وهو بطى السير وكان قوس المكث اكثر من يج فانه يرى وان كان اقل من ذلك لم ير واذا كان فى

الهلال من النور ثلثا اصبع لم يطمع فى رويته
* فان زاد على ثلثى اصبع وكان بهته سريعا وكان قوس المكث يد درجة فانه يرى وان كان اقل من ذلك لم ير
* وان كان فيه من النور ثلثا اصبع وهو بطى السير وكان قوس المكث يه درجة فانه يرى وان نقص من ذلك لم ير
وبالجملة فان سرعة سيره وزيادة نوره يعين على رويته وان قلت قوس المكث وان ابطا سيره وقلة نوره تمنع من رويته وان كانت قوس المكث كثيرة الا ان يبلغ قوس المكث يه درجة فيطلق عليه الروية ولا يعتد بقلة نوره ولا ببطء سيره ومن اجل اغفال ذلك يقع الخطأ دايما فى روية الهلال وجميع ما ذكرناه انما يكون مع صفاء الجو ونقايه وان اختلفت احوال الهوآء (!) كان ذلك عايقا عن الروية وبالله التوفيق

التحقيق : *١* في ل : البلد *٢-٢* في الهامش في ل

Acknowledgements

Microfilms of the main manuscripts used in this study were kindly provided by the Egyptian National Library in Cairo, the British Library in London, and the Bibliothèque Nationale in Paris. My research on lunar crescent visibility conducted in 1985 was facilitated by a grant from the U.S. National Science Foundation; this support is gratefully acknowledged. It is also a pleasure to thank my colleagues Professor Frans Bruin, American University of Beirut, and Dr Yasakatsu Maeyama, Frankfurt University, for their assistance.

REFERENCES

1. Namely, Ya'qūb ibn Ṭāriq (*fl.* Baghdad, c. 760) in Kennedy *et al.*, 157–63, now supplemented by Hogendijk 1; Thābit ibn Qurra (*fl.* Baghdad, c. 900) in Kennedy *et al.*, 140–3, and Morelon, pp. xciii–cxviii, 93–117, and 230–59; al-Battānī (*fl.* Raqqa, c. 910) in Nallino, i, 85–92 (Latin trans.) and 265–72 (comm.), and iii, 129–38 (Arabic text), and a new, English translation and commentary in Bruin; Maimonides (*fl.* Cordova then Cairo, c. 1175) in Neugebauer 1, 349–60, and 2; and various early tables for determining lunar crescent visibility examined in my contribution to the *Kennedy Festschrift*, now supplemented by Hogendijk 2.
2. I am currently preparing for publication surveys of lunar crescent visibility in later Egyptian / Syrian and Yemeni sources. An investigation of the legal texts on the subject is a formidable task for the future.
3. On Ibn Yūnus and his works see my article in *DSB*, xiv, 574–80, to which add the sources listed in *Cairo survey*, no. B59. The listings in Sezgin, v, 342–3, vi, 228–31, and vii, 173, confuse the *Ḥākimī Zīj* with the Cairo corpus of tables for timekeeping and include a spurious

astrological text. On Ibn Yūnus's milieu see the articles "Fāṭimids", "Fusṭāṭ", "al-Ḥākim", and "al-Ḳāhira" in EI_2.

A detailed analysis of the spherical astronomy in the *Ḥākimī Zīj* is contained inKing 1. In addition I have prepared a new translation of the observation accounts first published in Caussin as well as an analysis of the planetary astronomy in this work.

4. On astronomy in medieval Egypt and Yemen see King 3/III, and 2 (also 3/IV), respectively.
5. The titles of the 81 chapters of the *Ḥākimī Zīj* are listed in the introduction published in Caussin, 67–79. There was no chapter devoted entirely to lunar cresent visibility, but this topic was probably discussed in Chap. 76 entitled *Fī ẓuhūr al-kawākib wa-khtifā'ihā*, "On the first and last visibility of the planets [including the moon?]".
6. On the *Muṣṭalaḥ Zīj* (Kennedy, no. 47) see King 3/III, 535, and V, 135, and also *Cairo survey*, no. C12.
7. de Slane, 446.
8. *Cairo survey*, no. C12, p. 57a *sub* 2.1.15(c).
9. *Ibid.*, no. B59, p. 43b *sub* 2.1.6.
10. On Abu l-'Uqūl and his works see King 2, Section II.9 (pp. 30–32) and also 3/IV, 63. The *Mukhtār Zīj* is listed as Kennedy, no. 57.
11. Rieu, 525–6.
12. See Kennedy, 141–2, on planetary longitude determinations, and al-Bīrūnī, Section 321 on pp. 186–91 for an example of a medieval Islamic ephemeris.
13. See Kennedy, 143b, and as-Saleh in Kennedy *et al.*
14. Kennedy, 140b–141a.
15. *Ibid.*, 142b.
16. On the 'arc of light' see Kennedy *et al.*, 128. In Hogendijk 2 it is shown that the earliest Muslim astronomers used a more complicated formula than (2) in order to tabulate $L(\Delta\lambda,\beta)$.
17. On the shadow functions see Schoy and also Kennedy, 140a.
18. See King 1, *passim.*
19. See Kennedy, 140b, and the article "*Maṭāli'*" in EI_2.
20. Kennedy, 170b, and King 1, Part II, Section 14.
21. See Ilyas for a recent study of the problem of lunar crescent visibility by a modern astronomer.
22. See the references to Neugebauer's studies in ref. 1 above.
23. *Ibid.*
24. *Ibid.*
25. On al-Khalīlī see Suter, no. 418; *Cairo survey*, no. C37; and my article in *DSB*.

BIBLIOGRAPHICAL ABBREVIATIONS

al-Bīrūnī: R. R. Wright, trans., *The book of instruction in the elements of the art of astrology by ... al-Bīrūnī* (London, 1934).

Bruin: F. Bruin, "The first visibility of the lunar crescent", *Vistas in astronomy*, xxi (1979), 331–58.

Cairo catalogue and *survey*: D. A. King, *A catalogue of the scientific manuscripts in the Egyptian National Library*, 2 vols (Cairo, 1981–87), and *idem*, *A survey of the scientific manuscripts in the Egyptian National Library* (*Publications of the American Research Center in Egypt*) (Winona Lake, Indiana, 1987).

Caussin: Caussin de Perceval, "Le livre de la grande table Hakémite, observée par le Sheikh ... ebn Iounis ...", *Notices et extraits des manuscrits de la Bibliothèque Nationale*, 7 (An XII = 1804), 16–240. [The pagination in the separatum runs from 1 to 224.]

DSB: *Dictionary of scientific biography*, 14 vols and 3 supp. vols (New York, 1970–80).

EI_2: *Encyclopaedia of Islam*, 2nd ed., 5 vols to date (Leiden, 1960–).

Hogendijk 1: J. P. Hogendijk, "New light on the lunar crescent visibility table of Ya'qūb ibn Ṭāriq", to appear in *Journal of Near Eastern studies*.

Hogendijk 2: J. P. Hogendijk, "Three Islamic lunar crescent visibility tables", *Journal for the history of astronomy*, xix (1988), 29–44.

Ilyas: M. Ilyas, *A modern guide to the astronomical calculations of Islamic calendar, times & qibla* (Kuala Lumpur, 1984).

Kennedy: E. S. Kennedy, "A survey of Islamic astronomical tables", *Transactions of the American Philosophical Society*, n.s., xlvi:2 (1956), 123–77.

Kennedy *et al.*: E. S. Kennedy, Colleagues and Former Students, *Studies in the Islamic exact sciences* (Beirut, 1983). Includes reprints of: E. S. Kennedy, "The crescent visibility theory of Thābit bin Qurra", *Proceedings of the Mathematical and Physical Society of the United Arab Republic*, 1960, 71–74, repr. on pp. 140–3; E. S. Kennedy and M. Janjanian, "The crescent visibility table in Al-Khwārizmī's Zīj", *Centaurus*, xi (1965), 73–78, repr. on pp. 151–6; E. S. Kennedy, "The lunar visibility theory of Ya'qūb ibn Ṭāriq", *Journal of Near Eastern studies*, xxvii (1968), 126–32, repr. on pp. 157–63; and J. A. as-Saleh, "Solar and lunar distances and apparent velocities in the astronomical tables of Ḥabash al-Ḥāsib", *al-Abhath*, xxiii (1970), 129–77, repr. on pp. 204–52.

Kennedy Festschrift: D. A. King and G. Saliba (eds), *From deferent to equant : Studies in the history of science in the ancient and medieval Near East in honor of E. S. Kennedy*. *Annals of the New York Academy of Science*, d (=500) (1983). Includes: D. A. King, "Some early Islamic tables for lunar crescent visibility", on pp. 185–225.

King 1: D. A. King, *Spherical astronomy in medieval Islam: The Ḥākimī Zīj of Ibn Yūnus*, to appear.

King 2: D. A. King, *Mathematical astronomy in medieval Yemen: A bio-bibliographical survey* (*Publications of the American Research Center in Egypt*) (Malibu, California, 1983).

King 3: D. A. King, *Islamic mathematical astronomy* (London, 1986). Includes reprints of: III, "The astronomy of the Mamluks", *Isis*, lxxiv (1983), 531–55; IV, "Mathematical astronomy in medieval Yemen", *Arabian studies*, v (1979), 61–65; and V, "A double-argument table for the lunar equation attributed to Ibn Yūnus", *Centaurus*, xviii (1974), 129–46.

Morelon: R. Morelon, *Thābit ibn Qurra: Oeuvres d'astronomie* (Paris, 1987).

Nallino: C. A. Nallino, *al-Battani sive Albatenii opus astronomicum* (*Pubblicazioni del Reale Osservatorio de Brera in Milano*, xl), 3 vols (Milan and Rome, 1899–1907; reprinted Frankfurt, 1969).

Neugebauer 1: O. Neugebauer, "The astronomy of Maimonides and its sources", *Hebrew Union College annual*, xxii (1949), 321–63, reprinted in his *Astronomy and history: Selected essays* (New York, 1983), 381–423.

Neugebauer 2: O. Neugebauer, "Astronomical commentary", in S. Gandz, J. Obermann and O. Neugebauer, *The Code of Maimonides, Book 3, Treatise 8, Sanctification of the New Moon* (New Haven, Connecticut, 1956).

Renaud: H. P. J. Renaud, "Additions et corrections à Suter 'Die Mathematiker und Astronomen der Araber' ", *Isis*, xviii (1932), 166–83.

Rieu: C. Rieu, *Supplement to the Catalogue of the Arabic manuscripts in the British Museum* (London, 1894).

Schoy: C. Schoy, *Über den Gnomonschatten und die Schattentafeln der arabischen Astronomie* (Hanover, 1923).

Sezgin: F. Sezgin, *Geschichte des arabischen Schrifttums*, 10 vols to date (Leiden, 1967–).

de Slane: Mc Guckin de Slane, *Catalogue des manuscrits arabes* (Paris, 1883–95).

Suter: H. Suter, "Die Mathematiker und Astronomen der Araber und ihre Werke", *Abhandlungen zur Geschichte der mathematischen Wissenschaften*, x (1900), and "Nachträge und Berichtigungen", *ibid.*, xiv (1902), 157–85 (repr. Amsterdam, 1982, and again in his *Beiträge zur Geschichte der Mathematik und Astronomie im Islam*, 2 vols (Frankfurt, 1986)).

IV

LUNAR CRESCENT VISIBILITY PREDICTIONS IN MEDIEVAL ISLAMIC EPHEMERIDES

INTRODUCTION

Since the Islamic calendar is lunar and the beginnings of the months are regulated by the sightings of the lunar crescent, the determination of the possibility of crescent visibility has been a matter of concern to Muslims over many centuries. When the Muslims were pre-eminent in the various branches of science, Muslim astronomers devoted considerable attention to the complicated problem of predicting crescent visibility. They formulated conditions for ascertaining visibility and compiled tables for facilitating its determination. Some of these theoretical conditions and tables have been investigated in recent years, using the vast manuscript sources available for the study of the history of Islamic civilization.[1]

The determination of the possibility of visibility of the moon on a given evening at the beginning of a lunar month was one of the most complicated problems confronting medieval astronomers; indeed even modern astronomers have not produced an infallible theory of crescent visibility.[2] Fig. 1 shows the crescent moon above the western horizon at sunset shortly before visibility. The moon will be seen only if it is far enough away from the sun after conjunction and far enough above the horizon. I denote the solar and lunar longitudes by λ_s and λ_m, and the lunar latitude by β. In Fig. 1 the moon is shown at longitude difference $\triangle \lambda = \lambda_m - \lambda_s$ from the sun and with latitude β north of the ecliptic. Its altitude above the horizon is h. The apparent distance of the moon from the sun I denote by e, and the difference in setting times of the sun and moon over the local horizon (measured on the celestial equator) by s. Some astronomers computed the amount of light on the crescent, here denoted by L and defined by $L = e/15$. The units used for L were called *asabi*^{c}, «digits»: note that L = 12 digits when the moon is full at opposition, i.e. when e = 180°.

Values of $\triangle \lambda$ and β could be found from solar and lunar tables (see below); then values of *e*, *s* and *h* could be computed for the local latitude using approximate methods based on plane trigonometry or more sophisticated methods involving spherical trigonometry. Medieval visibility conditions mostly involved conditions on *s*, *s* and *e* (or *L*), or *s*, *e* and *h*. Each quantity would have to be greater or equal to a certain prescribed minimum value, in order that visibility be assured. Thus, for example, several Muslim astronomers proposed:

$$e = \sqrt{(\triangle \lambda)^2 + \beta^2} \qquad \text{and} \quad E = e/15.$$ [3]

The difference between the setting times of the sun and moon, *s*, can be computed using:

$$s = \rho\,(\lambda_m' + 180°) - \rho\,(\lambda_s + 180°),$$

where λ_m' is the longitude of the ecliptic which sets with the moon (i.e. the lunar longitude adjusted by an amount dependent on the lunar latitude and the terrestrial latitude, and $\rho\,(\lambda)$ are the «ascensions» *(maṭaliᶜ)* or time taken by the ecliptic arc λ to rise over the eastern horizon. The ascensions were a standard of function of classical and medieval astronomy, and tables of $\rho\,(\lambda)$ for different latitudes were available.[4]

The simple Indian condition $s > 12°$ was advocated in many early Islamic astronomical treatises, although already in the eighth century, Yaᶜqub ibn Tariq had proposed the joint conditions:

$s \geqslant 12°$ and $L > 0;45$ (i.e. $e > 3/4$ of $15°$), or

$s \geqslant 10°$ and $L > 1;0$ (i.e. $e > 15°$).[5]

Several other astronomers advocated conditions of the form:

$$\triangle \lambda + \mu\,\beta > f_n,$$

where $0 < \mu < 1$ is a constant dependent on terrestrial latitude and f_n is a series of limits for each zodiacal sign *n*, determined such that $s \geqslant 12°$.[6] Considerably more complicated conditions were prescribed by some of the major Muslim astronomers.[7]

Such conditions are usually elaborated in *zijes*, or astronomical handbooks with tables, of which close to 200 examples were compiled by Muslim astronomers between the eighth and nineteenth centuries.[8] But the practical application of these theories was not recorded in *zijes;* rather, it is documented in other sources, which have yet to be studied properly. The purpose of this paper is to present some newly discovered material on the subject.

ISLAMIC EPHEMERIDES

From the ninth century onwards, Muslim astronomers prepared ephemerides displaying daily positions of the sun, moon and five naked-eye planets for a given year. These tables, called *daftar al-sana* or *taqwim* in medieval Arabic, were computed for specific terrestrial longitudes. The celebrated eleventh-century scientist of Central Asia, al-Biruni, described such ephemerides based on the Persian (solar) calendar.[9] Ephemerides served only the year for which they were compiled and thus had a high rate of obsolescence; they are well named ephemeris, plural ephemerides. No examples survive from the earliest period of Islamic astronomy, but fragments of various late Fatimid (twelfth century Egyptian) ephemerides have survived in the Cairo Geniza and have recently been studied for the first time.[10] The earliest complete Islamic ephemerides are two of Yemeni provenance, which date from the fourteenth and early fifteenth centuries, but these have not been studied properly yet.[11] Numerous examples survive from the Ottoman and Safavid Empires, but none of these have been investigated either.[12]

Most medieval Islamic ephemerides compiled for the Hijra calendar contain information on the lunar crescent at the beginning of each Muslim month. Extracts from two facing pages from the earlier of the two Yemeni ephemerides mentioned above (which serves the year 727 Hijra = 1326/27) are shown in Plates 1 and 2. The tables on the right hand page (Plate 1) display the daily positions of the sun, moon and planets during the first few days of the month of Ramadan. At the top of the page there is written information on the time of the conjunction of the sun and moon. Numbers are expressed sexagesimally (to base 60) in the standard alphanumerical *(abjad)* notation of medieval Islamic astronomical works.[13] The information recorded for each day on the left hand page (Plate 2) includes statements about the position of the moon on the zodiac as well as its position relative to the planets and the corresponding predictions for events on earth *(salih* for «good news» and *fasid* for «bad news»). At the top of the page there is additional information about crescent visibility: the moon will be seen as a very thin crescent *(saghiran)* on the Tuesday evening in a certain lunar mansion; its size will be one digit (one twelfth of the area of the moon); its latitude will be so many degrees and minutes. The phrase *wa-yattafiq al-hisab wa-'l-ru'ya* means «there will be agreement between the prediction by calculation and what is actually seen,» but a cautionary *in sha' Allah* would have been appropriate.

Plate 3 displays an extract from an Egyptian ephemeris for the year 1023 Hijra (= 1614 / 15). Diagrams illustrating the direction of the crescent for each month of the year relative to the horizon of Cairo are drawn, in addition to such numerical information. Plate 4 displays a *ghurra-name* for the period 1086-88 Hijra (= 1675-78), in which only the days of the week when visibility will occur *(ghurra)* for each month are recorded.

The calculations underlying such numerical information are not generally recorded in ephemerides. However, two ordered sets of such calculations have recently been located, and to these we now turn.

THE TABLES IN THE FLORENCE MANUSCRIPT

Plate 5 shows a set of tables taken from MS Florence Laurenziana Or. 152, fol. 145v(?)[14] which display various functions required to determine visibility. Calculations are given for each month of the solar year beginning with the vernal equinox of 1460, corresponding to VI 865-V 866 Hijra. The table forms part of an ephemeris for that solar year, whose purpose was to ascertain the astrologically significant positions of the moon relative to the sun and five planets. The provenance of the ephemeris is not stated, but I suspect that it was compiled in what is now northern Iraq and north-western Iran. Unfortunately the history of astronomy in this region during the late medieval period is particularly obscure.[15]

The ten columns in the table bear the following titles:

shuhur al-ᶜArab: the names of the months of the Hijra calendar

shams al-ghurub: the longitude of the sun at sunset (given in zodiacal signs, degrees and minutes)

qamar al-ghurub: the longitude of the moon at (sun) set (given likewise)

al-buᶜd bi-'l-magharib: the distance between the two luminaries, measured in descensions *(magharib)*, i.e. on the celestial equator (though see below) (given in degrees and minutes). The descensions σ (λ) are defined in terms of the ascensions by the relation:

$$\sigma(\lambda) = \rho(\lambda + 180°).$$

ᶜard al-qamar: the lunar latitude (given in degrees and minutes)

qaws al-ru'ya: the arc of visibility (see below) (given in degrees and minutes)

asabiᶜ al-nur; the digits of light (see below)

jihatuhu ᶜinda 'l-ru'ya: the direction of the lunar latitude at visibility (*shamal* = north or *janub* = south of the ecliptic)

surat al-ru'ya: the orientation of the crescent, relative to the local horizon.

ma yura wa-ma la yura: «what will be seen and what will not be seen,» more precisely, a statement whether or not the crescent will be seen on the first day of the civil month.

The apparent sophistication of the table is dispelled by closer examination. Indeed the table was clearly prepared by an incompetent. The «descensional distance» between the sun and moon is no more than their longitudinal difference, $\Delta\lambda$. The «arc of visibility» is no more than the quantity $\Delta\lambda + \beta$, and the «digits of light» are 1/15 of this. Thus, the compiler has mislabelled the longitude difference with a name that would lead one to suspect that he was tabulating the difference in setting times, and he has used a very crude approximation for $e(\Delta\lambda, \beta)$. Finally, he has imposed a very crude condition for visibility, quite independent of the local latitude, namely, $L > 0;48$, equivalent to $e > 12°$.[16] The condition:

$$\Delta\lambda + \beta > 12°$$

is attested in only one other source known to me, namely, an (Iraqi?) *zij* based on the work of the twelfth-century Iraqi astronomer al-Fahhad.[17]

If the predictions for sensitive cases in the table were correct, it was only by chance. It was such astronomers who gave the profession a bad name and who caused the legal scholars, who were responsible for the actual regulation of the calendar, to disregard their pronouncements, a tradition which has persisted up to the present day.

THE TABLES IN THE CAIRO MANUSCRIPT

Plates 6-8 show a set of similar tables preserved in MS Cairo Dar al-Kutub *sinaʿa* 166, 2, fols. 40r-41r.[18] The tables are probably of Egyptian provenance, and serve each month of the six-year period 1125-30 Hijra (= 1713-18). The thirteen columns in the tables bear the following titles:

asma' al-ahilla: the names of the crescents (= the names of the Muslim months)

ʿalamat layali 'l-ghurra: the signa (numbers 1-7 representing the days of the week) corresponding to the evenings of first visibility

ʿadad al-ayyam hisaban: the day of the civil month (according to the standard scheme of alternating 29 and 30 days for the Muslim months with occasional leap years) for which visibility is to be determined (1 or 2)

muqawwam qamar al-ru'ya: the lunar longitude at the time of the first visibility (given in signs, degrees and minutes)

al-ʿard: the lunar latitude (given in degrees and minutes)

jihat al-ʿard: the direction of the lunar longitude (*sh* = *shamal* = north or *j* = *janub* = south of the ecliptic), and whether it is increasing (*d* = *za'id*) or decreasing (*t* = *habit*). Thus *jt* = southerly decreasing, etc.

qaws al-nur: the arc of light, i.e. the apparent angular distance between the sun and the moon, *e* (given in degrees and minutes)

daqa'iq al-nur: the minutes of light, *L*

qaws al-ru'ya irtifaʿuhu: the altitude corresponding to the arc of visibility, i.e. the altitude of the moon at first visibility, *h* (given in degrees)

qaws al-makth: the arc of tarrying, i.e. the difference in setting times of the sun and moon over the local horizon, *s* (given in degrees and minutes)

al-manzila: the lunar mansion corresponding to the lunar longitude, together with a number (from 1 to 13) representing the moon's position within the mansion

al-sifa: the orientation of the crescent relative to the local horizon (*b* = *muntasib* = erect, i.e. roughly perpendicular to the horizon; *f* = *munharif* = inclined; *y* = *mustawi* = straight, i.e. roughly parallel to the horizon)

hukm *al-ru'ya:* the verdict concerning visibility: the crescent will be seen clearly *(yura zahiran* or *bayyinan);* it will be seen faintly *(khafiyan);* it will probably be seen *(ghaliban);* visibility will be difficult (*ᶜasurat al-ru'ya* [text has *ᶜsr* but the subject is feminine]); and the crescent will not be seen at all *(la yura abadan).*

There are several obvious copyist's mistakes (*e.g.* *L* and *h* in IX 1126 and *h* in XI 1126) and some of the predictions are muddled: when conditions are excellent, it is said that visibility is only probable (e.g. III 1125). Also it is not clear how the quantity *h* was computed.

Notice also that the solar longitude at visibility is not given, so that it is not feasible to attempt to reconstruct the calculations. One could, of course, compute the actual values of the various quantities tabulated using modern tables, but I have considered this beyond the call of duty. If the tables were prepared specifically for Cairo, they might have been computed using solar and lunar ephemerides based on the recension by the fifteenth-century Egyptian astronomer Ibn Abi 'l-Fath al-Sufi of the *Zij-i Sultani* of Ulugh Beg of Samarqand, which was the most popular *zij* in Egypt during the Ottoman period.[19] On the other hand, they may have been computed using ephemerides generated by the auxiliary tables associated with the fifteenth-century Egyptian astronomer Ibn al-Majdi.[20]

Since the tables shown in Plates 6-8 display values of the minutes of light *(L),* the difference in setting times of the two luminaries *(s),* and the altitude of the crescent *(h),* it is reasonable to suppose that the conditions used to determine visibility involved these three quantities. There is, as yet, no published material on the various visibility theories used by Mamluk and Ottoman astronomers which are described in their numerous surviving writings, but conditions involving these three quantities *e, s* and *h* were indeed recorded by a series of such astronomers.[21] However, these conditions predict only whether or not the crescent will be seen, not how it will be seen. Conditions covering different stages of visibility (lifted from the *Zij-i Ilkhani* compiled at the Observatory of Maragha in the mid-thirteenth century) are recorded in the *Zij-i Sultani* and Egyptian (Arabic) versions thereof, but these involve only the quantities $\triangle \lambda$ and *s.*[22] There seems to be little point in trying to determine the conditions for visibility which might underlie our tables until the contemporary Egyptian materials on crescent visibility theory have been properly investigated.

ACKNOWLEDGEMENTS

It is a pleasure to thank the authorities of the Egyptian National Library in Cairo for unlimited access to their enormous holdings of Islamic scientific manuscripts and for permission to use the photographs illustrating this paper. Likewise, I am grateful to the authorities of the Biblioteca Medicea Laurenziana in Florence for the facilities afforded to me within the Library and for permission to use the photograph used as Plate 5. My research on lunar crescent visibility theory conducted at New York University was supported by a grant from the National Science Foundation, Washington, D.C. This support is gratefully acknowledged. It is also a pleasure to acknowledge my gratitude to my colleagues, Prof. E.S. Kennedy and Dr. Jan Hogendijk, for their comments on a preliminary draft of this paper.

BIBLIOGRAPHY AND ABBREVIATIONS

al-Biruni, *Tafhim:* R.R. Wright, trans., *The Book of Instruction in the Elements of the Art of Astrology by... al-Biruni* (London, 1934).

Cairo Survey: D.A. King, *A Survey of the Scientific Manuscripts in the Egyptian National Library,* Publications of the American Research Center in Egypt (Winona Lake, Ind., 1986).

EI^2: The Encyclopedia of Islam, 4 vols. to date (Leiden, 1960-).

Goldstein & Pingree: B.R. Goldstein and D. Pingree, «Astrological Almanacs from the Cairo Geniza,» *Journal of Near Eastern Studies,* 38 (1979), pp. 153-175 and 231-256, and *Journal of the American Oriental Society,* 103 (1983), pp. 673-690.

Ilyas: M. Ilyas, *A Modern Guide to Astronomical Calculations of Islamic Calendar, Times and Qibla* (Kuala Lumpur, 1985).

Kennedy *et al.,* Studies: E.S. Kennedy, Colleagues and Former Students, *Studies in the Islamic Exact Sciences,* Publications of the American University of Beirut (Beirut, 1983).

Kennedy Festschrift: D.A. King and G. Saliba, eds., *From Deferent to Equant: Studies in the History of Science in the Near East in Honor of E.S. Kennedy, Annals of the New York Academy of Science,* 1986.

Kennedy, *Zijes:* E.S. Kennedy, «A Survey of Islamic Astronomical Tables,» *Transactions of the American Philosophical Society,* 46:2 (1956).

King, *IMA:* D.A. King, *Islamic Mathematical Astronomy* (London, 1986).

King, *LCV:* D.A. King, «Some Early Islamic Tables for Determining Lunar Crescent Visibility,» in *Kennedy Festschrift.*

King, *MAY:* D.A. King, *Mathematical Astronomy in Medieval Yemen: A Biobibliographical Survey,* Publications of the American Research Center in Egypt (Malibu, Calif 1983).

Menage: V.L. Menage, «The Beginnings of Ottoman Historiography,» in B. Lewis and P.M. Holt, eds., *Historians of the Middle East* (London, 1962).

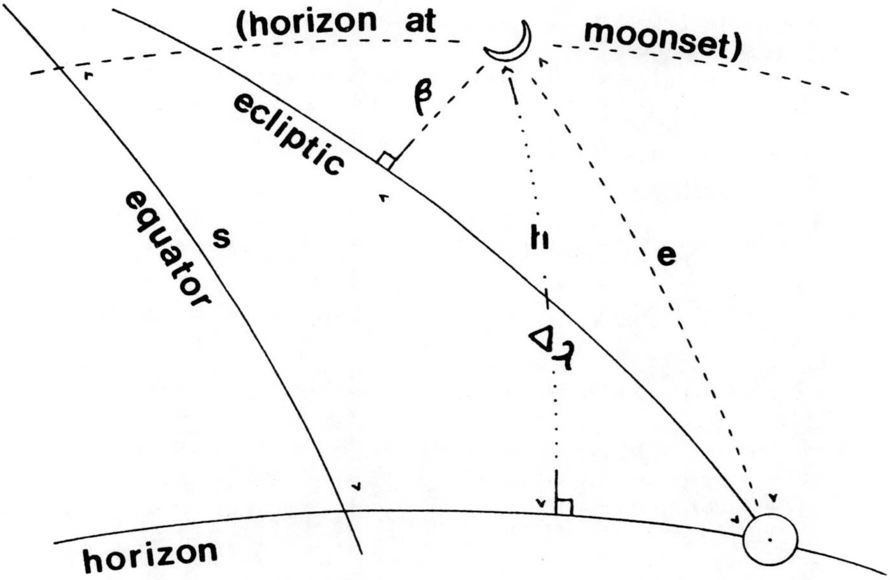

Figure 1: The sun and moon at sunset on the evening of first visibility, which occurs shortly after sunset. The difference in ecliptic longitudes is shown as $\Delta \lambda$ and the lunar latitude as β . Most medieval visibility predictions involved the apparent angular separation of the two luminaries, *e*, and the difference in setting times of the sun and moon, *s*, measured on the celestial equator. Some also involved the altitude of the moon at sunset, *h*, or at the moment of visibility. Most medieval theories asserted that only if *s*, or each of *s* and *e*, or each of *s*, *e* and *h*, were greater than certain prescribed minimum values, would the crescent be seen.

Plate 1: An extract from the fourteenth-century Yemeni ephemeris in MS Cairo Dar al-Kutub *miqat* 817, 1, fol. 77v.

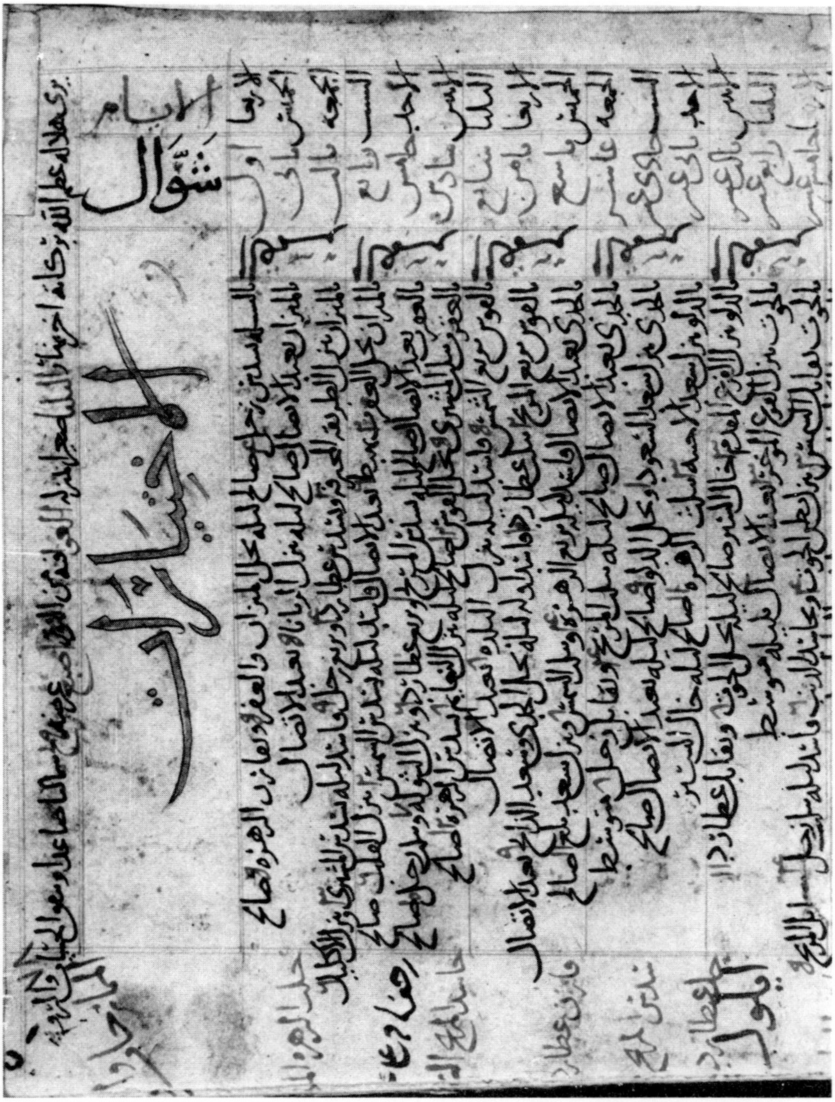

Plate 2: An extract from the astrological information recorded opposite the solar, lunar and planetary ephemerides shown in Plate 1. Taken from MS Cairo Dar al-Kutub ***miqat*** 817, 1, fol. 78r.

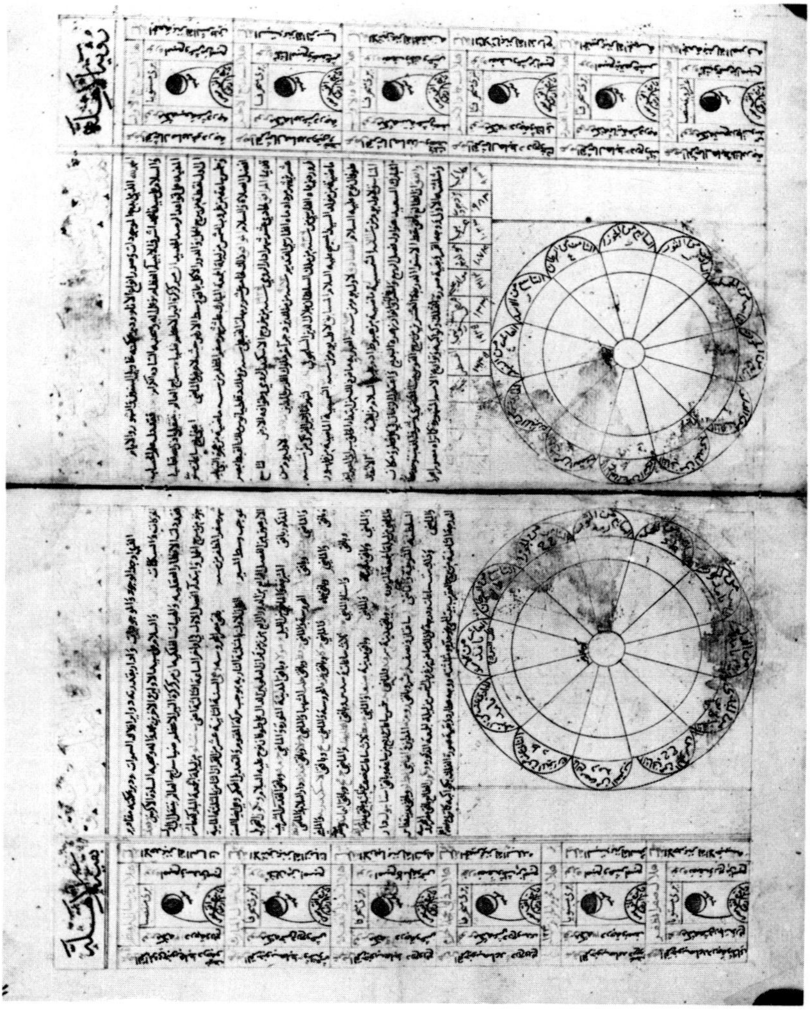

Plate 3: The introductory pages of a seventeenth-century Egyptian ephemeris showing orientations of the crescents for each month relative to the horizon of Cairo. Taken from MS Cairo Dar al-Kutub *miqat* 141, 3, fols. 21v-22r.

Plate 4: A Persian *ghurrat-name* for the years 1086-88 Hijra (= 1675-78). Taken from MS Cairo Tal^cat *majami*^c 298, 7, fol. 55r.

Plate 5: Calculations for predicting crescent visibility for each month of the solar year corresponding to 865/66 Hijra (= 1460/61), taken from MS Florence Or. 152, fol. 45v.

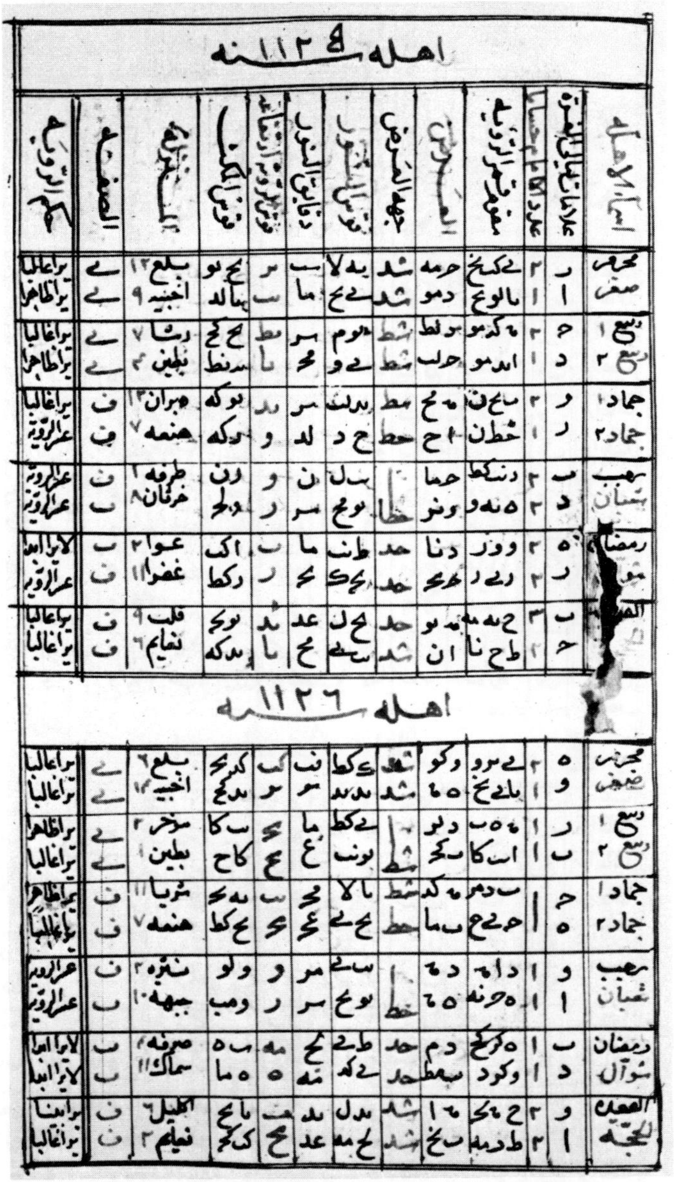

Plates 6-8: Calculations for predicting crescent visibility for each month of the years 1125-30 Hijra (= 1713-18), taken from MS Cairo Dar al-Kutub *sinaᶜa* 166, 2, fols. 40r-41r.

اهله سنه ١١٢٧

اسماء الاشهر	علامات اوائل الشهور	عدد الايام بحساب	تقويم قمر الزيج	[illegible]	جهة العرض	قوس النور	دقائق النور	[illegible]	تقويم الكلب	المنزل	الصفة	حكم الرؤية
محرم	ب	٢	ظ كركه	دكا	شد	ب كا	نا	كح	نه لو	بلدة ١١	ف	ترا غالبا
صفر	د	٢	نادس	دنو	شط	لط نا	عو	كا	كح كح	سعود ١١	ى	ترا غالبا
ربيع ١	ه	٢	ما كح مد	دو	شط	كح لو	نه	نه	ند لو	مقدم ٧	ى	ترا غالبا
ربيع ٢	و	١	ما كد ما	ب كط	شط	ح ب	له	ى	ى كه	رشا ٧	ف	ترا خفيا
جماد ١	ا	٢	ب اح	ما ند	سط	نو لا	ع	نه	ك ط	شرطا ٧	ف	ترا غالبا
جماد ٢	ب	١	ف كه ما	به كح	سط	كح ك	كح	ى	سلط	هنعه ٤	ف	ترا بينا
رجب	ح	١	ح نو لط	د نو	جط	ح نو	لو	د	دو	ذراع ١	ف	لا يرا ابدا
شعبان	ه	١	دكا لح	نو نط	جد	ند ند	س	و	ولو	طرف ١١	ف	عسر الرؤية
رمضان	و	١	ه ب ب	دكه	جد	ح مو	لح	ب	ب ط	خراتان ٦	ف	لا يرا ابدا
شوال	ا	١	وه ح	مت نو	جد	ما نو	مه	ه	ه لا	عوا ٣	ب	لا يرا ابدا
العقدة	ح	٢	رح كح	نه لو	شد	كح لا	ند	ط	نار	زمانا ٧	ف	ترا [illegible]
الحجة	ه	٢	ح كا ا	خ ح	شد	نو كه	سو	نه	بر نا	شوله ١	ف	[illegible] غالبا

اهله سنه ١١٢٨

اسماء الاشهر	علامات اوائل الشهور	عدد الايام بحساب	تقويم قمر الزيج	[illegible]	جهة العرض	قوس النور	دقائق النور	[illegible]	تقويم الكلب	المنزل	الصفة	حكم الرؤية
محرم	و	٢	ط س كح	د ل	شد	ط ى	لو	ط	ما لا	نعايم ٩	ف	ترا خفيا
صفر	ا	٢	ى نو مو	د نه	شد	كح	نت	نه	نه كد	بلع ٦	ى	ترا غالبا
ربيع ١	ح	٣	ما كح مط	ح نو	شط	كح ن	عه	كا	كا نو	مقدم ٤	ى	ترا غالبا
ربيع ٢	د	٢	ما نو ب	اكو	شط	ب له	كح	ند	ند ح	رشا ١	ى	ترا غالبا
جماد ١	ر	٢	اى لب	ما مو	جط	ح د	لب	ر	ح مد	شرطين ١٢	ب	عسر الرؤية
جماد ٢	ا	٢	ب نط ى	و مو	جط	نر خ	عب	كح	نر ل	دبران ١٣	ف	ترا غالبا
رجب	ا	٢	ح ب ب	دح	جط	كح كو	نو	ط	ى ما	هقعه ١	ف	ترا خفيا
شعبان	ب	١	ده ب	دنر	جد	ى كح	ما	د	ب مح	نثره ٧	ف	لا يرا ابدا
رمضان	د	٣	ه ما لح	ح كح	جد	نو كد	سو	ح	ح مح	خرتان ٥	ب	عسر الرؤية
شوال	ه	١	وح نو	امو	جد	ط ما	ما	د	ه كط	صرفه ١٣	ب	لا يرا ابدا
العقدة	ر	٢	رو نه	ا نه	شد	كح ل	نو	ى	ما ط	غفر ٨	ف	ترا خفيا
الحجة	ب	٢	ح ط م	ح مو	شد	نو لا	سو	نه	نو كح	قلب ٣	ف	ترا غالبا

اهله سنه ١١٢٩

أسماء الأهلة	علامات ليالي الغرة	عدد ايامها بحساب	مقوم قمر الرؤية	المعرض	جهة العرض	قوس النور	دقائق النور	قوس الرؤية ارتفاعه	قوس المكث	المنزل	الصفة	حكم الرؤية
محرم صفر	ح ه	١ ١	ط مه ک ے و س	د مد د مط	شد شط	ح نو ے یح	لب مه	ح یا	ے یح یب	شوله ١٠ ذابح ٤	ف ے	عمر الرؤية يرا ظاهرا
ربيع ١ ربيع ٢	ز ا	٢ ١	ما و مه ما یح د	حط الحج	شط	یح ک ه یح	یح کا	نه و	نه ما ه مد	سعود ١٢ مقدم ٨	ے ے	يرا غالبا لا يرا ابدا
جماد ١ جماد ٢	ح د	٢ ١	اه مه الحج له	اب ح لط	حط	ب م ح نه	یح لو	یا و	ند ا رح	شرطين ٦ ثريا ٤	ے ے	يرا ظاهر عمر الرؤية
رجب شعبان	و ز	٢ ١	حونا دهب	ه مه دلط	حط حد	یح یح ع مو	عح نو	ب ز	نه لح زم	هقعه ٤ نثره ١	ف ف	يرا غالبا عمر الرؤية
رمضان [illegible]	ا ح	١ ١	دیح نه ومه ب	ح کد ما که	حد حد	طم ز مه	لط یح	د ے	حوط ما مه	طرفه ١٣ صرفه	ف ب	لا يرا ابدا يرا ظاهرا
القعدة الحجة	د و	١ ١	وکحب وکرما	ام در	سد سد	نا و نوا	مد سر	٥ نه	ے کح نوط	سماك ٧ الاكليل ٤	ب ف	عمر الرؤية يرا غالبا

اهله سنه ١١٣٠

أسماء الأهلة	علامات ليالي الغرة	عدد ايامها بحساب	مقوم قمر الرؤية	المعرض	جهة العرض	قوس النور	دقائق النور	قوس الرؤية ارتفاعه	قوس المكث	المنزل	الصفة	حكم الرؤية
محرم صفر	ز ب	١ ١	ح یح مر ط کا مط	د ند دح	شد شد	طح ما نا	لد مه	ط س	س ح یح ر	قلب ١٢ بلده ٦	ف ف	يرا بينا يرا غالبا
ربيع ١ ربيع ٢	د و	٢ ٢	ے کدم ما کو مح	ب مت مه ے	شط	ب کس نه ٤	مط بس	یح نه	ند ے ز مه	سعود ١ مقدم ٨	ے ے	يرا غالبا يرا غالبا
جماد ١ جماد ٢	ز ب	٢ ٢	مه نط مت اکه کط	ب ط دکر	حط	ح نه نه م	لد س	ز یا	ح ٥ نه کد	رشا ٤ ثريا ١	ے ے	عمر الرؤية يرا غالبا
رجب شعبان	ح ه	٢ ٢	ب نو یح حکه لا	دنط دو	حط حد	طب زخ	یح ع	٥ یا	ومط ندد	دبران ١٠ ذراع ٩	ف ف	لا يرا ابدا يرا غالبا
رمضان شوال	و ز	٢ ١	دیح د ه ما لح	ب که مه کو	حد حد	یح نه طب	یح لو	ز ٥	ط ے ٥ مه	طرفه ٧ خرتان ٤	ف ب	عمر الرؤية لا يرا ابدا
القعدة الحجة	ب ح	٢ ١	ونط بر زس ب	ب مط دنو	سد سد	مزنو یاند	ع یح	ب ے	یح نو سکط	سماك ٣ زبانا ١	ب ف	يرا غالبا يرا بينا

NOTES

1. Most modern studies of lunar crescent visibility were conducted by Prof. E.S. Kennedy of the American University of Beirut and his students. See, for example, the various monographs reprinted in Kennedy *et al.*, *Studies*, pp. 140-143 and 151-163, and my contribution to the *Kennedy Festschrift* (listed as King, *LCV* and containing an extensive bibliography).
2. A recent work on the subject by a Muslim astronomer, the first of its kind, is listed as Ilyas.
3. See, for example, Kennedy *et al.*, *Studies*, p. 159.
4. See Kennedy, *Zijes*, p. 140 and the article *Matali*ᶜ in *EI*².
5. Numerous tables based on the condition $s \geqslant 12^{\circ}$ are investigated in King, *LCV*. For Yaᶜqub ibn Tariq's conditions, see Kennedy *et al.*, *Studies*, p. 129.
6. See King, *LCV*, Section 7.
   See, for example, Kennedy *et al.*, *Studies*, pp. 71-74 (Thabit ibn Qurra) and my forthcoming study of the theory of Ibn Yunus.
8. See Kennedy, *Zijes*, for a survey of some 125 *zijes* and accounts of the contents of some of the most important ones.
9. al-Biruni, *Tafhim*, Section 321, pp. 186-191.
10. See Goldstein & Pingree.
11. See *Cairo Survey*, No. E11, and King, *MAY*, pp. 33 and 39 on these two ephemerides.
12. See *Cairo Survey*, Nos. D209 (Egypt), G112 (Iran), and H78 (Turkey), and King, *MAY*, p. 13 (Yemen). See also Menage, p. 170 on Ottoman almanacs.
13. See R.A.K. Irani, «Arabic Numeral Forms,» repr. in Kennedy *et al.*, *Studies*, pp. 710-721.
14. The facing page is numbered 46, but my notes on this manuscript indicate that the ephemeris is contained in fols. 145r-155r.
15. It was in this region that the *Shamil Zij* (Kennedy, *Zijes*, No. 29; see also Nos. 40, 56 and 73) was used. This work, extant in several copies, is currently being investigated by my colleague Dr. Jan Hogendijk.
16. When L = 0;47 (line 8), the prediction is negative; when L = 0;51 (line 1), it is positive. Notice that in the entry for Shawwal 865 (line 5), L = 0;48, but the word *la* = «not» in the prediction has been crossed out.
17. See King, *LCV*, Section 7 (f). On the *Zijes* of al-Fahhad, see Kennedy, *Zijes*, p. 176 (index).
18. See *Cairo Survey*, No. D226. The other work contained in this manuscript is a treatise on mechanical clocks by the sixteenth-century Istanbul astronomer Taqi 'l-Din: see *Cairo Survey*, No. H12, sub 7.1.6.
19. See Kennedy, *Zijes*, Nos. 12 and 107, and D.A. King, «The Astronomy of the Mamluks,» repr. in *idem*, *IMA*, III, esp. p. 536.
20. See E.S. Kennedy and D.A. King, «Ibn al-Majdi's Tables for Calculating Ephemerides,» repr. in King, *IMA*, VI.
21. Kushyar ibn Labban (fl. Iran, ca. 1000; see Kennedy, *Zijes*, Nos. 7 and 9) advocated the following condition in his *Zij:*

$$e \geqslant 10^{\circ},\ s \geqslant 8^{\circ},\ \text{and}\ h \geqslant 6^{\circ}$$

His conditions are repeated in the twelfth-century Egyptian *Mustalah Zij* (Kennedy, *Zijes*, No. 47; King, *IMA*, III, pp. 535-536). The fourteenth-century Syrian astronomer Ibn al-Shatir (Kennedy, *Zijes*, No. 11; King, *IMA*, p. 536) misquoted Kushyar's conditions as:

$$e \geqslant 10^\circ,\ s \geqslant 12^\circ, \text{ and } h \geqslant 8^\circ,$$

and noted that «another astronomer» had prescribed:

$$e \geqslant 10^\circ,\ s \geqslant 8^\circ, \text{ and } h \geqslant 7^\circ.$$

Other Mamluk and Ottoman astronomers such as al-Khalili, Ibn al-Majdi and Taqi 'l-Din proposed similar conditions.

22. The conditions are that provided $\triangle \lambda \geqslant 10^\circ$, then when:

$$10^\circ \leqslant s \leqslant 12^\circ,\ 12^\circ \leqslant s \leqslant 14^\circ, \text{ and } s > 14^\circ,$$

the crescent will be seen faintly, moderately clearly, and clearly, respectively.

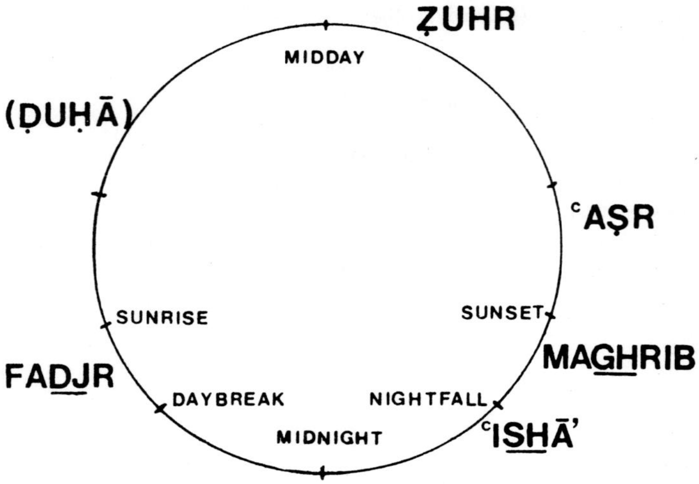

Fig. 1. The Islamic day.

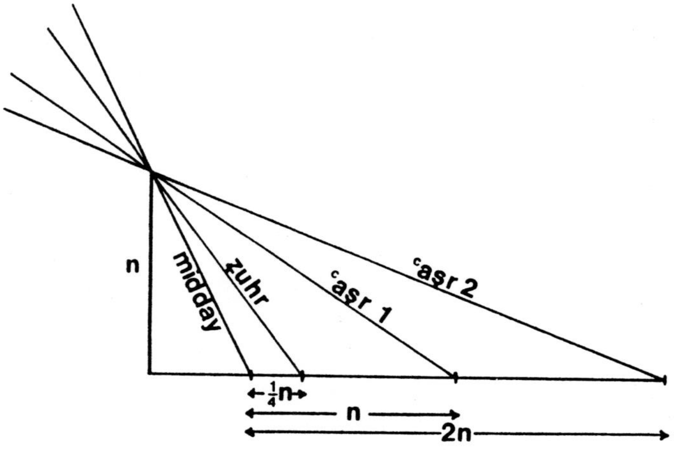

Fig. 2. Shadow increases at the *ẓuhr* (Andalusī/Maghribī definition) and at the beginning and end of the *ᶜaṣr*.

MĪḲĀT: ASTRONOMICAL TIMEKEEPING

MĪḲĀT (A., *mifʿāl* form from *w-ḳ-t*, plural *mawāḳīt*) appointed or exact time. In this sense the term occurs several times in the Ḳurʾān (II, 185; VII, 138, 139, 154; XXVI, 37; XLIV, 40; LVI, 50; LXXVIII, 17).

Astronomical aspects. *ʿIlm al-mīḳāt* is the science of astronomical timekeeping by the sun and stars and the determination of the times (*mawāḳīt*) of the five prayers. Since the limits of the permitted intervals for the prayers are defined in terms of the apparent position of the sun in the sky relative to the local horizon, their times vary throughout the year and are dependent upon the terrestrial latitude. When reckoned in terms of a meridian other than the local meridian, the times of prayer are also dependent upon terrestrial longitude.

The definitions of the times of prayer outlined in the Ḳurʾān and *ḥadīth* [see 1, Legal aspects] were standardized in the 2nd/8th century and have been used ever since. According to these standard definitions, the Islamic day and the interval for the *maghrib* prayer begin when the disc of the sun has set over the horizon. The intervals for the *ʿishāʾ* and *fadjr* prayers begin at nightfall and daybreak. The permitted time for the *ẓuhr* begins either when the sun has crossed the meridian, or when the shadow of any object has been observed to increase, or, in mediaeval Andalusian and Maghribī practice, when the shadow of any vertical object or gnomon has increased over its midday minimum by one-quarter of the length of the object. The interval for the *ʿaṣr* begins when the shadow increase equals the length of the gnomon and ends either when the shadow increase is twice the length of the gnomon or at sunset. See Figs. 1 and 2.

The names of the prayers are derived from the names of the corresponding seasonal hours [see SĀʿA] in pre-Islamic classical Arabic, the seasonal hours (*al-sāʿāt al-zamāniyya*) being one-twelfth divisions of the day and night. The definitions of the times of the daylight prayers in terms of shadow increases (as opposed to shadow lengths in the *ḥadīth*) represent a practical means of regulating the prayers in terms of the seasonal hours.

In some circles, a sixth prayer, the *ḍuḥā*, was performed at the same time before midday as the *ʿaṣr* was performed after midday. These definitions of the *ḍuḥā, ẓuhr*, and *ʿaṣr* correspond to the third, sixth and ninth seasonal hours, the links being provided by an approximate Indian formula relating shadow increases to the seasonal hours (see below). The Umayyad Caliph ʿUmar b. ʿAbd al-ʿAzīz [*q.v.*] is reported to have used a (Graeco-Roman) sundial (marked with the seasonal hours) for regulating his prayers. The early ʿAbbāsid astronomer al-Kh^w^ārazmī [*q.v.*] was still toying with different definitions of the *ẓuhr* intended to associate it with the sixth and the seventh seasonal hours.

The regulation of the times of prayer was conducted at two different levels. At the popular level, the simple techniques of folk astronomy were used. Muslim astronomers, on the other hand, used sophisticated tables and instruments for timekeeping.

As we see from treatises on folk astronomy and on the sacred law, in popular practice the daylight prayers were regulated by simple arithmetical shadow schemes of the kind also attested in earlier Hellenistic and Byzantine folk astronomy. The night prayers were regulated by observation of the lunar mansions [see MANĀZIL]. As the 7th/13th century Yemeni legal scholar al-Aṣbaḥī wrote in his treatise on folk astronomy: "The times of prayer are not to be found by the degrees (marked) on an astrolabe and not by calculation using the science of the astronomers; they are to be found only by direct observation ... The astronomers took their knowledge from Euclid and the *Sindhind*, and from Aristotle and other philosophers; all of them were infidels.''

Some twenty different shadow schemes have been located in the Arabic sources. In most cases, they are not the result of any careful observations. Usually a single one-digit value for the midday shadow of a man 7 *ḳadams* (''feet'') tall is given for each month of the year. One such scheme, attested in several sources, is (starting with the value for January):

9 7 5 3 2 1 1 2 4 5 (or 6) 8 10.

The corresponding values for the shadow length at the beginning of the *ʿaṣr* prayer are 7 units more for each month.

Other arithmetical schemes are sometimes presented in order to find the shadow length at each seasonal hour of day. The most popular formula advocated in order to find the increase (Δs) of the shadow over its midday minimum at T (<6) seasonal hours after sunrise or before sunset is:

$$T = 6\ n\ /\ (\Delta s + n),$$

where n is the length of the gnomon. This is the formula which was first used to establish the values $\Delta s = n$ for the 3rd and the 9th seasonal hours of daylight, the beginnings of the *ḍuḥā* and *ʿaṣr* ($T = 3$), and $\Delta s = 2n$ for the 10th hour, sometimes taken as the end of the *ʿaṣr* ($T = 2$).

The Muslim astronomers had a passion for compiling tables, and some of these were specifically for the purpose of timekeeping. Al-K͟hʷārazmī prepared the first known tables for regulating the times of the daylight prayers. Computed for the latitude of Bag͟hdād, his tables display the shadow lengths at the *ẓuhr* and the beginning and end of the *ʿaṣr*, with values for each 6° of solar longitude (corresponding roughly to each six days of the year): see Pl. IX, 1. He also compiled some simple tables displaying the time of day in seasonal hours in terms of the observed solar altitude.

The 3rd/9th century astronomer ʿAlī b. Amād͟jūr compiled a more extensive table for timekeeping based on a simple approximate formula which serves all latitudes. The underlying formula is

$$T = 1/15 \text{ arc sin } (\sin h \,/\, \sin H),$$

where h is the observed altitude and H is the meridian altitude. (Note that $T = 0$ when $h = 0$ and $T = 6$ when $h = H$, as required for the cases when the sun is on the horizon and the meridian; in fact this formula is accurate only when the sun is at the equinoxes.) Ibn Amā̲d̲jūr simply tabulated $T(h, H)$ for each degree of both arguments ($h \leq H$).

In astronomical handbooks [see ZĪ̲D̲J̲] from the 3rd/9th century onwards, we find descriptions of an accurate method for finding $T(h, H)$. These involve the semi-diurnal arc D and the use of the versed sine function (vers $\theta = 1 - \cos \theta$). The standard mediaeval formula is (in modern notation):

$$\text{vers } (D - T) = \text{vers } D - \sin h \text{ vers } D \,/\, \sin H,$$

which can be derived with facility by means of an analemma construction applied to the celestial sphere. This formula was adopted by Muslim astronomers from Indian sources.

The modern formula for the hour-angle t ($= D - T$) can also be derived by these procedures. It is:

$$\cos t = (\sin h - \sin \delta \sin \varphi) \,/\, (\cos \delta \cos \varphi),$$

where δ is the solar declination and φ is the local latitude (note that $H = 90° - \varphi + \delta$), and is used in this form by later Muslim astronomers (see al-K̲h̲alīlī, below).

Likewise, from the 3rd/9th century onwards, we find descriptions of how to find the time of day or night using an analogue computer such as an astrolabe [see ASṬURLĀB] or using a calculating device such as a sine quadrant [see RUBᶜ]. In the first case, there is no need to know the formula; in the second case, one uses the formula to compute specific examples.

ᶜAlī b. Amā̲d̲jūr also compiled a table of $T(h, H)$ for Bag̲h̲dād based on an accurate trigonometric formula. Some prayer-tables for Bag̲h̲dād preserved in the *Zī̲d̲j* of the 7th/13th century astronomer al-Ḳāsim b. Maḥfūẓ al-Bag̲h̲dādī, which display, for example, the duration of twilight for each day of the year, were

1. Al-Khwārazmī's prayer-tables for latitude 33° (Baghdād). For each 6° of solar longitude the lengths of the shadow of a gnomon of length 12 units are tabulated for the beginning of the *ẓuhr* and for the beginning and end of the *ʿaṣr*. Note that some entries are written in Hindu-Arabic numerals [see ḤISĀB] and others in the more usual alpha-numerical system [see ABDJAD]. Taken from Ms Berlin Ahlwardt 5793 (Landberg 56), fol. 94a, with kind permission of the Deutsche Staatsbibliothek, Preußischer Kulturbesitz, Orientabteilung.

PLATE IX

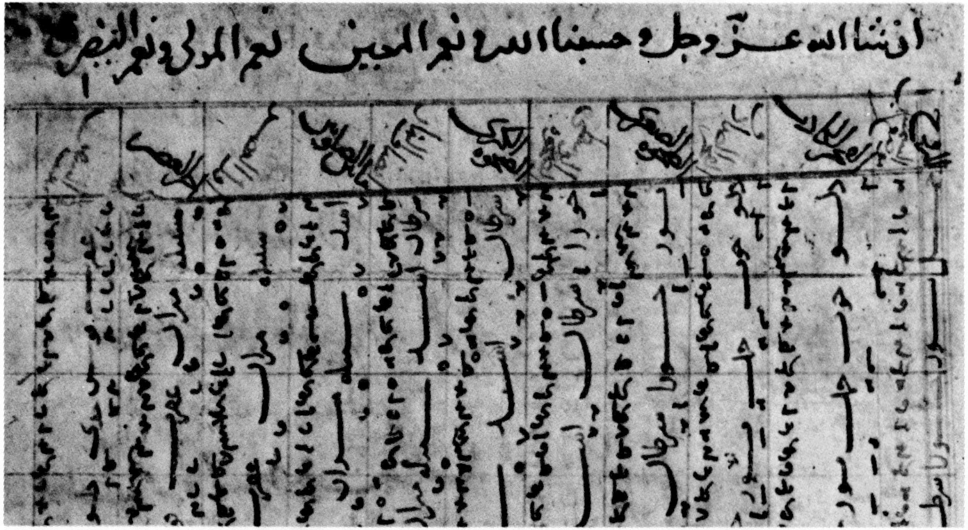

2. An extract from some anonymous prayer-tables for an unspecified latitude (about 40°). The function tabulated is the ascendant at ten different times of the day and night, and values are given for each 6° of solar longitude. The times served are dusk (*mughīb al-shafaḳ*) and daybreak (*al-fadjr al-ṣādiḳ* and *al-fadjr al-kādhib*) and the seven moments which divide the length of daylight into eight equal parts. The table is evidence of a survival of the early tradition of associating the *ḍuḥā* and the *ʿaṣr* with the seasonal hours. Taken from Ms Cambridge U.L. O.1, fol. 179a, with kind permission of Cambridge University Library.

probably another early ʿAbbāsid production. Quantitative estimates of the angle of depression of the sun at nightfall and daybreak occur in the *Zīdj* of the 3rd/9th-century astronomer Ḥabash al-Ḥāsib [*q.v.*]. Quantitative estimates of the effect of refraction at the horizon were first made in the 4th/10th century by the Cairene astronomer Ibn Yūnus [*q.v.*]. Isolated tables displaying the altitudes of the sun at the *ẓuhr* and *ʿaṣr* prayers and the duration of morning and evening twilight occur in several early mediaeval Islamic astronomical works, usually of the genre known as *zīdj*. Several early examples of extensive tables for reckoning time by day from the solar altitude or for reckoning time of night from altitudes of certain prominent fixed stars have come to light. All of these tables were computed for a specific locality, and to use any of them one needed an instrument, such as an astrolabe, to measure celestial altitudes or the passage of time. There is no evidence that these tables were widely used.

In practice, at least before the 7th/13th century, the regulation of the prayer-times was the duty of the *muʾadhdhin* or muezzin [see ADHĀN and MASDJID]. The muezzins were appointed for the excellence of their voices and their character, and they needed to be proficient only in the rudiments of folk astronomy. They had to know the shadows at the *ẓuhr* and the *ʿaṣr* for each month, and which lunar mansion was rising at daybreak and setting at nightfall, information which was conveniently expressed in the form of mnemonics. They did not need astronomical tables or instruments. The necessary techniques are outlined in the chapters on prayer in the books of sacred law [see FIḲH], and the qualifications of the muezzin are sometimes detailed in works on public order [see ḤISBA and IḤTISĀB]. See, for example, the discussion of their responsibilities by Ibn al-Ukhuwwa [*q.v.*] (ed. R. Levy, text, 176-8, tr. 63-5) and the criticism of their mentality by the scientist al-Bīrūnī [*q.v.*] (E.S. Kennedy, *Shadows*, i, 75-6, 226-30, and ii, 28-9, 142-3).

In the 7th/13th century, there occurred a new

development whose origins are obscure. In Egypt we find the first mention of the institution of the *muwaḳḳit* [*q.v.*], a professional astronomer associated with a religious institution, whose primary responsibility was the regulation of the times of prayer. Simultaneously, there appeared astronomers with the epithet *mīḳātī* who specialized in spherical astronomy and astronomical timekeeping (*ʿilm al-mīḳāt*), but who were not necessarily associated with any religious institution.

In Cairo later in that century, a *mīḳātī* named Abū ʿAlī al-Marrāku<u>sh</u>ī [*q.v.*] compiled a compendium of spherical astronomy and instruments from earlier sources which was to set the tone of *ʿilm al-mīḳāt* for several centuries. His treatise, appropriately entitled *<u>Dj</u>āmiʿ al-mabādiʾ wa-'l-ghāyāt fī ʿilm al-mīḳāt* ("An A to Z of astronomical timekeeping"), was investigated by the Sédillots *père et fils* in the 19th century.

Al-Marrāku<u>sh</u>ī's contemporary, <u>Sh</u>ihāb al-Dīn al-Maḳsī, compiled a set of tables displaying the time since sunrise as a function of solar altitude *h* and solar longitude λ for the latitude of Cairo (based on an earlier set by the 4th/10th century astronomer Ibn Yūnus [*q.v.*]), which in the 8th/14th century was expanded and developed into a corpus of tables covering some 200 manuscript folios and containing over 30,000 entries. The Cairo corpus of tables for timekeeping was used for several centuries and survives in numerous copies, no two of which contain the same tables. Besides tables displaying the time since sunrise / hour-angle (time remaining until midday) / solar azimuth for each degree of solar altitude, which make up the bulk of the corpus (see Pl. X, 3), there are other tables displaying the solar altitude and hour-angle at the *ʿaṣr*, the solar altitude and hour-angle when the sun is in the direction of the *ḳibla* [*q.v.*], and the duration of morning and evening twilight. In some later copies of the corpus, there are tables for regulating the time when the lamps on minarets during Ramaḍān should be extinguished (*ṭafy al-ḳanādīl*) and when the muezzin should pronounce a

3. An extract from the corpus of tables used for timekeeping in mediaeval Cairo. These tables display the time since sunrise (*al-dāʾir*), the hour-angle (*faḍl al-dāʾir*) and the solar azimuth (*al-samt*) as a function of solar altitude (here 15°) and each degree of solar longitude (*al-ʿadad*). Time is reckoned in equatorial degrees (1° = 4 minutes since 360° = 24 hours). Taken from Ms Cairo Dār al-Kutub *mīqāt* 690, fols. 15b-16a, with kind permission of the Egyptian National Library.

PLATE X

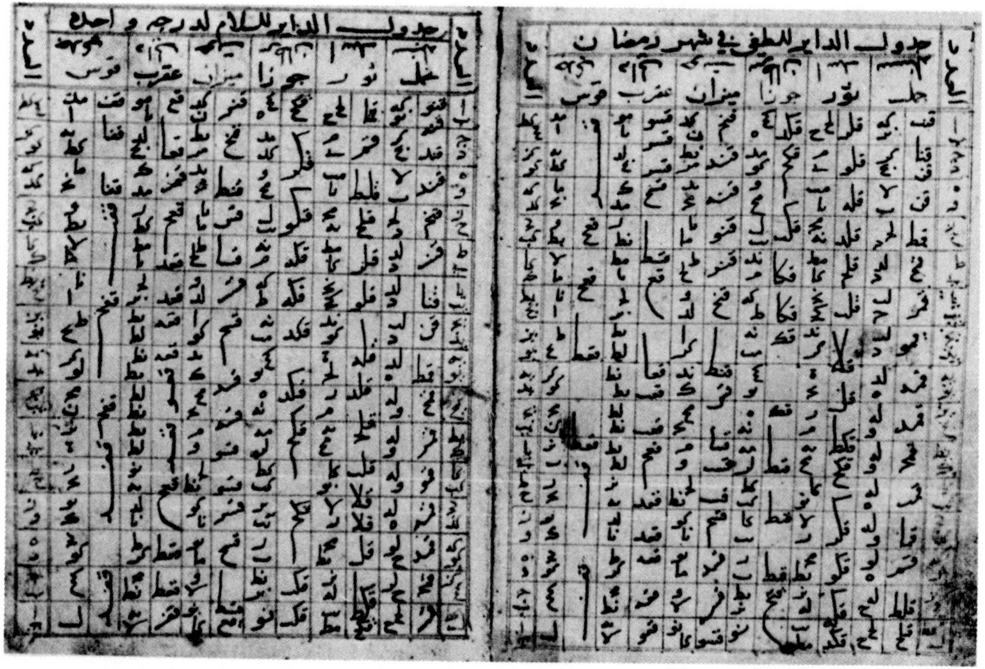

4. Another extract from the tables in the Cairo corpus displaying the times since sunrise up to the prescribed moments for the *muʾadhdhin* to extinguish the lamps on the minarets during Ramaḍān and to announce a benediction on the Prophet Muḥammad. Taken from Ms Cairo Dār al-Kutub *mīḳāt* 954,2, fols. 3b-4a, with kind permission of the Egyptian National Library.

blessing (*salām*) on the Prophet Muḥammad: see Pl. X, 4. In some copies, early and late, there is a table for orienting the large ventilators (*bādahandj* [see BĀDGĪR in Suppl.]) on Cairo rooftops so that their south-eastern sides face winter sunrise [see MAKKA. 4. As the centre of the world].

Al-Maḳsī also compiled an extensive treatise on sundial theory, including tables of coordinates for marking the curves on horizontal sundials for different latitudes and vertical sundials for Cairo with any inclination to the local meridian [see MIZWALA]. The latter were particularly useful for constructing sundials on the walls of mosques in Cairo, and the special curves for the *ẓuhr* and *ʿaṣr* enabled the faithful to see how much time remained until the muezzin would announce the call to prayer.

A contemporary of al-Marrākushī and of al-Maḳsī named Nadjm al-Dīn al-Miṣrī was clearly disturbed by the fact that the available tables for timekeeping served only specific latitudes and that they were either for timekeeping by day of for timekeeping by night. He compiled a table for timekeeping which served all latitudes and could be used for timekeeping by the sun or by the stars. In his table, one feeds in the instantaneous altitude of the sun or any star, as well as the meridian latitude and half arc of visibility of the sun or of that star, and simply reads off the time since rising of the sun or of the star. The table contains over 250,000 entries and survives in a unique manuscript probably copied by Nadjm al-Dīn himself.

Another region of the Islamic world in which the writings of al-Marrākushī and the output of the early Cairo *muwaḳḳit*s were influential was the Yemen. Under the Rasūlids [*q.v.*], mathematical astronomy was practiced and patronised. In particular, the Sultan al-Ashraf (1295-6) compiled a treatise on instrumentation inspired by that of al-Marrākushī. The Yemeni astronomer Abu 'l-ʿUḳūl, who worked for the Sultan al-Muʾayyad in Taʿizz, compiled a corpus of tables for timekeeping by day and night which is the largest such corpus compiled by any

Muslim astronomer, containing over 100,000 entries.

In Cairo in the 8th/14th century there were several *muwaḳḳit*s producing works of scientific merit but the major scene of *ʿilm al-mīḳāt* was Syria. The Aleppo astronomer Ibn al-Sarrā<u>dj</u>, who is known to have visited Egypt, devised a series of special quadrants and universal astrolabes and trigonometric grids; his works represent the culmination of the Islamic achievement in astronomical instrumentation.

Two other major Syrian astronomers, al-Mizzī and Ibn al-<u>Sh</u>āṭir, studied astronomy in Egypt under Ibn al-Akfānī [*q.v.*] and others. Al-Mizzī returned to Syria and compiled a set of hour-angle tables and prayer-tables for Damascus modelled after the Cairo corpus. Ibn al-<u>Sh</u>āṭir compiled some prayer-tables for an unspecified locality with latitude 34° (the new Mamlūk city of Tripoli?). Al-Mizzī also compiled various treatises on instruments, but Ibn al-<u>Sh</u>āṭir turned his attention to theoretical astronomy and devised a new series of models which represent the culmination of the Islamic endeavour in this field [see ʿILM AL-HAYʾA]. This notwithstanding, he also devised the most splendid sundial known from the Islamic Middle Ages, the fragments of which are preserved in the garden of the Archaeological Museum in Damascus; a replica made in the late 19th century by the last Syrian *muwaḳḳit* of any reknown, al-Ṭanṭāwī, is still *in situ* on the main minaret of the Umayyad Mosque [see MIZWALA].

It was a colleague of al-Mizzī and Ibn al-<u>Sh</u>āṭir named <u>Sh</u>ams al-Dīn al-<u>Kh</u>alīlī who made the most significant advances in *ʿilm al-mīḳāt*. Al-<u>Kh</u>alīlī recomputed the tables of al-Mizzī for the new parameters (local altitude and obliquity of the ecliptic [see MAYL]) derived by Ibn al-<u>Sh</u>āṭir. His corpus of tables for timekeeping by the sun (*i.e.* functions $t(h,\lambda)$ and $T(h,\lambda)$) and for regulating the times of prayer for Damascus was used there until the 13th/19th century: see Pl. XI, 5. In addition, he compiled some tables of auxiliary trigonometric functions for solving all of the standard problems of spherical astronomy for any latitude. The functions tabulated are:

$$f(\varphi,\theta) = R \text{ Sin } \theta \,/\, \text{Cos } \varphi,$$

$$g(\varphi,\theta) = \text{Sin } \theta \text{ Tan } \varphi \,/\, R, \text{ and}$$

$$K(x,y) = \text{arc Cos } (R\, x \,/\, y),$$

where the trigonometric functions are to base $R = 60$. Values are given to two sexagesimal digits and are invariably computed accurately. Using these tables, the hour-angle can be found with a minimum of calculation. Al-Khalīlī describes a procedure equivalent to:

$$t(h,\ \delta,\ \varphi) = K\{\,[f(\varphi,h)\ -\ g(\varphi,\delta)],\ \delta\,\}$$

which is mathematically equivalent to the modern formula noted above. Al-Khalīlī also compiled a table displaying the *ḳibla* or local direction of Mecca as a function of terrestrial longitude and latitude [see ḲIBLA]. These tables represent the culmination of the Muslim achievement in the computation of tables for spherical astronomy.

Some of the activities of the Damascus school became known in Tunis in the 8th-9th/14th-15th centuries. Extensive auxiliary tables and prayer-tables for the latitude of Tunis were compiled there by astronomers whose names are not known to us. Prayer-tables were also prepared for various latitudes in the Maghrib.

More significant was the influence of the Cairo and Damascus schools on the development of *ʿilm al-mīḳāt* in Ottoman Turkey. Cairene and Damascene astronomers of the 8th/14th century had already prepared sets of prayer-tables for the latitude of Istanbul, as well as for Jerusalem, Mecca, and Aleppo, but several new sets of tables were prepared by Ottoman astronomers for Istanbul and elsewhere in Turkey after the model of the corpora for Cairo and Damascus. Prayer-tables for Istanbul are contained in the very popular almanac [see RŪZNĀMA] of the 9th/15th century Ṣūfī Shaykh Wefāʾ (see Pl. IX, 2), and in the less widely distributed almanac of the 10th/16th century scholar Darendeli. These remained in use until the 19th century.

Large sets of tables for timekeeping by the sun

and/or stars were prepared for Istanbul and for Edirne. One set for the sun was compiled by Taḳī 'l-Dīn b. Maʿrūf [*q.v.*], director of the short-lived Istanbul Observatory in the late 10th/16th century. In the 12th/18th century, the architect Ṣāliḥ Efendī produced an enormous corpus of tables for timekeeping which was very popular amongst the *muwaḳḳit*s of Istanbul. A feature distinguishing some of these Ottoman tables from the earlier Egyptian and Syrian tables is that values of the time of day are based on the convention that sunset is 12 o'clock. This convention, inspired by the fact that the Islamic day begins at sunset (because the calendar is lunar and the months begin with the sighting of the crescent shortly after sunset), has the disadvantage that clocks registering "Turkish" time need to be adjusted by a few minutes each few days. Prayer-tables based on this convention were compiled all over the Ottoman Empire and beyond; examples have been found in the manuscript sources for localities as far apart as Algiers and Yarkand, and Crete and Ṣanʿāʾ: see Pl. XI, 6.

In the 19th/20th centuries, the times of prayer were and are tabulated in annual almanacs, wall calendars and pocket diaries, and the times for each day are listed in newspapers. In Ramaḍān, special sets for the whole month are distributed. These are called *imsākiyya*s and indicate in addition to the times of prayer, the time of the early morning meal (*suḥūr*) and the time before daybreak—called the *imsāk*—when the fast should begin. Modern tables are usually prepared by the local survey department or observatory or some other agency approved by the religious authorities (see Fig. 3). Recently, electronic clocks and watches have appeared on the market which are programmed to beep at the prayer times for different localities, and to pronounce a recorded prayer-call [see ADHĀN].

Bibliography: On the definition of the times of prayer as they appear in the astronomical sources, see E. Wiedemann and J. Frank, *Die Gebetszeiten im Islam*, in *SBPMSE*, lviii (1926), 1-32, repr. in E. Wiedemann, *Aufsätze zur arabischen*

PLATE XI

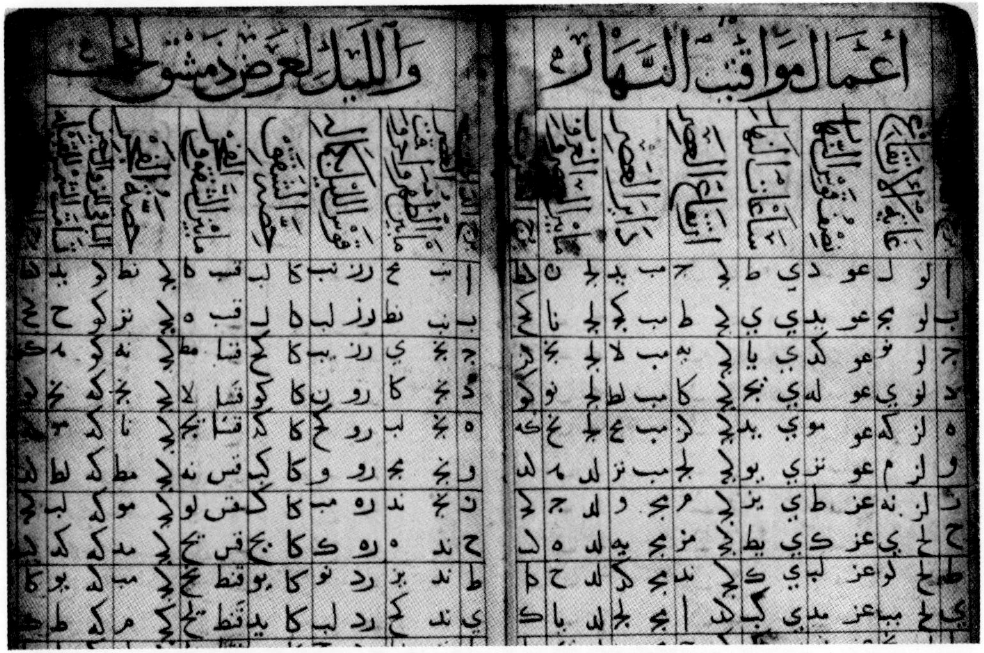

5. An extract from al-Khalīlī's corpus of tables for Damascus showing some 12 functions for regulating the times of prayer. Values are displayed for each degree of solar longitude (this extract serves the first few degrees of Aquarius). Taken from Ms Paris B.N. ar. 2558, fols. 10b-11a, with kind permission of the Bibliothèque Nationale.

PLATE XI

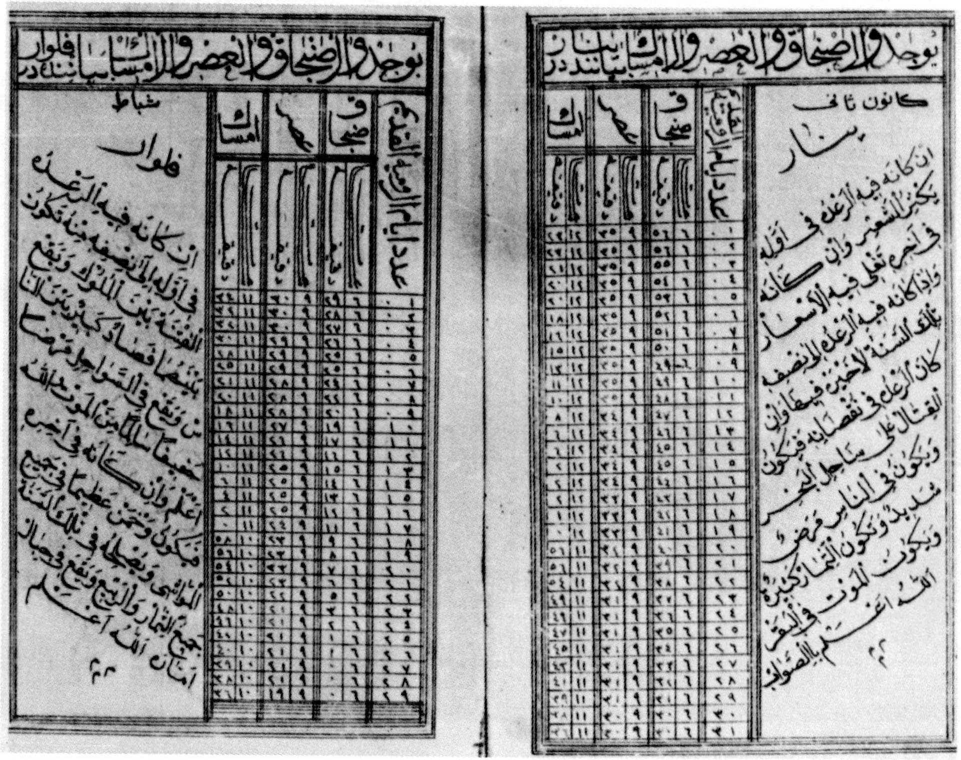

6. An extract from some Ottoman prayer-tables for the latitude of Algiers. For each day of each month (here February) the times of the *ḍuḥā* (here called *ṣandjāk*), the ʿ*aṣr* and the *imsāk* are tabulated. The times are expressed in hours and minutes and are reckoned according to the Ottoman convention that sunset is 12 o'clock. Taken from Ms Cairo Ṭalʿat *falak turkī* 9, fol. 41a, with kind permission of the Egyptian National Library.

Wissenschaftsgeschichte, Hildesheim 1970, ii, 757-88. For al-Bīrūnī's detailed discussion, see E.S. Kennedy, *The Exhaustive Treatise on Shadows by ... al-Bīrūnī: translation and commentary*, Aleppo 1976, i, 210-30, ii, 132-43, and idem *et alii, Studies in the Islamic exact sciences*, Beirut 1983, 299-310. On the formulae for timekeeping used by the Muslim astronomers, see the papers by M.-L. Davidian, N. Nadir and B.R. Goldstein repr. in Kennedy *et alii, Studies*, 274-96, and the studies listed below. On the detailed treatment of spherical astronomy by Ibn Yūnus and al-Bīrūnī, see, respectively, D.A. King, *Spherical astronomy in medieval Islam: the* Ḥākimī Zīj *of Ibn Yūnus* (forthcoming), and M.-Th. Debarnot, *La trigonométrie sphérique chez les Arabes de l'est à la fin du Xᵉ siècle*, Damascus 1985. On al-Marrākushī and his treatise, and the rise of the institution of the *muwaḳḳit*, see King, *The astronomy of the Mamluks*, in *Isis*, lxxiv (1983), 531-55, esp. 539-40 and 534-5, repr. in idem, *Islamic mathematical astronomy* (hereafter *IMA*), London 1986, III. On the earliest known tables for regulating the prayer times and reckoning time of day from solar altitude, see King, *Al-Khwārizmī and new trends in mathematical astronomy in the ninth century*, in *Occasional Papers on the Near East* (New York University, Hagop Kevorkian Center for Near Eastern Studies), ii (1983), 7-11. On the corpora of tables for Cairo, Taᶜizz, Damascus and Jerusalem, Tunis and Istanbul, see respectively, King, *Ibn Yūnus'* Very Useful Tables *for reckoning time by the sun*, in *Archive for history of exact sciences*, x (1973), 342-94, repr. in idem, *IMA*, IX; *Mathematical astronomy in medieval Yemen ...*, Malibu, Calif. 1983, 31-2; *Astronomical timekeeping in fourteenth-century Syria*, in *Procs. of the First International Symposium for the History of Arabic Science (Aleppo, 1976)*, ii, 75-84 and plates, repr. in idem, *IMA*, X; (with E.S. Kennedy) *Indian astronomy in fourteenth-century Fez ...*, in *Journal for the History of Arabic Science*, vi (1982), 3-45, repr. in idem, *IMA*, VIII, esp. 8-9; and *Astronomical*

الأربعاء

ميلادية	هجرية	شمسية
١٩٧٤	١٣٩٤	١٣٩٠
مايو	ربيع ثاني	نيسان
١	٩	١٨

الأوقات

فجر	٥٦–٩	للتوقيت العربية السعيدة	عصر	١٣–٨
شروق	٦ –١١		مغرب	١٢
ظهر	٣٨–٥		عشاء	١

البرج	منزلة الشمس	معالم الزراعة
الثور	البطين	غروب كامد
٩	٨	٨

زيارتك لمكتبة الجيل الجديد بصنعاء جولة في عالم جديد

a

الاربعاء ١١ جادى الاول ٢١ مايو ايار ٣٠ الثور

عرض	فجر	اشراق	ظهر	عصر	عشاء
مكة	٩,٢٢	١٠٫٤٩	٥,٢٧	٨٫٤٢	١٫٣٠

الخميس ١٢ جادى الاول ٢٢ مايو ايار ٣١ الثور

المدينة	٩,٧	١٠٫٣٨	٥٫٢٢	٨٫٤٧	١٫٣٠
نجد	٩٫١١	١٠٫٤٠	٥٫٢٣	٨٫٤٥	١٫٣٠

استخفافاً بحقهن كان له عند الله عهد –

b

كل عام وانتم بخير
امساكية
شهر رمضان المبارك
لعام ١٣٩٥ هـ.

f

الأيام	رمضان	الطول	الامساك افرنجي س د	الامساك عربي س د	الفجر افرنجي س د	الفجر عربي س د	الشمس افرنجي س د	الشمس عربي س د	الظهر افرنجي س د	الظهر عربي س د	العصر افرنجي س د	العصر عربي س د	المغرب س	المغرب د	العشاء افرنجي س د	العشاء عربي س د
السبت	١	٦	٣ ٢٣	٩ ١٦	٣ ٤٣	٩ ٣٦	٥ ١٦	١١ ٩	١١ ٤٠	٥ ٣٤	٣ ١٦	٩ ٩	٦	٧	٧ ٣٠	١ ٢٣
الاحد	٢	٧	٢٣	١٨	٤٣	٣٨	١٦	١١	٤٠	٣٥	١٤	٩		٥	٢٨	٢٣
الاثنين	٣	٨	٢٤	٢٠	٤٤	٤٠	١٧	١٣	٤٠	٣٦	١٤	١٠		٤	٢٦	٢٣
الثلاثاء	٤	٩	٢٥	٢٢	٤٥	٤٢	١٨	١٥	٤٠	٣٧	١٤	١١		٣	٢٥	٢٣

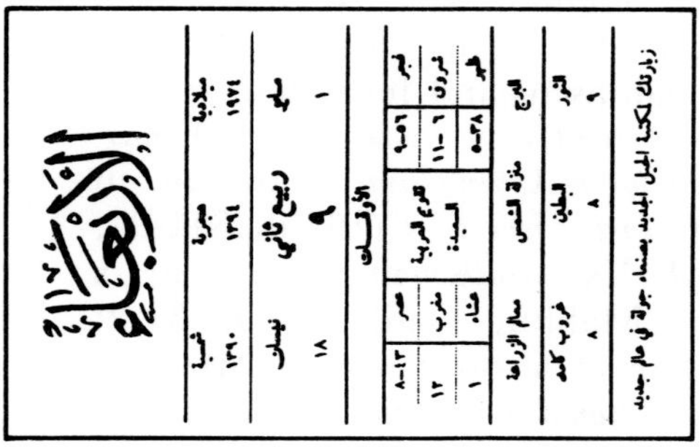

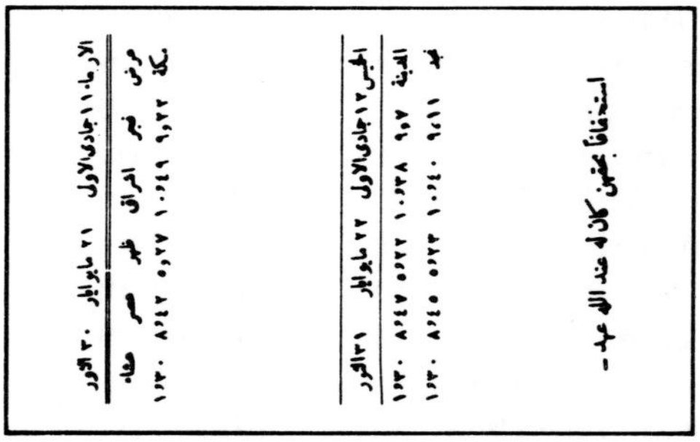

c

٨) شعبان (برائت) آیی

٢٩ کون (١ــ١٥ گهچه)

1984 نچی میلادی ییلی — ١٤٠٤ نچی هجری سنه

نماز وقتلری تاشکینت وقتی ایله

شعبان	هفته کونلری	Mali	صبح تانگ س	صبح تانگ م	طلوع کون چیقیش س	طلوع کون چیقیش م	قیام توش س	قیام توش م	عصر نماز دیگر س	عصر نماز دیگر م	مغرب کون باتیش س	مغرب کون باتیش م	عشاء خفتن س	عشاء خفتن م
١	سیهشنبه	1	٤	٤٨	٦	٢١	١	٢٠	٦	٢١	٨	٢٠	٩	٥١
٢	چهارشنبه	2	٤	٤٦	٦	١٩	١	٢٠	٦	٢٢	٨	٢٢	٩	٥٢
٣	پنجشنبه	3	٤	٤٦	٦	١٨	١	٢٠	٦	٢٣	٨	٢٣	٩	٥٣
٤	جمعه	4	٤	٤٣	٦	١٧	١	٢٠	٦	٢٤	٨	٢٤	٩	٥٣
٥	شنبه	5	٤	٤١	٦	١٦	١	١٩	٦	٢٥	٨	٢٥	٩	٥٥
٦	یکشنبه	6	٤	٣٩	٦	١٥	١	١٩	٦	٢٦	٨	٢٦	٩	٥٦
٧	دوشنبه	7	٤	٣٨	٦	١٣	١	١٩	٦	٢٧	٨	٢٧	٩	٥٧
٨	سیهشنبه	8	٤	٣٧	٦	١٢	١	١٩	٦	٢٨	٨	٢٨	٩	٥٨
٩	چهارشنبه	9	٤	٣٦	٦	١١	١	١٩	٦	٢٩	٨	٢٩	٩	٥٩
١٠	پنجشنبه	10	٤	٣٥	٦	٩	١	١٩	٦	٣٠	٨	٣٠	١٠	٠

d

e

Fig. 3. Extracts from some modern tables for regulating the times of prayer for different localities. The examples shown serve (a) Yemen; (b) Mecca, Medina and Najd; (c) Tunis; (d) Tashkent; (e) Istanbul, Ankara, Izmir and Erzerum; and (f) Nābulus (for Ramadan, 1395/1975).

timekeeping in Ottoman Turkey, in *Procs. of the International Symposium on the Observatories in Islam (Istanbul, 1977)*, 245-69, repr. in idem, *IMA*, XII. Three sets of auxiliary tables are analysed in R.A.K. Irani, *The* Jadwal al-taqwīm *of Ḥabash al-Ḥāsib*, unpubl. Master's diss., American University of Beirut 1956; C. Jensen, *Abū Naṣr's approach to spherical astronomy as developed in his treatise* The Table of Minutes, in *Centaurus*, xvi (1972), 1-19; and King, *Al-Khalīlī's auxiliary tables for solving problems of spherical astronomy*, in *Journal for the History of Astronomy*, iv (1973), 99-110, repr. in idem, *IMA*, XI. See further King, *Studies in astronomical timekeeping in Islam, I. A survey of tables for reckoning time by the sun and stars*; *II. A survey of tables for regulating the times of prayer; III. A survey of shadow schemes for simple time-reckoning; IV. On the origin of the prayers in Islam;* and *V. On the role of the Muezzin and the Muwaqqit in medieval Islamic society* (forthcoming, I-II with Springer-Verlag, New York, III-IV in *Oriens*). On the Ottoman convention of reckoning sunset as 12 o'clock, see J. Würschmidt, *Die Zeitrechnung im Osmanischen Reich*, in *Deutsche optische Wochenschrift* (1917), 88-100. On the *muwaḳḳit-khānas*, the buildings adjacent to the major Ottoman mosques which were used by the *muwaḳḳits*, see A.S. Ünver, *Osmanlı Türklerinde ilim tarihinde muvakkithaneler*, in *Atatürk konferensları*, v (1975), 217-57. See also the *Bibl.* to MIZWALA.

VI

Universal Solutions in Islamic Astronomy

What I mean by universal solutions [1] are instruments, mathematical procedures and tables which aim at providing solutions for all terrestrial latitudes to those problems of spherical astronomy in which latitude is involved. I am fully aware that I am imposing a modern categorization on the medieval source material, but it seems to me useful to consider these solutions together and thereby to try to add a new dimension to our understanding of the achievements of the Muslim scientists.

The expression 'universal solution' is not a particularly happy one, and I use it for lack of anything better. In our sources, the notion is usually but not always restricted to the oekoumene; I have resisted the temptation to use the term oecumenical in my title even though Asger might have enjoyed it. To illustrate the problem of terminology, we note that, for example, all trigonometric solutions are general and thus universal, but these I shall not treat. Given the formula, calculations still have to be performed for different latitudes. Instruments designed to facilitate such computations I shall discuss. Tables of various spherical astronomical functions computed for each of the climates of classical geography are in a sense not universal, but these will be considered. Tables which serve each degree of latitude between the Yemen and the Bosphorus I regard as universal in that, as remarked by Abū Naṣr (Central Asia, *ca.* 1000), there was no-one beyond these bounds who was interested in spherical astronomy anyway [2]. I hope that the reader will gain a clearer notion of precisely what I mean by universal solutions as he or she reads on.

The problems of spherical astronomy essentially involve converting coordinates from one orthogonal coordinate system to another. Of particular interest to Muslim astronomers was the determination of time from an observed solar or stellar altitude, that is, converting from a horizon- to an equator-based coordinate system. This and other problems can be solved using either an analogue computer or trigonometry. With the former, the instantaneous celestial configuration relative to the horizon is represented in three or two dimensions; then the time is read from a scale. Otherwise the three-dimensional problem is solved by spherical trigonometry, or is reduced to two dimensions and solved by plane trigonometry; then the resulting formula is applied to the numerical data, either to calculate the solution by means of trigonometric tables or to derive it without computation by means of a trigonometric grid.

This one aspect of Muslim scientific creativity that I shall describe was a direct continuation of various Greek developments along the same lines. One example of a 'universal solution' inherited by the Muslims from the Greeks is the planisphaeric astrolabe with plates for each of the seven climates or for each few degrees of latitude. The astrolabe can be made to simulate the apparent daily rotation of the heavens over the horizon of any locality. The underlying theory of stereographic projection, together with a numerical investigation of the size and position of the various markings on such an instrument, is presented in Ptolemy's *Planisphaerium* [3]. Other examples are Ptolemy's tables in *Almagest* II.8 and II.13 of oblique ascensions and the angles between the ecliptic and the horizon and meridian for each of the climates [4]; his universal dial for coordinate conversion and tables for this purpose (and hence also for constructing sundials) described in the *Analemma* [5]; and the tables for planetary visibility and parallax in the *Handy Tables* [6]. Of these works, only the *Analemma* was not translated into Arabic (contrary to what is often maintained in the secondary literature).

The main centres of the activity I shall describe were Baghdad in the ninth century, Andalusia in the eleventh, and from the thirteenth to the fifteenth, Egypt and Syria [7]. Some of the tables occur in the astronomical handbooks called *zīj*es [8], but many are found in other compilations [9]. Most of the material I survey here is described in greater detail in other papers, not all published yet.

1. TABLES FOR ASTROLABE CONSTRUCTION

A more accurate means of making astrolabes than geometric construction is to compute the radii of the various circles and the positions of their centres relative to the centre of the instrument. Al-Khwārizmī (Baghdad, *ca.* 825) devised an auxiliary function for generating values of such radii and centre-distances. (His method was, of course, equivalent to that of Ptolemy.) His contemporary al-Farghānī went so far as actually to tabulate both of these functions for each degree of latitude from 15° to 50° and each degree of altitude for the almucantars and each degree of azimuth for the azimuth circles, thus achieving a universal solution to the problem of astrolabe construction.

Later Muslim astronomers, such as the Yemeni Sultan al-Ashraf (*ca.* 1295), compiled astrolabe tables for a set of latitudes of specific localities. A 14C Egyptian astronomer completed al-Farghānī's tables so that they served each degree of terrestrial latitude from 1° to 90°. Astrolabe tables were apparently widely used in medieval Egypt and Syria, if not elsewhere. In Iran, a grid called *dastūr* which displayed graphically values of the auxiliary function tabulated by al-Farghānī seems to have been more popular [10].

2. TABLES FOR SUNDIAL CONSTRUCTION

The tables for coordinate conversion in Ptolemy's *Analemma* could be used to contruct markings for the seasonal hours on different kinds of planar sundials for all climates. Although the Muslims, as far as we know, were not familiar with this work of Ptolemy, they did develop more practical tables for sundial construction. Al-Khwārizmī compiled a set of tables for marking the curves on horizontal sundials for a series of latitudes. His tables display radial coordinates for the intersections of the hour lines with the solstitial curves for each of these latitudes. Al-Khwārizmī also presented a table for marking a polar sundial.

Al-Farghānī wrote on sundial theory, but unfortunately his treatise is lost. Thābit ibn Qurra (Baghdad, *ca.* 875) produced a treatise on sundial theory (without tables) along the same lines as Ptolemy's *Analemma,* apparently independently. His contemporary Ibn al-Adamī compiled a set of tables to facilitate the marking of curves on vertical sundials inclined at any angle to the meridian for any latitude. Later astronomers compiled sundial tables for latitudes of specific localities such as Cairo, Damascus and Jerusalem. Such tables remained in use for centuries in Egypt and Syria [11].

3. THE SPHERICAL ASTROLABE

The spherical astrolabe was apparently developed from the planisphaeric astrolabe. It is first described in a unique manuscript preserved in Istanbul and attributed to Ḥabash (Baghdad, *ca.* 850). The instrument consists of two parts: (1) a globe divided into two hemispheres by a circle representing the horizon of any locality, the upper hemisphere marked with almucantar circles, a meridian and azimuth circles, and the lower with markings for the seasonal hours; and (2) a rete with ecliptic and star pointers, attached to the globe by pins fitted into holes on the meridian. The ensemble thus constitutes a universal analogue computer. Several later descriptions survive, as well as one complete 15C instrument [12].

4. THE UNIVERSAL ASTROLABE AND QUADRANT

Another universal solution which was first developed in 9C Baghdad was the universal plate *(ṣafīḥa āfāqīya),* a single astrolabe plate displaying stereographic projections of horizons for each few degrees of latitude. Ḥabash devised such a plate for use with an astrolabe rete, and various later astronomers in the Islamic East wrote on its use [13]. This tradition has yet to be investigated properly.

More familiar to historians of science is the universal astrolabe devised in 11C Toledo by ᶜAlī ibn Khalaf al-Shakkāz. In this, known only from the 13C *Libros del saber,* the plate bears so-called *shakkāzīya* markings based on a stereographic projection for latitude zero. The rete consists of a semicircle of *shakkāzīya* markings and a semi-circle comprising an ecliptic and star pointers. Clearly, al-Shakkāz' astrolabe is extremely useful for solving problems of spherical astronomy, but the instrument no longer simulates the apparent daily rotation of the heavens over any horizon. His better-known contemporary al-Zarqāllu replaced the rete by an alidade fitted with a movable perpendicular cursor to produce a universal plate. On another plate called *zarqāllīya* for converting ecliptic and equatorial coordinates he introduced a second set of *shakkāzīya* markings inclined to the first at an angle equal to the obliquity [14].

The universal astrolabe of ᶜAlī ibn Khalaf was reinvented by Ibn al-Sarrāj (Aleppo, *ca.* 1325). A second, more sophisticated version, designed and constructed by Ibn al-Sarrāj himself, is now housed in the Benaki Museum in Athens. A treatise by al-Wafā'ī (Cairo, *ca.* 1450), who owned the very instrument now preserved in Athens, enables us to understand what Ibn al-Sarrāj intended to achieve with it. This splendid astrolabe could be used universally in five different ways. First and second, it could be used in the same way as the astrolabe of al-Shakkāz exploiting the features of either of the two halves of the rete; third, the rete could be used with a series of quarter plates inscribed with almucantar markings for each

3° of latitude; fourth, the rete could be used with a set of horizons for each 3° of latitude; and fifth, there is on the back a highly ingenious trigonometric grid and alidade (see below) [15].

Jamāl al-Dīn al-Māridīnī (Cairo, *ca.* 1400) devised a quadrant based on the same principle as the universal astrolabe. Single *shakkāzīya* quadrants were also used by Egyptian and Syrian astronomers, but with these only approximate solutions could be attained [16].

5. QUADRANTS AND TRIGONOMETRIC GRIDS

Al-Khwārizmī or one of this contemporaries developed a sine quadrant with a set of lines drawn parallel to one edge. This could be used with a thread attached at the centre and a movable marker to perform trigonometric calculations graphically. Later, grids with two orthogonal sets of parallels were devised for the same purpose [17]. The horary quadrant with a movable cursor for timekeeping by the sun for any latitude is likewise first attested in a treatise from 9C Baghdad, a copy of which is now preserved in Cairo. Some later horary grids were designed for specific latitudes, but others were intended to be used for any latitude, the procedure being approximate. Such sine and universal horary quadrants are often found on the backs of Islamic astrolabes [18].

Several Egyptian and Syrian astronomers devised different kinds of trigonometric grids, of which the most ingenious (if least practical) is the one on the back of the astrolabe of Ibn al-Sarrāj. Such grids could be used to yield numerical solutions to all of the standard problems of spherical astronomy for any latitude, although only the simpler sine quadrant can be said to have been widely used for such purposes [19].

6. COMPENDIA

Ibn al-Shāṭir (Damascus, *ca.* 1350) devised a compendium, that is, a multi-functional instrument, in the form of a box with a hinged lid and various movable parts. Some parts of this "jewel-box" made by Ibn al-Shāṭir himself survive in Aleppo. A treatise preserved in Berlin explains the way in which the ensemble is to be used. The box contained a magnetic compass for aligning it in the cardinal directions and a polar sundial on a plate which could be attached to the lid and erected in a plane perpendicular to the celestial equator for any of a series of latitudes. Also, sights could be attached to the lid to measure the hour-angle at any of these latitudes, and a set of horizons engraved on the lid enabled the user to compute the instantaneous configuration of the ecliptic relative to the appropriate horizon. In addition, the plate bore markings indicating the qiblas or local directions of Mecca for a series of localities.

Al-Wafā'ī (Cairo, *ca.* 1450) devised an equatorial semi-circle with a sighting device, inclined at an adjustable angle to a base that could be oriented in the cardinal directions by means of a compass. Several examples of this instrument and copies of al-Wafā'ī's treatise on its use are available [20].

Both of these instruments were the precursors of the simpler qibla boxes which were widely used in Ottoman times. Any influence on the development of compendia in Europa, if such there was, has yet to be established.

7. TABLES OF SPHERICAL ASTRONOMICAL FUNCTIONS

Several Muslim astronomers from the 10C onwards presented in their *zījes* extensive sets of oblique ascension tables for each degree of terrestrial latitude (up to about 40°) [21]. Other spherical astronomical functions were generally tabulated only for the latitude of the locality where the *zīj* was compiled, although sometimes auxiliary tables were presented with which tables of such functions could be conveniently generated for all latitudes [22].

For example, from the 9C onwards we find tables of tan $\delta(\lambda)$, where δ is the solar declination and λ the solar longitude. With such a table, the equation of half daylight $d\,(\lambda, \phi)$ for latitude ϕ is given by the relation:

$$\sin d\,(\lambda, \phi) \;=\; \tan \delta(\lambda) \cdot \tan \phi\;,$$

and the oblique ascensions $\rho(\lambda, \phi)$ can then be found from the right ascensions $\alpha(\lambda)$ using

$$\rho(\lambda, \phi) \;=\; \alpha(\lambda) \;-\; d(\lambda,\phi)\;.$$

8. TABLES FOR TIMEKEEPING

Muslim astronomers compiled numerous tables for timekeeping [23]. Al-Khwārizmī compiled a table serving all latitudes and displaying the solar altitude as a function of the meridian altitude and the time since sunrise in seasonal hours. Ibn Amājūr (Baghdad, *ca.* 900) compiled a table intended to serve all latitudes displaying the time in seasonal hours as a function of the observed altitude and the meridian altitude. The formula underlying these two tables was necessarily approximate. However, Ibn Amājūr also compiled a table of time since sunrise for the latitude of Baghdad as a function of the same arguments based on an exact formula. Both of his tables are extant in a unique manuscript preserved in Paris.

Various Muslim astronomers compiled similar and still more sophisticated tables for timekeeping by the sun and stars for specific latitudes, often containing tens of thousands of entries. One, compiled by Najm al-Dīn al-Miṣrī (Cairo, *ca.* 1250), was different from all the others in that it was not only universal but could be used for timekeeping by the sun or by any star. The two halves of the unique manuscript are preserved in Cairo and Oxford. In this table, one enters with the meridian altitude, the observed altitude, and the half-arc of visibility for the celestial body in question, and reads off the time since rising. There are over a quarter of a million entries, a feature of the table which would not attract many copyists; indeed there is even reason to suppose that the unique manuscript is in the hand of Najm al-Dīn himself and that the table was never copied thereafter.

9. AUXILIARY TABLES

Ḥabash devised and tabulated a set of trigonometric functions with no direct astronomical significance, but such that successive applications thereof would yield the solutions of some of the standard problems of spherical astronomy [24]. Several later astronomers tried their hand at compiling such auxiliary tables, and the ultimate solution was achieved by al-Khalīlī (Damascus, *ca.* 1360).

Al-Khalīlī tabulated three double-argument functions for each degree/integer of both arguments, which we denote (with the standard capital notation to represent medieval functions to bases other than unity [here R = 60]) as follows:

$$f(\phi,\theta) = R \text{ Sin } \theta / \text{Cos } \phi \, ,$$
$$g(\phi,\theta) = \text{Sin } \theta \text{ Tan } \phi / R \, ,$$
$$K(x,y) = \text{arc Cos } (x\, R / \text{Cos } y) \, .$$

The first two functions are tabulated for each degree of latitude from 1° to 55°. With these tables, the most complicated problems of spherical astronomy reduce to simple arithmetical operations. Thus, for example, if one wishes to determine the hour angle t or the azimuth a (measured from the meridian) for latitude when a celestial body with declination has altitude h, one simply performs the following:

$$t = K\{[f(\phi,h) - g(\phi,\delta)], \delta\}, \quad \text{and}$$
$$a = K\{[g(\phi,h) - f(\phi,\delta)], h\}.$$

Al-Khalīlī's tables were known in Egypt and Turkey as well as in Syria, and they survive in several copies. A few later astronomers compiled auxiliary tables, but none was as sophisticated or as useful as those of al-Khalīlī. Indeed his tables represent the ultimate numerical solution to problems of spherical astronomy in the medieval context. He even described how to use the tables for converting ecliptic and equatorial coordinates.

10. TABLES FOR DETERMINING THE TIMES OF PRAYER

The times of the five daily prayers in Islam are astronomically defined, the daytime prayers in terms of shadow lengths and the nighttime ones in terms of twilight phenomena. Already al-Khwārizmī prepared a set of tables for regulating the times of the daytime prayers, computed for the latitude of Baghdad. In the succeeding centuries highly sophisticated tables for regulating the daytime and nighttime prayers were computed for various urban centres in the Muslim world [25].

Al-Wafā'ī is supposedly the compiler of an extensive set of prayer-tables, extant in a unique manuscript in Princeton, with tables for each degree of latitude from 15° to 40°, another universal solution. Also Ibn Abi 'l-Khayr al-Ḥusnī (Cairo, *ca.* 1550) compiled a table displaying the duration of twilight at the equinoxes and solstices for each half climate, extant in a unique manuscript in Cairo. Certain Egyptian astronomers attributed to their illustrious predecessor Ibn Yūnus (*ca.* 990) – correctly or not, we do not know – a theory for the effect of refraction on the arc of daylight, with numerical values for the equinoxes and solstices at Cairo. ᶜAbd al-Qādir al-Minūfī (Cairo, *ca.* 1550) compiled a table purporting to display the correction for refraction at the horizon for each degree of terrestrial latitude, extant in a unique manuscript in Istanbul [26].

11. TABLES FOR DETERMINING LUNAR AND PLANETARY VISIBILITY

The Muslim calendar is lunar and is regulated by the sighting of the lunar crescent. Numerous Muslim astronomers adopted the simple Indian condition that visibility occurs if the difference in setting times between the sun and moon is 12 equatorial degrees. Al-Khwārizmī compiled a table displaying the minimum ecliptic distance between the two luminaries to ensure visibility according to this Indian condition. Values were computed for each zodiacal sign and were specifically for the latitude of Baghdad. Later more complicated theories were formulated and other lunar visibility tables were compiled, mostly for specific latitudes [27].

One astronomer, however, probably in Andalusia during the eleventh or twelfth century but possibly earlier, compiled a table seemingly based on Indian theory for each of the climates, another universal solution. This is available in several extremely corrupt copies mainly of Maghribi and Egyptian provenance.

The *Zīj*es of Ḥabash and al-Battānī (Raqqa, *ca.* 910) contain planetary visibility tables for each of the climates borrowed from the *Handy Tables*. The *Zīj* of al-Khāzinī (Merv, *ca.* 1125), however, has a different set based on an Indian theory and computed by one of his predecessors [28].

12. TABLES FOR PARALLAX

Theon's tables of parallax components for each of the climates were incorporated into several Islamic *zīj*es. The corresponding tables in the *Zīj* of Ulugh Beg (Samarqand, *ca.* 1425), however, are computed for his newly-found value of the obliquity and for each 5° of latitude from 25° to 50° [29].

13. THE DETERMINATION OF THE QIBLA

The first mathematical solutions to the problem of finding the qibla or local direction of Mecca, apparently dating from the early 9C, were cartographic, and hence approximate. Exact solutions derived by means of an analemma or by spherical trigonometry were formulated in Baghdad within a few decades [30]. An unidentified 9C or early-10C Baghdad astronomer prepared a table displaying the qibla as an angle counted from the meridian for each degree (up to 20°) of two arguments which are stated to be longitude and latitude difference from Mecca. The entries in the table were based on an accurate procedure, but for the stated arguments the entries are approximate. (The second argument must be slightly modified in order that the table yield exact qibla values.)

Ibn al-Haytham (Cairo, *ca.* 1025) authored three treatises on the qibla. In one he solved the problem geometrically by means of an analemma, in another by spherical trigonometry considering 16 possible cases. In the third, known only from a reference in his autobiography, he tabulated the qibla as a function of latitude and longitude differences. Various other qibla tables, based on either approximate or exact formulae, were compiled over the centuries (including one for longitudes up to 60° on either side of the meridian of Mecca). But it was al-Khalīlī again who produced the ultimate qibla table. Based on an exact formula, and with arguments simply latitude and longitude, his table contains over 3,000 entries, each accurately computed or in error by at most 1′ or 2′. It is available to us in three manuscripts and was apparently not widely known in later astronomical circles.

NOTES

[1] This is a modified version of a paper presented in 1981 at the XVth International Congress of the History of Science in Bucharest. It is a pleasure to acknowledge my debt to my colleagues J.-P. Hogendijk and E. S. Kennedy for their comments on the penultimate version of the paper.
[2] See [Jensen 1971, 5] and [King 1986, XI, 108, n. 19].
[3] On the astrolabe in the Islamic context see the article *"Asṭurlāb"* by W. Hartner in *EI*, 2nd. ed. On the theory of stereographic projection see [Neugebauer 1975, II, 857-879].
[4] [Neugebauer 1975, I, 34-37 and 45-52].
[5] *ibid.*, II, 839-856.
[6] *ibid.*, II, 1017-1025 and 990-999.
[7] A more detailed survey of Mamluk (13C to 15C Egyptian and Syrian) universal solutions is to appear in the *Winder Festschrift*.
[8] Some 125 *zīj*es are surveyed in [Kennedy 1956].
[9] The categories of tables not contained in *zīj*es are briefly described in [King 1986, II].
[10] On the tables of al-Khwārizmī, al-Farghānī and al-Ashraf see [King 1983, 22-27; 1986, II, 53-55; 1987, II, 108-111]. A survey of all known Islamic astrolabe tables is in preparation. On the *dastūr* see, for example, [Michel 1947, 47-48].
[11] On al-Khwārizmī's tables see [King 1983, 17-22]. Thābit's treatise was published with German translation by K. Garbers and commentary by P. Luckey (see [Sezgin 1978, 168, no. 9]). A survey of Islamic sundial tables is in preparation; see [King 1986, II, 51-53] (and the works of C. Schoy there cited), and [King 1987, II, 112, and XVI].
[12] [Seemann 1925] is a survey of Islamic writings on the spherical astrolabe. Ḥabaṣh's treatise is listed in [Sezgin 1978, 175 (no. 4)]; in *ibid.*, 181, Qusṭā ibn Lūqā's treatise – if such it was – is confused with his writings on the celestial globe. On the surviving instrument see [Maddison 1962].
[13] The information on Ḥabash is taken from W. H. Morley in [Gunther 1932, I, 7]. The treatise on the universal plate by al-Sijzī (*ca.* 975) is listed in [Sezgin 1978, 226 (no. 6)].
[14] For more details see [King 1987, VII], and the various textual studies of J. Millás Vallicrosa there cited. Two recent contributions on the same subject are [Puig 1986a and 1986b].
[15] See [King 1987, IX].
[16] The single and double *shakkāzīya* quadrants are discussed in [Samsó and Catalá 1971] and [King 1987, X].
[17] On the sine quadrant see [Schmalzl 1929, 71-99], and [King 1983, 28-31; 1986, XIII, 111-115].
[18] On the horary quadrant see [Lorch 1981] and [North 1981]. The early treatise is listed in [*Cairo Survey*, no. B105 (see also no. F6)].
[19] On some by Ibn al-Sarrāj see [King 1987, IX]. On two by Ibn al-Shāṭir see [Schmalzl 1929, pp. 100-108], reprinted in [Kennedy and Ghanem 1976, 27-35].
[20] On the compendia of Ibn al-Shāṭir and al-Wafā'ī see [King 1987, XII], and [Tekeli 1960], [Brice, Imber and Lorch 1976], and [Dizer 1977], respectively. Some qibla boxes are featured in [King 1987, XIV].
[21] [Kennedy, 1956, 140-141 and 170].
[22] All such auxiliary tables are surveyed in [*SATMI*, I, Sections 6-8]. Some examples of tables of tan $\delta(\lambda)$ are discussed in [Kennedy *et al.* 1983, 125-127].

[23] All known Islamic tables for timekeeping are surveyed in [*SATMI*, I]. See [King 1986, II, 44-46], and also the article "*Mīḳāt* (astronomical aspects)" in *EI*, 2nd. ed., for a survey, and [King 1986, IX, X, and XII] on the corpuses for Cairo, Damascus and Jerusalem, and Istanbul. On the table of Najm al-Dīn ([*Cairo Survey*, no. C16]) see [King 1986, II, 44-45, and III, 540-541].

[24] All known Islamic auxiliary tables are surveyed in [*SATMI*, I, Section 9]. Detailed studies of individual tables are [Irani 1956] (Ḥabash); [Jensen 1971] (Abū Naṣr); and [King 1986, XI] (al-Khalīlī).

[25] See the article "*Mīḳāt*" (astronomical aspects)" in *EI*, 2nd. ed., for an overview. All known Islamic prayer-tables are surveyed in [*SATMI*, II]. See also [King 1983, 7-11], and various studies on the prayer-tables used in Cairo, Damascus, and Istanbul reprinted in [King 1986, IX, X, and XII].

[26] See the study cited in note 7 for further details. On the theory of refraction attributed to Ibn Yūnus see [King 1986, IX, 373-376].

[27] For al-Khwārizmī's table see [King 1983, 4-6]. (The one investigated in [Kennedy *et al.* 1983, 151-156], is not by al-Khwārizmī.) The anonymous Andalusian table, and several others, are investigated in my contribution to the *Kennedy Festschrift*.

[28] [Kennedy *et al.* 1983, 137-140].

[29] *ibid.*, 41.

[30] See the articles "*Ḳibla* (astronomical aspects)" and "Makka (centre of the world)" in *EI*, 2nd. ed., for an overview of the qibla problem. Studies of individual qibla tables are [King 1983, 12-16] (Abbasid) and [Lorch 1980] (al-Khāzinī), both superseded by [King 1987b] (miscellaneous), and [King 1986, XIII] (al-Khalīlī). Ibn al-Haytham's treatise on the solution by spherical trigonometry (listed in [Sezgin 1974, 368, no. 16] and [Sezgin 1978, 259, no. 19]) has recently been investigated by A. Dallal of Columbia University.

BIBLIOGRAPHY

Brice, W., Imber, C. and Lorch, R. 1976. The Dā'ire-yi Mu'addel of Seydī Alī Re'īs. *Seminar on Early Islamic Science* (The University of Manchester), no. 1.

Cairo Survey: D. A. King, *A Survey of the Scientific Manuscripts in the Egyptian National Library,* (Publications of the American Research Center in Egypt), Winona Lake, Ind.: Eisenbrauns, 1986.

Dizer, M. 1977. The Dā'irat al Mu^{c}addil in Kandilli Observatory .... *Journal for the History of Arabic Science* 1, 257-260 and 3 plates.

EI, 2nd. ed.: *The Encyclopaedia of Islam*. 2nd ed., 5 vols. to date. Leiden: E. J. Brill, 1960 onwards.

Gunther, R. T. 1932. *The Astrolabes of the World.* 2 vols. Oxford: University Press, repr. London: The Holland Press, 1976.

Irani, R. A. K. 1956. *The Jadwal al-Taqwīm of Ḥabash al-Ḥāsib.* Unpublished master's thesis, The American University of Beirut.

Jensen, C. 1971. Abū Naṣr Manṣūr's Approach to Spherical Astronomy as Developed in His Treatise *The Table of Minutes. Centaurus* 16, 1-19.

Kennedy, E. S. 1956. A Survey of Islamic Astronomical Tables. *Transactions of the American Philosophical Society*. N.S., 46: 2, 121-177.

Kennedy, E. S. 1960. *The Planetary Equatorium of Jamshīd Ghiyāth al-Dīn al-Kāshī.* Princeton, N.J.: Princeton University Press.

Kennedy, E. S. and Ghanem, I. 1976. *The Life and Work of Ibn al-Shāṭir: an Arab Astronomer of the Fourteenth Century.* Aleppo: Institute for the History of Arabic Science.

Kennedy, E. S., Colleagues and Former Students. 1983. *Studies in the Islamic Exact Sciences.* Beirut: American University Press.

Kennedy Festschrift: King, D. A. and Saliba, G., eds.1987. *From Deferent to Equant: A Volume of Studies in the History of Science in the Near East Dedicated to E. S. Kennedy. Annals of the New York Academy of Sciences 500.*

King, D. A. 1983. al-Khwārizmī and New Trends in Mathematical Astronomy in the Ninth Century. *Occasional Papers on the Near East* (New York University, Hagop Kevorkian Center for Near Eastern Studies), no. 2.

King, D. A. 1986. *Islamic Mathematical Astronomy.* London: Variorum Reprints.

King, D. A. 1987. *Islamic Astronomical Instruments.* London: Variorum Reprints.

King, D. A. 1987b. The Earliest Islamic Mathematical Methods and Tables for Finding the Direction of Mecca. *Zeitschrift für Geschichte der Arabisch-Islamischen Wissenschaften.* To appear in vol. 3.

Lorch, R. 1980. The Qibla-Table Attributed to al-Khāzinī. *Journal for the History of Arabic Science* 4, 259-264.

Lorch, R. 1981. A Note on the Horary Quadrant. *Journal for the History of Arabic Science* 5, 115-120.

Maddison, F. 1962. A 15th Century Islamic Spherical Astrolabe. *Physis* 4, 101-109.

Michel, H. 1947. *Traité de l'astrolabe.* Paris: Gauthier-Villars.

Neugebauer, O. 1975. *A History of Ancient Mathematical Astronomy.* 3 vols. Berlin, *etc.:* Springer Verlag.

North, J. 1981. Astrolabes and the Hour-Line Ritual. *Journal for the History of Arabic Science* 5, 113-114.

Puig Aguilar, R. 1986a. *Ibn al-Naqqāš al-Zarqālluh (Azarquiel): Al-Šakkāziyya,* edición, traducción y estudio. Barcelona: Universidad de Barcelona, Instituto "Millás Vallicrosa" de Historia de la Ciencia Árabe.

Puig Aguilar, R. 1986b. Concerning the Ṣafīḥa Shakkāzīya. *Zeitschrift für Geschichte der Arabisch-Islamischen Wissenschaften* 2, 123-139.

Samsó, J. and Catalá, M. 1971. Un instrumento astronómico de raigambre Zarqāli: el cuadrante šakkāzī de Ibn Ṭībughā. *Memorias de la Real Academia de Buenas Letras de Barcelona* 13, 5-31.

SATMI: King, D. A. *Studies in Astronomical Timekeeping in Medieval Islam.* New York, *etc.:* Springer-Verlag, in press.

Schmalzl, P. 1929. *Zur Geschichte des Quadranten bei den Arabern.* Munich: Saliesianische Offizin.

Seemann, H. 1925. Das kugelförmige Astrolab nach den Mitteilungen von Alfons X. von Kastilien und den vorhandenen arabischen Quellen. *Abhandlungen zur Geschichte der Naturwissenschaften und der Medizin (Erlangen)* 8.

Sezgin, F. 1974 and 1978. *Geschichte des arabischen Schrifttums,* V: Mathematik, and VI: Astronomie. Leiden: E.J. Brill.

Tekeli, S. 1960. (The) Equatorial Armilla of ... al-Wafai and the Torquetum. *Ankara Üniversitesi Dil ve Tarih – Coğrafya Fakültesi Dergisi* 18, 227-259.

Winder Festschrift: Kazemi, F. and McChesney, R. B., eds. *Essays in Honor of R. Bayly Winder.* To appear.

VII

Universal Solutions to Problems of Spherical Astronomy from Mamluk Egypt and Syria

Medieval Islamic mathematical astronomy was divided into three main branches: planetary astronomy, spherical astronomy, and mathematical astrology. Spherical astronomy, with which we are concerned here, is the study of problems relating to the apparent daily rotation of the celestial sphere. The main problems of this discipline that confronted the medieval astronomer were the determination of the time and azimuth of a celestial body from an observed altitude, given its declination (or distance from the celestial equator), and the local terrestrial latitude (measured by the altitude of the celestial pole above the local horizon).

Timekeeping was important to the Muslim astronomers not only because of their need to accurately record and predict celestial phenomena but also because the five daily prayers in Islam are astronomically defined. Another problem of major concern was the prediction of the appearance of the lunar crescent, upon which depends the regulation of the lunar calendar. Yet another was the determination of the *qibla,* or direction of Mecca from any locality; the qibla is the direction of prayer and of various other ritual acts. The problem of finding the qibla from the appropriate geographical coordinates belongs properly to mathematical geography, but medieval scientists often treated it as a problem of spherical geometry: it is then required to find the azimuth of the zenith of Mecca on the local horizon.

The medieval Muslim astronomers had at their disposal five main practical means of obtaining the solutions to problems of spherical astronomy:

1. Observational instruments (such as an equatorial armilla) which could be aligned so that the time could be read from the instrument without any calculation whatsoever.
2. Analogue computer devices (such as an astrolabe) which could be made to simulate an instantaneous celestial configuration relative to the local horizon and provide an instant solution.
3. Geometric solutions (such as the analemma) from which solutions to specific numerical examples could be found by two-dimensional constructions.
4. Trigonometric solutions, derived either from geometric solutions or by the application of spherical trigonometry. Tables of the sine/cosine and tangent/cotangent functions were available for calculating the solutions to specific numerical examples.
5. Trigonometric computational devices (such as a sine quadrant) for deriving without calculation the solution to a specific numerical example, given the appropriate formula.

Tables or instruments providing solutions to problems in spherical astronomy for all latitudes rather than one specific latitude will hereafter be labeled *universal.*

* * *

There was considerable interest in astronomy in medieval Egypt and Syria during two centuries of Mamluk rule, especially from the middle of the thirteenth century to the middle of the fifteenth. This Egyptian-Syrian tradition began in thirteenth-century Cairo, reached its culmination in fourteenth-century Aleppo and Damascus, and continued with occasional innovations in fifteenth-century Cairo.

Some of the activity of the Egyptian and Syrian astronomers was in the field of planetary astronomy, both theoretical, in the form of modifications to the planetary models of Ptolemy, and practical, in the form of new *zijes* or astronomical handbooks compiled for Cairo and Damascus. These achievements do not concern us here. However notable advances were also made by the Egyptian and Syrian astronomers in the field of spherical astronomy, particularly astronomical timekeeping. It has long been clear that these astronomers were mostly *muwaqqits*— astronomers associated with mosques and responsible for regulating the astronomically defined times of Muslim prayer. Until recently, their known works, mainly treatises on the use of the sine and almucantar quadrants, have tended to create an impression among historians of science that their contribution to the development of science was negligible. My recently published essay on Mamluk astronomy illustrates the new Mamluk material investigated for the first time during the past ten years which attests to the sophistication of their activities.[1]

My purpose here is to draw attention to one particular aspect of the Mamluk scientific tradition—the interest in instruments, computational devices, and mathematical tables, which facilitate the universal solution of problems in spherical astronomy.

In another study of a more general nature, I note Greek achievements in the field of universal solutions and survey the Muslim activity in this area (notably in ninth-century Baghdad, eleventh-century Andalusia, and Mamluk Egypt and Syria).[2] These include the spherical astrolabe;[3] modifications and additions to the astrolabe, such as the plate of horizons *(safiha afaqiya)*, the universal astrolabe of ʿAli ibn Khalaf, the universal plate known as the *shakkaziya*, and the universal plate of al-Zarqallu;[4] trigonometric grids such as the sine or sinical *(mujayyab)* quadrant, invented in ninth-century Baghdad;[5] al-Farghani's tables of coordinates for marking astrolabe plates for each degree of latitude;[6] al-Khwarizmi's tables of coordinates for marking sundials for a whole range of latitudes;[7] parallax, visibility, and ascension tables, each based on Hellenistic models;[8] auxiliary tables for solving particular problems of spherical astronomy;[9] tables for timekeeping based on an approximate Indian formula independent of local latitude and hence considered universal;[10] and tables displaying the qibla for each degree of terrestrial latitude and longitude.[11]

The material here shows that a succession of highly competent astronomers in Mamluk Egypt and Syria turned their attention to universal solutions and also came up with some impressive instruments and tables. I shall discuss each of these Mamluk astronomers in chronological order, indicating their contribution to universal solutions. The reader should bear in mind that only one aspect of Mamluk astronomy is being considered; many important Mamluk astronomers are not mentioned because they did not, as far as we know, write on this one aspect of spherical astronomy. Also, even for those who did, I have refrained from mentioning their productions on universal instruments which are well-known, such as the planispheric astrolabe with a complete set of plates, the spherical astrolabe, and the sine quadrant.

* * *

In what follows I shall assume that the reader is familiar with the basic problems of medieval spherical astronomy[12] and their solution by means of an astrolabe or sine quadrant.[13] I shall use the following notation freely:

ϕ	local latitude
λ	solar longitude
δ	celestial declination
h	instantaneous altitude
H	meridian altitude
D	half arc of visibility
t	hour-angle
T	time since rising
a	azimuth (measured from the prime vertical)
L	local longitude
q	qibla (measured from the meridian)

I shall also refer to the medieval trigonometric functions using the standard capital notation, thus:

$$\text{Sin}\,\theta = R\sin\theta;\ \text{Cos}\,\theta = R\cos\theta;\ \text{Cot}\,\theta = R\cot\theta;$$
$$\text{Vers}\,\theta = R\,\text{vers}\,\theta = R\,(1 - \cos\theta),\ \text{etc.}$$

where R is generally 60.[14] As far as trigonometric tables are concerned, it should be borne in mind that tables of the sine and cotangent functions with values to three sexagesimal digits for each minute of argument were standard equipment of the Mamluk astronomers.[15]

1. *al-Marrakushi*

Abu ʿAli al-Marrakushi was an astronomer of Moroccan origin who worked in Cairo about the year 1280.[16] He compiled a compendium of spherical astronomy and astronomical instruments, which is perhaps the single most valuable source for the history of Islamic astronomical instrumentation. In this work, which exists in several manuscripts copied in Egypt, Syria, and Turkey, al-Marrakushi presents a detailed discussion of spherical astronomy and then deals first with different kinds of plane sundials, and then other instruments such as the armillary sphere, the planispheric astrolabe, the universal plate known as the *shakkaziya*, the trigonometric grid called *mujayyab*, and a variety of quadrants for determining time from solar altitude.

The first half of al-Marrakushi's treatise, dealing with spherical astronomy and sundials, has been translated by J. J. Sédillot, and the second half, dealing with instruments, has been summarized in a rather chaotic fashion by L.A. Sédillot.[17] al-Marrakushi's sundial theory has been studied by C. Schoy.[18]

al-Marrakushi was not concerned with the kind of solutions I label universal, but those Mamluk astronomers who were knew of his work. For the modern reader who wishes to obtain an overview of spherical astronomy in thirteenth-century Egypt the treatise of al-Marrakushi, as presented by the Sédillots and Schoy, is the best introduction. It is a sad reflection on the state of the field that this work has been neglected for over a century. Certainly it merits publication with a new translation and analysis.

2. *al-Maqsi*

In the late thirteenth century, the Egyptian astronomer Shihab al-Din al-Maqsi[19] compiled a set of tables of the time since sunrise for each degree of solar altitude and solar longitude, with values computed for the latitude of Cairo, taken correctly as 30°. There are indications that al-Maqsi based these tables of $T(h, \lambda)$ on some earlier tables by the tenth-century Egyptian astronomer Ibn Yunus, and the precise nature of al-Maqsi's own contribution is obscure.[20] Certainly the format of these tables is that of the solar azimuth tables of Ibn Yunus, which display the function $a(h, \lambda)$ for each degree of both arguments, likewise for the latitude of Cairo.

al-Maqsi's tables of $T(h, \lambda)$, together with a related set displaying the hour-angle $t(h, \lambda)$ and Ibn Yunus' tables of $a(h, \lambda)$, formed a major part of the corpus of tables for timekeeping that was used in Cairo from the thirteenth century onward. The other tables in the corpus display the astronomically-defined intervals during which the five daily prayers may be performed. The functions tabulated include the duration of morning and evening twilight for regulating the prayers at daybreak and nightfall, and the times between sunrise and midday, midday and mid-afternoon (defined in terms of shadow lengths), and mid-afternoon and sunset, for regulating the prayers at mid-day, mid-afternoon, and sunset. Each of these functions is displayed for each degree of solar longitude.

These tables and others based on them were used in Cairo until the nineteenth century. The corpus contains about 200 pages of tables with more than 30,000 entries,[21] each computed to degrees and minutes. Although the corpus survives in over two dozen manuscripts, no two contain exactly the same tables, and not one contains a reliable statement of the authorship of each of the tables.

al-Maqsi's tables of the time since sunrise are the earliest known tables for timekeeping computed for arguments h and λ. All earlier Islamic tables displaying the time of day (or night) as a function of solar (or stellar) altitude have the meridian altitude as the second argument.[22] No such tables of $T(h,H)$ are known to have been compiled for Cairo. However, it was surely al-Maqsi's tables of $T(h, \lambda)$ for timekeeping by the sun and the Islamic tradition of compiling tables of $T(h,H)$ for specific latitudes which inspired Najm al-Din al-Misri (no. 3) to compile a universal table displaying the time since rising of the sun or any star for all latitudes.

al-Maqsi also wrote a treatise on sundials, including an introduction on sundial theory and tables for constructing horizontal sundials for Mecca, Cairo, Jerusalem, and Damascus, as well as more than 100 tables specifically for Cairo for marking sundials in various planes inclined to the meridian and to the prime vertical and skew to both.[23] These tables are considerably more extensive than those of al-Marrakushi.

al-Maqsi included tables for constructing sundials in the plane of the celestial equator and the perpendicular plane, and pointed to their universal application. The former have the disadvantage that they serve only northern or southern declinations. The latter, known as polar sundials, were discussed by later astronomers such as Ibn al-Sarraj, Ibn al-Shatir, and al-Tizini (nos. 5, 8, and 13), and a unique Islamic example of such a sundial, made in the year 1786/87, is preserved in Acre.[24]

3. *Najm al-Din al-Misri*

The Egyptian astronomer, Najm al-Din al-Misri,[25] a contemporary of al-Maqsi,[26] decided to make a table for timekeeping which would serve all latitudes and could be used for timekeeping by both the sun and the stars.[27] He tabulated the time since rising as a function of the observed altitude and two variables which depend on the local latitude and celestial declination—the meridian altitude and half arc of visibility. Thus the function tabulated is $T(H,h,D)$— see plate 1. Values are given for each degree of both arguments, and number over 250,000. The computation of the entries is not difficult by virtue of the choice of argument, and the errors that I have investigated in spot-checks amount to no more than $\pm 0;15°$, that is, one minute of time. The underlying formula for the hour-angle t $(= D - T)$ is:

$$\text{Vers } t = (1 - \text{Sin } h \,/\, \text{Sin } H) \text{ Vers } D.$$

The table is a highly ingenious extension of the Islamic tables for timekeeping for specific latitudes. It is extant in a unique contemporary manuscript copy, the two halves of which are preserved in Oxford and Cairo. This manuscript is probably in the hand of Najm al-Din himself. I find it somewhat surprising that Najm al-Din's enormous table is not mentioned in later astronomical sources, and think that it is safe to assume that the table was scarcely used. With over 250,000 entries, the table would be unlikely to attract many copyists, and besides, tables for daytime timekeeping were already available for Cairo (see no. 2).

4. *al-Abhari*

In the year 1320 the Syrian astronomer Amin al-Din al-Abhari compiled a treatise on instruments, discussing four instruments in par-

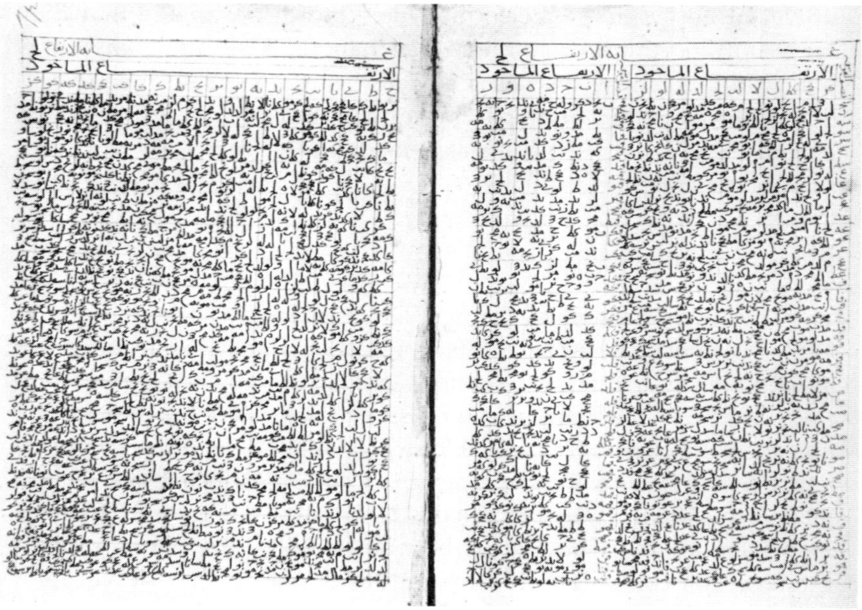

1. An extract from Najm al-Din al-Misri's enormous table for time-keeping by the sun and stars. The extract serves:

$H = 79°$ and $h = 32°, 33°, \ldots, 70°$ (entered horizontally),
and $D = 121°, 122°, \ldots, 150°$ (entered vertically).

Taken from MS Oxford Bodleian Marsh 672, fols. 33v–34r, courtesy of the Bodleian Library. [Replacement photo courtesy of the Egyptian National Library].

ticular—the *shakkaziya* plate, the armillary sphere, the circular *mujayyab* trigonometric grid, and the *afaqiya* plate of horizons.[28] These instruments are all universal—they are not restricted to a specific latitude—and it seems likely that it was this feature which led al-Abhari to treat them together. His treatise was not widely known, and survives in only two manuscripts, now preserved in El Escorial and Gotha.

5. *Ibn al-Sarraj*

Ibn al-Sarraj worked in Aleppo in the first half of the fourteenth century.[29] He wrote on mathematics and astronomy, but his main interest was in astronomical instruments. He was a first-class instrument maker, and displayed considerable ingenuity in the design of new varieties of instruments.

Ibn al-Sarraj re-invented the universal astrolabe of the eleventh-century Andalusian astronomer ʿAli ibn Khalaf.[30] There is no evidence that ʿAli ibn Khalaf's instrument or the treatise he wrote on its use was known outside Andalusia except in Europe. Ibn al-Sarraj certainly knew of the single *shakkaziya* plate, and may have known of the universal astrolabe of the Moroccan astronomer Ibn Baso,[31] who was a contemporary of al-Marrakushi. However, Ibn al-Sarraj probably hit upon the idea independently, and wrote a treatise on the use of the instrument which he called *Sarrajiya,* after himself. In this astrolabe, described in a unique manuscript preserved in Cairo, a half *shakkaziya* grid could rotate over a whole *shakkaziya* plate, and the other half of the rete consisted of a mixed north-south astrolabe projection of the ecliptic and position of selected stars, with which one could solve problems of spherical astronomy for all latitudes. Another anonymous Syrian or Egyptian wrote a treatise on the same instrument, calling it *al-asturlab al-mughni,* "the astrolabe which does not need [any plates]."

A universal astrolabe made by Ibn al-Sarraj himself in 1329 survives in the Benaki Museum in Athens. However, only the front of this astrolabe and the lower part of the back correspond to the instrument that Ibn al-Sarraj described in this treatise. The surviving astrolabe of Ibn al-Sarraj contains a set of plates for each three degrees of latitude which could be used with the half rete bearing the ecliptic and star pointers. The almucantars for each latitude are folded over so that each set takes up a quarter of each plate. Ibn al-Sarraj's instrument is a universal astrolabe which can be used universally in five ways; as such it ranks as one of the most interesting, if not the most interesting, of medieval astronomical instruments—see plate 2. Ibn al-Sarraj did not write on the use of this particular instrument, but the Egyptian astronomer al-Wafaʾi (no. 14), who owned the astrolabe about a century after Ibn al-Sarraj made it, did write a treatise on its use, extant in two copies in Princeton and Dublin.

Fortunately, al-Wafaʾi included an account of the operations that can be performed with the alidade and the mysterious grid on the upper back of Ibn al-Sarraj's astrolabe. Ibn al-Sarraj devised a number

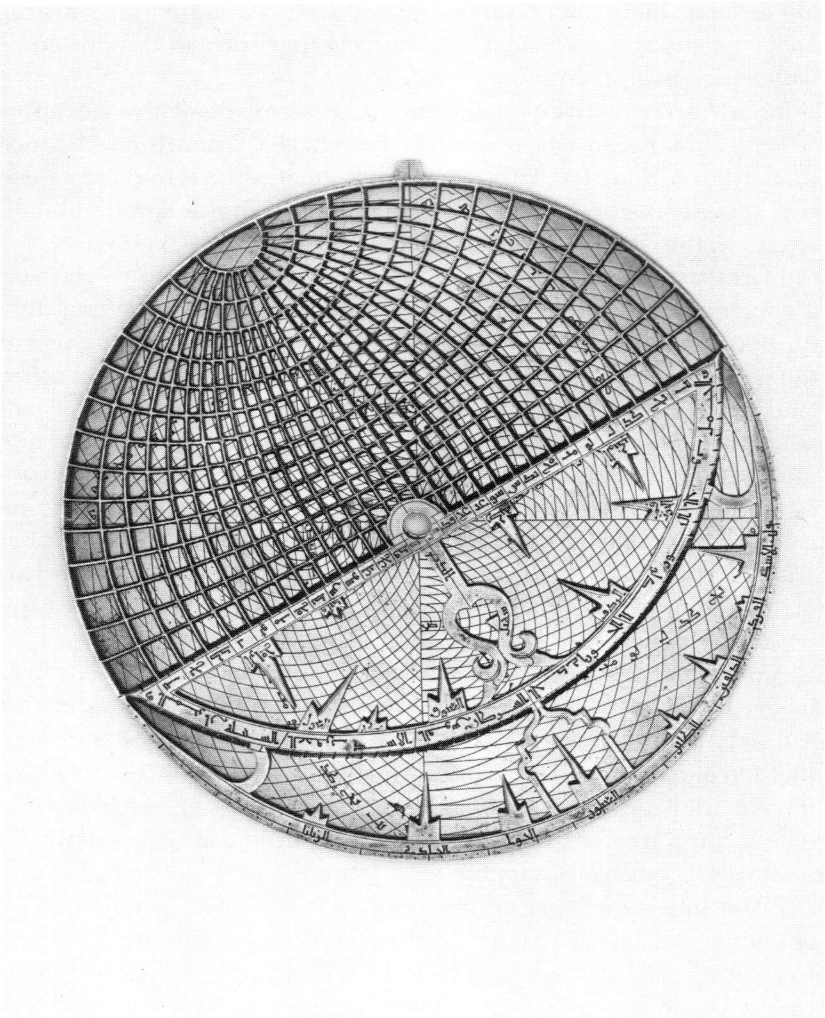

2. The front of Ibn al-Sarraj's universal astrolabe. Courtesy of the Benaki Museum, Athens.

of other grids in the shape of semicircles and quadrants, the purpose of which was to solve the standard problems of spherical astronomy for all latitudes. With the grid on the back of his astrolabe and the special alidade one can very simply effect a transformation of the form $(a,b) \rightarrow (c,d)$, where

$$c = \text{arc sin } (\sin a \sin b) \text{ and } d = \text{arc cos } (\sin a \cos b),$$

an operation which has several applications in spherical astronomy.

Another of Ibn al-Sarraj's inventions consists of a semicircular plate with scales along the diameter and around the circumference and two threads with movable markers attached at both ends of the diameter. The semicircle needs no other markings: Ibn al-Sarraj called it *al-jayb al-gha'ib*, "the trigonometric grid without any grid." A related instrument in the form of a quadrant and again without any markings was called by Ibn al-Sarraj *al-rubʿ al-jamiʿ*, "the universal quadrant." Yet another quadrant called *al-mujannah*, "the winged quadrant," bore two families of arcs of circles. All of these instruments can be used, with the kind of ingenuity which Ibn al-Sarraj possessed and which is in evidence in all his treatises on their use, to solve all of the standard problems of spherical astronomy for any latitude.

6. *Tibugha al-Baklamshi*

A contemporary Syrian scholar, Tibugha al-Baklamshi,[32] appears to have been the first to note that the problems of spherical astronomy can be solved using just one quadrant of *shakkaziya* markings with a thread attached at the center. Tibugha wrote a treatise on the use of such an instrument in which he claimed to be its inventor. One disadvantage of this instrument is that the determination of time from celestial altitude is approximate.[33]

Various Syrian and Egyptian astronomers in the next century, such as Ibn Tibugha (no. 11), al-Shadhili, and Abu Tahir, wrote treatises on the use of the single *shakkaziya* quadrants.[34]

7. *al-Bakhaniqi*

A fourteenth-century Egyptian astronomer called al-Bakhaniqi[35] took the tables of al-Farghani (*fl. ca.* 825) displaying coordinates for marking astrolabe plates for each degree of latitude from 15° to 50°[36] and extended them to cover the complete range 0° to 90°. He did this for a Yemeni ruler who had probably noticed that al-Farghani's tables did not serve latitudes in the Yemen. His tables survive in a unique manuscript now in Dublin.

Another set of tables compiled by al-Bakhaniqi, which survives in a unique manuscript in Cairo, displays the hour-angle at the beginning of the mid-afternoon prayer for each 10° of solar longitude at each degree of terrestrial latitude from 10° to 50°.

8. Ibn al-Shatir

Ibn al-Shatir was the head of the astronomers associated with the Umayyad Mosque in Damascus in the mid-fourteenth century and worked mainly in planetary astronomy.[37] His contribution to planetary theory is now well-known[38] although none of his treatises or his new *zij* for Damascus has been published. Ibn al-Shatir was inevitably also concerned with spherical astronomy and timekeeping; indeed, he constructed the magnificent sundial on the minaret of the Umayyad Mosque in Damascus.[39] He devised various instruments for solving problems in spherical astronomy for all latitudes; however, he faced the problem that Ibn al-Sarraj had already almost exhausted the possibilities.

One of Ibn al-Shatir's inventions was a compendium consisting essentially of two universal sundials, the first a horizontal sundial for the equator which could be inclined to serve as a sundial in any latitude, and the second an equatorial plate which could be inclined in the plane of the celestial equator in any latitude and used to measure the hour-angle of the sun by day or the stars by night. Ibn al-Shatir called this instrument *sanduq al-yawaqit*, "box of sapphires," and wrote a treatise on its use, as did the later Egyptian astronomer al-Sufi (no. 17). Both of these treatises are preserved in a unique manuscript now in Berlin. Ibn al-Shatir devised other different trigonometric grids,[40] including one in the form of an equilateral triangle which he labeled *al-rubʿ al-tamm*, "the complete quadrant"— (see plate 4). another labeled *al-rubʿ al-ʿAlaʾi* in the form of a quadrant with an orthogonal grid of markings and one family of parallel diagonals, and yet another in the form of a square.[41] For each he prepared a treatise showing how it could be used to solve problems of trigonometry and spherical astronomy.

Ibn al-Shatir also designed a circular plate, one side of which displayed two orthogonal grids superimposed at an angle equal to the obliquity of the ecliptic. The second side consisted of a fixed stereographic projection of the celestial equator and solstitial day-circles and ecliptic over which could rotate a grid of horizons for all latitudes: an

3. The trigonometric grid on the back of the Ibn al-Shatir's universal astrolabe. Photo courtesy of the late M. Alain Brieux, Paris.

4. The "complete" quadrant invented as a universal trigonometric grid by Ibn al-Shatir. Photo O. Gingerich, courtesy of the Egyptian National Library.

astrolabe in reverse. An example of this instrument, made by Ibn al-Shatir himself and unfortunately now missing the rete and alidade, is preserved in Cairo (see plate 3), His treatise on its uses survives in manuscript form in Leningrad, Aleppo, Cairo, and Princeton.

Ibn al-Shatir, like Ibn al-Sarraj, did not, as far as we know, prepare any universal auxiliary tables for solving the problems of spherical astronomy.

9. *al-Khalili*

Shams al-Din al-Khalili[42] was a *muwaqqit* at the Umayyad Mosque and a contemporary of Ibn al-Shatir. He made singularly impressive contributions to spherical astronomy, mainly in the form of astronomical tables.

al-Khalili compiled a corpus of tables for timekeeping for Damascus,[43] including tables of $t(h,\lambda)$ for each degree of both arguments and tables for regulating the times of prayer, all computed specifically for Damascus. A few years previously some prayer-tables for Damascus had been compiled by al-Mizzi[44], but al-Khalili used the parameters of Ibn al-Shatir. al-Khalili's tables were used in Damascus until the nineteenth century and exist in about two dozen manuscripts, the best of which is preserved in Paris.

al-Khalili also devised a set of tables for solving all the standard problems of spherical astronomy for any latitude.[45] The functions tabulated are:

$$f_\phi(\phi) = R \text{ Sin } \theta \,/\, \text{Cos } \phi$$
$$g_\phi(\theta) = \text{Sin } \theta \text{ Tan } \phi \,/\, R, \text{ and}$$
$$G(x,y) = \text{arc Cos } (x R \,/\, \text{Cos } y)$$

and values are given for each degree or unit of argument. There are more than 13,000 entries in the complete set (see plate 5). With these functions the determination of the hour-angle reduces to an application of the formula:

$$t(h,\delta,\phi) = G\{[f_\phi(h) - g_\phi(\delta)], \delta\}$$

which is easily shown to be equivalent to the modern formula

$$t = \text{arc cos } \{(\sin h - \sin \delta \sin \phi) \,/\, \cos \delta \cos \phi\}.$$

al-Khalili's *Universal Table*, called in Arabic *al-jadwal al-afaqi*, exists in several copies of the Damascus corpus and was known to later astronomers in Syria, Egypt, and Turkey, and even in the Maghrib.

al-Khalili compiled another set of auxiliary tables for timekeeping by the sun which was less well known. It survives in a unique manuscript preserved in Dublin.[46] The functions tabulated are:

$$B(\phi,\lambda) = \text{Cos } \delta(\lambda) \text{ Cos } \phi \,/\, 2R \text{ and}$$
$$V(x,y) = \text{arc Vers } (R y \,/\, 2x),$$

with which one finds $t(h,\lambda,\phi)$ by first computing

$$X(h,\lambda,\phi) = \text{Sin } H(\lambda\phi) - \text{Sin } h,$$

5. Two subtables in al-Khalili's universal auxiliary tables displaying values of his first two functions for latitudes 19° and 20°. Taken from MS Paris B.N. ar. 2558, opening no. 73, courtesy of the Bibliothèque Nationale.

and then using:

$$t(h,\lambda,\phi) = V\{B(\phi,\lambda),\ H(h,\lambda,\phi)\}.$$

al-Khalili's *pièce de résistance* was a table displaying the qibla for each degree of terrestrial latitude and longitude.[47] Qibla tables based on approximate formulae are attested in several earlier Islamic sources,[48] but that of al-Khalili is based on an accurate formula, and the entries, more than 3,000, are all correctly computed to within a minute or two of arc. It is not yet clear how al-Khalili achieved this degree of accuracy. The modern formula for the qibla in terms of the local latitude and longitude (ϕ, L) and the latitude and longitude of Mecca (ϕ_M, L_M) is:

$$q = \text{arc cot}\{\ (\sin\phi \cos\Delta L - \cos\phi \tan\phi_M) \ /\ \sin\Delta L\ \}$$

(where $\Delta L = L \sim L_M$), so that the computation of a single qibla value is not a trivial exercise. al-Khalili's qibla table does not appear to have been widely used in later centuries. It is not known to have been mentioned by any later astronomer, and is extant now in only three of the known manuscripts of the Damascus corpus preserved in Paris and Berlin.

10. *Ibn al-Ghazuli*

Ibn al-Ghazuli was an Egyptian astronomer contemporary with Ibn al-Shatir and al-Khalili (nos. 8 and 9).[49] He wrote on the *jayb gha'ib* of Ibn al-Sarraj, and on the use of an octant which he devised, consisting of a half quadrant of *mujayyab* markings and a half quadrant of almucantar markings (see plate 6).

11. *Ibn Tibugha*

ʿAli ibn Tibugha was probably the son of Tibugha al-Baklamshi (no.6) and was also an astronomer.[50] Like Ibn al-Sarraj (no. 5) he worked in Aleppo rather than Damascus. Ibn Tibugha was the author of a treatise on the *shakkaziya* quadrant, and in his treatise on the sine and almucantar quadrants he presented a table displaying the qibla for each degree of latitude and longitude. This table is based on an accurate formula but is independent of that of al-Khalili (no. 9). The

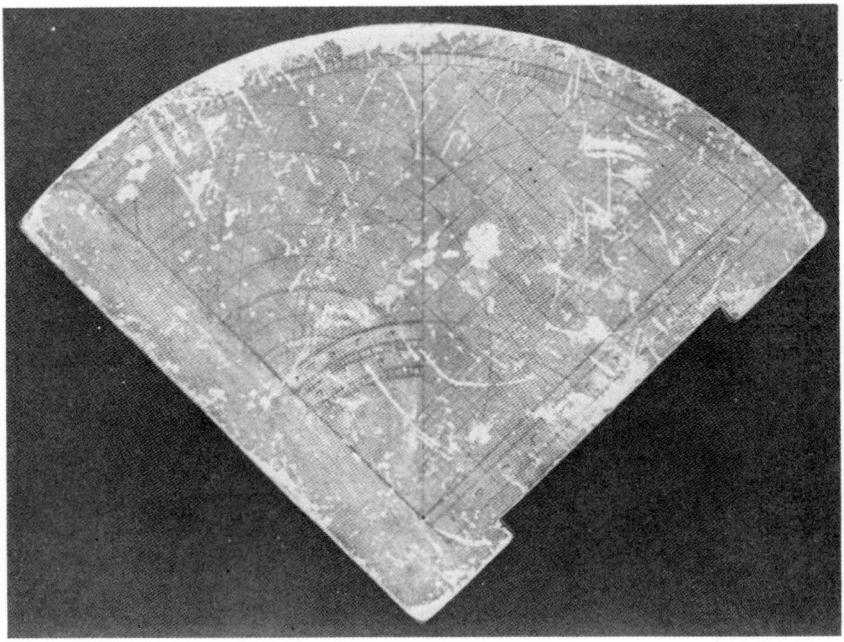

6. The unique example of an octant as devised by Ibn al-Ghazuli, showing almucantar and sinical markings on the two halves of a quadrant. Photo O. Gingerich, courtesy of the Egyptian National Library.

values serve a smaller range of arguments, but are likewise very carefully computed.

12. *al-Maridini*

The late-fourteenth-century astronomer Jamal al-Din al-Maridini was born in Syria but later moved to Egypt.[51] He was the author of several treatises on instruments and has often been confused with his grandson, the prolific Cairo astronomer Sibt al-Maridini.[52] al-Maridini wrote a treatise on the use of a double *shakkaziya* quadrant in which a *shakkaziya* grid rotates over a solid *shakkaziya* grid and can be used to solve all the problems of spherical astronomy for any latitude (see plate

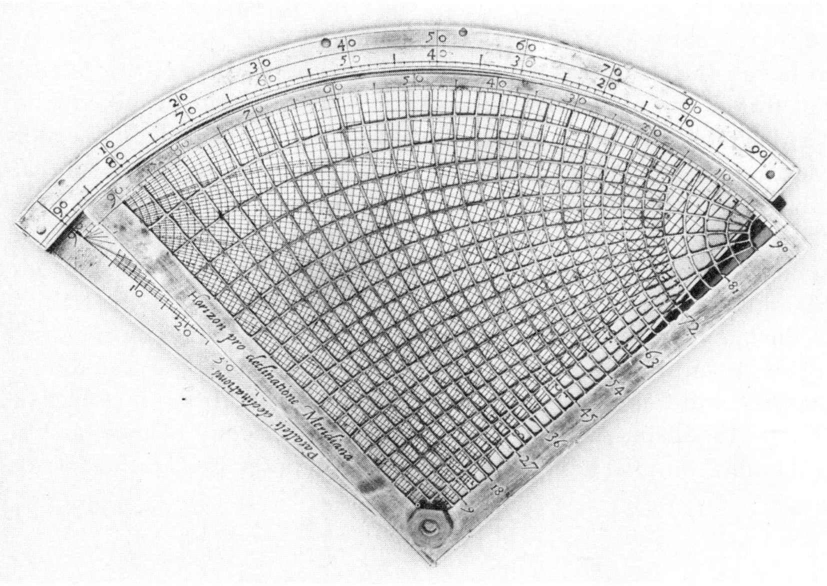

7. A European version of the double *shakkaziya* quadrant first described by al-Maridini. Courtesy R. and M. Webster, Adler Planetarium, Chicago, Ill.

7).[53] The inspiration for such a quadrant was surely the universal astrolabe of Ibn al-Sarraj (no. 5). al-Maridini's treatise is extant in three manuscripts in Damascus and Istanbul. He also devised a set of very simple and somewhat impractical auxiliary tables for solving problems of spherical astronomy,[54] which are extant in two manuscripts in Paris and Cairo.

13. *Miscellaneous (Sundial Tables)*

Since universal tables were compiled by Mamluk astronomers for astrolabe construction, regulation of the prayer times, timekeeping by the sun and stars, and solving problems of spherical astronomy, it is somewhat surprising that there is no set of tables of coordinates for marking sundials for all latitudes. However, one such set had been compiled by al-Khwarizmi in the early ninth century and this was known in early Mamluk Syria.[55] Also Ibn al-Sarraj (no. 5) is known to have compiled an extensive work on sundial theory with tables, but this is unfortunately no longer extant. As we have noted, the sundial tables compiled by al-Maqsi (no. 2) include isolated tables for Mecca, Jerusalem, and Damascus. All later Mamluk Egyptian sundial tables were computed specifically for Cairo, by Ibn Sudun, Ibn al-Muhallabi, and Sibt al-Maridini,[56] and for other localities in Egypt such as Alexandria, Damietta, and Assiut. The only known set from Mamluk Syria was compiled by the Damascus astronomer al-Tizini[57] in the late fifteenth century and contains extensive tables for marking inclined vertical hour-angle dials for the latitudes of Damascus, Aleppo, and Cairo, the last mentioned being lifted from the tables of Ibn al-Muhallabi. As far as I know, the only attempt at universality in Mamluk sundial tables is a small table prepared by al-Sufi (see no. 17).

14. al-Wafa'i

The Egyptian astronomer ʿIzz al-Din al-Wafa'i lived in the early fifteenth century and was a *muwaqqit* at the Mosque of al-Mu'ayyad in Cairo.[58] He was the author of several treatises, mostly dealing with instruments. He wrote on universal instruments such as the octant of Ibn al-Ghazuli (no. 10) and the complete quadrant of Ibn al-Shatir (no. 8). Another of his treatises deals with an instrument of his own invention, an equatorial armilla in which a graduated semicircle affixed to a horizontal plate could be oriented in the plane of the celestial equator, and used to measure the hour-angle of the sun or any star in any latitude (see plate 8);[59] this instrument was called *da'irat al-muʿaddil* and became rather popular in later Egypt, Syria, and Turkey. Several examples survive, as do numerous copies of al-Wafa'i's treatise and various later treatises on the instrument. In an earlier version of this instrument, called *al-muqawwar,* al-Wafa'i put all of the component parts in a box built on the base: this confirms my impression that he was influenced by Ibn al-Shatir's "box of sapphires."[60]

al-Wafa'i became the owner of the universal astrolabe of Ibn al-Sarraj (no. 5) and wrote a treatise on the use of all its component parts, extant in two manuscripts in Dublin and Princeton. His mastery of the various functions of the instrument is impressive.[61]

8. A late Ottoman version of the *da' irat al-mu 'addil* invented by al-Wafa'i. (Replacement photo courtesy of Christie's of London.)

al-Wafa'i also compiled a set of auxiliary tables for solving the problems of spherical astronomy for a range of latitudes.[62] The functions tabulated are:

$$G(\phi,\theta) = R^3 / \mathrm{Cos}\phi\ \mathrm{Cos}\theta, \text{ and}$$
$$V(x) = \text{arc Vers}(x)$$

To find $t(h,\Delta,\phi)$ one first computes the quantity:

$$X(h,\delta,\phi) = \mathrm{Sin}\ H(\delta,\phi) - \mathrm{Sin}\ h$$

and then uses the tables to find:

$$H(h,\delta,\phi) = V\,\{H(h,\delta,\phi) \bullet G(\phi,\delta)\}.$$

al-Wafa'i's tables survive in a unique manuscript in the Vatican Library and were apparently not widely known. A commentary on the tables was written by the contemporary Cairo astronomer al-Karadisi, and this is likewise extant in a unique manuscript.[63]

By al-Wafa'i's time, sets of prayer-tables were available for Cairo, Damascus, and Jerusalem. al-Wafa'i is stated to be the author of a set of prayer-tables including these three sets as well as those for each degree of latitude from 15° (Yemen) to 41° (Istanbul).[64] These universal prayer-tables survive in a unique manuscript in Princeton.

15. *Ibn al-Mushrif*

Abu Bakr Ibn al-Mushrif was an astronomer in Cairo in the early fifteenth century.[65] He compiled a set of auxiliary tables[66] displaying the functions:

$$B(\phi,\delta) = \mathrm{Cos}\phi\ \mathrm{Cos}\delta\ /\ R$$
$$X(H,h) = \mathrm{Sin}\ H - \mathrm{Sin}\ h$$
$$V(x,y) = \mathrm{arc\ Vers}\ (R\ y\ /\ x)$$

With these the hour-angle can be found using:

$$t(h,\delta,\phi) = V\{B(\phi,\delta)\ ,\ X(H,h)\}$$

Rescensions of these tables were prepared by Sudun al-Bashtaki[67] and Ibn Abi'l-Fath al-Sufi (no. 17). The original tables and each of the rescensions survive in unique manuscripts, all preserved in Cairo.

16. *al-Aqfahsi*

The mid-fifteenth-century Cairo astronomer al-Aqfahsi[68] compiled a set of tables for constructing astrolabe plates for an extensive range of *southern* latitudes. These tables are extant in a unique manuscript in Berlin.

17. *al-Sufi*

Ibn Abi 'l-Fath al-Sufi was one of the leading astronomers in Cairo in the second half of the fifteenth century.[69] His major work was a *zij* for Cairo based mainly on the *zij* of Ulugh Beg for Samarqand.[70] His contributions to spherical astronomy include an extensive set of auxiliary tables for solving problems of spherical astronomy,[71] extant in a unique manuscript preserved in Oxford. These tables display the functions:

$$C(\theta_1,\theta_2) = \mathrm{Sin}\theta_1 \ \mathrm{Sin}\theta_2 \ / \ R$$
$$B(\theta_1, \theta_2) = \mathrm{Cos}\theta_1 \ \mathrm{Cos}\theta_2 \ / \ R, \text{ and}$$
$$S(x,y) = \mathrm{arc\ Cos}\ (Rx \ / \ y),$$

with which, for example,

$$t(h,\delta,\psi) = S\,\{[\mathrm{Sin}\ h - C(\phi,\delta), B(\phi,\delta)]\}.$$

They represent no improvement upon the more versatile tables of al-Khalili and Ibn Mushrif (nos. 9 and 15).

al-Sufi wrote several short treatises on sundial theory, one of which contains a table for facilitating the construction of hour-angle lines on any vertical dial.[72] The table displays the solar azimuth at the equinoxes as a function of time, for all latitudes. Using these azimuth values, one can determine the orientation of the hour-angle lines on a vertical dial inclined at any angle to the meridian for any latitude. The function tabulated by al-Sufi is:

$$a\ (\mathrm{Sin}\phi, T) = \mathrm{arc\ Tan}\ (\mathrm{Sin}\phi\ \mathrm{Tan}\ T \ / \ R)$$

Tables of a (T) for $\phi = 30°$ (Cairo) exist in earlier Egyptian sources.

Amongst al-Sufi's numerous works on instruments are treatises on the *mujannah* quadrant of Ibn al-Sarraj, and universal ones such as Ibn al-Shatir's "box of sapphires"[73] and each of al-Wafa'i's two equatorial armillas.

18. Ibn Abi 'l-Khayr al-Husni

Muhammad ibn Abi'l-Khayr al-Husni worked in Cairo in the sixteenth century.[74] He was the author of several astronomical works, including a set of tables displaying the duration of morning and evening twilight at the equinoxes and solstices for each half climate. These tables are extant in a unique manuscript in Cairo.[75]

19. *al-Minufi*

One last example of the Egyptian astronomers' penchant for universal solutions is a table prepared in the year 1567 by the Cairo astronomer ʿAbd al-Qadir al-Minufi.[76] This table displays the amount to be added to the solar day-circle to account for the effect of refraction at the horizon. Values of this function ΔD are displayed for each zodiacal sign and each degree of terrestrial latitude from 1° to 60°, and are based on the notion that ΔD increases linearly with latitude and assumes the values for latitude 30° (Cairo) attributed to Ibn Yunus.[77] al-Minufi's table survives in a unique manuscript in Istanbul but the underlying theory is attested in several later writings of Syrian and Turkish provenance.[78]

20. *Addenda*

In 1960 the late D. de Solla Price thought that he had found the origin of the European universal rectilinear dial for timekeeping known as * the *navicula Venetiis,* "little ship of Venice", in an Islamic instrument also shaped like a ship.[79] The latter device he found described in the treatise of al-Marrakushi,[80] but it is merely one of the many kinds of astrolabe retes devised by Muslim astronomers mainly in the tenth century. As far as I can tell, European universal dials have no precedent in the Muslim world.

In 1966 E. Poulle cited a fifteenth-century European report that the instrument known as the sexagenarium was of contemporary Egyptian provenance.[81] The sexagenarium is a quadrant with a trigonometric grid on one side, and a series of circular scales for computing mean longitudes and anomalies for the sun, moon, and planets on the other. The sine quadrant serves not only to provide numerical solutions to problems of spherical astronomy but also to compute equations of the sun, moon, and planets. (The latter function serves only to challenge the wits of the user, because auxiliary tables for computing the equations obviate the need for such tedious calculations.) While the sine quadrant and associated instructions for performing computations in spherical astronomy were well-known in fifteenth-century Cairo,[82] no such planetary dials are known from there. However, such a dial was described by al-Kashi (Samarqand,

* For "little ship of Venice", read "little ship of the Venetians",

ca. 1425)[83] although, as far as I know, he did not write on the sine quadrant and his planetary dial was not used by later Muslim astronomers.

Conclusion

The astronomers of Egypt and Syria during the Ottoman period, that is from the sixteenth century onward, did not share the interests of their Mamluk predecessors in universal solutions. To be sure, the manuscripts of the Mamluk works were copied during this period and presumably used, but I know of no serious activity in this area. Rather, the Ottoman astronomers in Egypt and Syria, and also Istanbul, which became the center of astronomical timekeeping in the sixteenth century, preferred to compile tables for specific latitudes and use instruments with the same limitations. Interest in astronomy did not subside in Ottoman times, but creative activity did. By the seventeenth century, the candle was burning low, and another one was burning more brightly in the West.

Acknowledgments

The materials for this paper were gathered during my research on the history of Islamic science conducted at the American Research Center in Egypt during the period 1972–79. This activity was sponsored mainly by the Smithsonian Institution and National Science Foundation, Washington, D.C. (1972–79), the American Philosophical Society (1972–74), and the Ford Foundation (1976–79). This support is gratefully acknowledged.

An abstract of this paper was presented at the 188th Annual Meeting of the American Oriental Society, held at Toronto in April 1978.

It is a pleasure to thank my colleagues Dr. Jan Hogendijk of the Mathematical Institute in Utrecht and Dr. Richard Lorch of the Bavarian Academy of Sciences in Munich for their suggestions on a preliminary version of this paper, without in any way implicating them in any remaining defects.

Notes

1. A survey of Mamluk astronomy is *King* 1. The main bio-bibliographical reference work for the astronomy of this period is *Suter*, with additional

sources listed in *Brockelmann* and *Cairo Survey*. Instrument makers are listed in *Mayer*. All of the tables for spherical astronomy mentioned in this study are analyzed in my forthcoming survey of astronomical timekeeping in medieval Islam (hereafter *SATMI*).

I purposely refrain from referring to the earlier tradition in Egyptian astronomy exemplified by the works of the late-tenth-century Cairo astronomer Ibn Yunus (on whom see the article in *DSB*), or the astronomers of the twelfth-century Afdal / Bata'ihi observatory (on which see *Sayili*, pp. 167–75), or the astronomers in thirteenth-century Cairo who compiled the astronomical handbook known as the *Mustalah Zij* (on this see *Kennedy* 1, no. 47, and *King* 1, pp. 535–36), which remained popular in Egypt until it was eventually replaced by Egyptian recensions of the *Zijes* of Ibn al-Shatir and Ulugh Beg (see *Kennedy* 1, nos. 11 and 16 and *King* 1, p. 536).

Likewise, I refrain from any discussion of the earlier tradition of astronomy in Syria. After the activity of Habash (*ca.* 830; see the article by W. Hartner in *DSB*), there is a large gap in our knowledge of mathematical astronomy in Syria until the mid-thirteenth century when there appeared Ibn al-Lubudi (*Suter*, no. 365 and *Kennedy 1*, nos. 86 and 87), who prepared two zijes, neither of which is extant, and Muhyi 'l-Din al-Maghribi (*Suter*, no. 376 and *Kennedy 1*, nos. 41 and 108), who compiled a *zij* for Damascus before taking off to work at the observatory in Maragha (see *King* 1, pp. 533–34). Thereafter there is another smaller gap until, almost as it were out of the blue, Ibn al-Sarraj appeared in Aleppo in early fourteenth century, followed by a whole school of other highly competent astronomers in Damascus. The main link between the traditions in Mamluk Syria and Egypt is the astronomical activity in the latter in the late thirteenth and early fourteenth century, which was such that individuals such as Ibn al-Shatir and al-Mizzi both went to Cairo and Alexandria to study astronomy.

2. See *King* 6.
3. For a survey of Islamic writings on the spherical astrolabe, see *Seemann* (now outdated).
4. See, for example, *Millás* 1 and 2. The term *shakkaziya* refers to the kind of grid generally associated with Gemma Frisius (on the use of which see *Waters* **et al.**, pp. 32–39). On the origin of the term see *King* 4, VII.
5. On the quadrant in Islam see *Schmalzl* and *King* 2, pp. 28–31.
6. On these see *King* 4, II pp. 53–55 and 2, p. 25.
7. On these see *King* 2, pp. 17–22.
8. On these see E.S. Kennedy, "Parallax Theory in Islamic Astronomy," repr. in *Kennedy* **et al.**, pp. 164–84; *Aaboe;* and *Kennedy* 1, pp. 140, 143, and 144.
9. Three examples are analyzed in *Irani* (Habash), *Jensen* (Abu Nasr), and *King* 4, XI (al-Khalili). Several more are discussed in *SATMI*, I, Sections 9.1 to 9.13.

10. Several such tables are discussed in *SATMI*, I, Section 2.5.
11. Several examples are noted in *King* 4, XIII 120–22.
12. See *King* 7 (Ibn Yunus), *Debarnot* (al-Biruni), and *Kennedy* 3 (al-Kashi).
13. See *Hartner*, pp. 287–318 (in part repr. from his article *"Asturlab"* in EI_2), *Waters et al.*, and *Schmalzl*.
14. See *Kennedy* 1, pp. 139–40.
15. See further *Cairo Survey*, nos. C72 and C137.
16. On al-Marrakushi see *Suter*, no. 363, *Cairo Survey*, no. C17, and the article in EI_2. He is usually described as a Morrocan because of his name. However, whatever his origin, he worked in Cairo, and all of his numerical examples and his tables involving terrestrial latitude are for the latitude of Cairo.
17. See *Sédillot-père* and *Sédillot-fils*. It appears that neither of the Sédillots realized that al-Marrakushi was writing in Cairo. Sédillot-père also misdated al-Marrakushi to 660 Hijra (= 1261/62) even though al-Marrakushi's solar tables and star catalogue are for 680 Hijra.
18. See *Schoy* 1. Schoy observed that al-Marrakushi was using the latitude of Cairo in his numerical examples.
19. On al-Maqsi see *Suter*, no. 383, and *Cairo Survey*, no. C15. See also *King* 1, p. 540.
20. These tables, and indeed the entire Cairo corpus of tables for timekeeping, are in several of the available manuscripts attributed to Ibn Yunus (on whom see the article in *DSB*). The corpus is analyzed in *King* 4, IX. The attribution of each of the various tables in the corpus is discussed in *SATMI*, II, Part 5.
21. Tables of the functions T, h, or a require about 10,000 entries to display values for each degree of both arguments h and λ. Tables of $T(h,H)$ with values for each degree of both arguments $(h < H)$ for a given latitude require about 3,000 entries to display values for each degree of both arguments.
22. On these tables see *SATMI*, I, Section 2.
23. See further *King* 4, II, pp. 51–53, and XVI, Sections 3 and 4.
24. Published in *Michel & Ben-Eli*.
25. On Najm al-Din al-Misri see *Suter*, no. 460 (confused) and *Cairo Survey*, no. C16. See also *King* 1, pp. 540–41.
26. The relationship between the three contemporary astronomers, al-Marrakushi, al-Maqsi, and Najm al-Din al-Misri, is by no means clear. Najm al-Din mentions al-Maqsi in the introduction to his tables, but neither al-Maqsi nor al-Marrakushi mention each other.
27. A more detailed analysis is given in *SATMI*, I, Section 2.6.1.
28. On Amin al-Din al-Abhari see *Suter*, no. 393. He is not to be confused with Athir al-Din al-Abhari (*Suter*, no. 364).
29. On Ibn al-Sarraj see *Suter*, no. 508, *Mayer*, p. 34, and *Cairo Survey*, no. C26. See also *King* 1, pp. 544–45. All of his instruments are discussed in detail in *King* 3, which includes texts and translations of all the related treatises.

30. On this instrument see note 4 above. ʿAli ibn Khalaf and his instrument are ignored in most general accounts of the history of the astrolabe.
31. On this see further *Renaud.*
32. On Tibugha see *Brockelmann,* 2: 168–169, and SII, p. 167, and *Cairo Survey,* no. C53.
33. This problem is investigated in *Samsó.*
34. On these astronomers see *Cairo Survey,* nos. C54, C55, and C56.
35. On al-Bakhaniqi see *MAY,* pp. 34–35, and *Cairo Survey,* no. C28.
36. On al-Farghani see the article in *DSB* by A.I. Sabra, and *Sezgin,*V, pp. 259–60 and 6, pp. 149–51, where these tables are not mentioned. I am currently collecting material for a survey of all available Islamic tables for making astrolabe plates (see *King* 4, II, pp. 53–55, for a brief account and 2, pp. 23–27, for the earliest examples).
37. On Ibn al-Shatir see the article in *DSB* and the references there cited, to which add *Kennedy & Ghanem* and *Cairo Survey,* no. C30.
38. See various articles by Kennedy, Roberts and Abboud reprinted in both *Kennedy & Ghanem* and *Kennedy et al.,* pp. 50–107.
39. On this sundial see *Janin.*
40. This instrument is analyzed in detail in *King* 4, XII.
41. On the "complete" and ʿAlaʾi quadrants see *Schmalzl,* pp. 100–108, repr. in *Kennedy & Ghanem,* pp. 27–35.
42. On al-Khalili *(Suter,* no. 418) see the article in *DSB* (supplement), and *Cairo Survey,* no. C32. See also *King* 1, pp. 542–43.
43. The Damascus corpus is analyzed in *SATMI,* II, Part X. See already *King* 4, X.
44. On al-Mizzi see *Suter,* no. 406, *Mayer,* pp. 61–62, and *Cairo Survey,* no. C34. On his tables see the references cited in the preceding note.
45. Analyzed in *King* 4, XI. See also *SATMI,* II, Section 10.4.
46. Analyzed in *SATMI,* II, Section 9.6
47. Analyzed in *King* 4, XIII.
48. See *ibid.,* pp. 120–22.
49. On Ibn al-Ghazuli see *Suter,* no. 412, *Mayer,* pp. 66–67, and *Cairo Survey,* no. C33.
50. On Ibn Tibugha see *Cairo Survey,* no. C54. The Cairo manuscript of the treatise used in *Samsó & Catalá* is the only evidence for the existence of a treatise on the *shakkaziya* quadrant by Ibn Tibugha. Samsó and Catalá wrongly associated the author with Tibugha al-Baklamshi and his grandson Ahmad ibn Rajab ibn Tibugha the Egyptian astronomer known as Ibn al-Majdi (*Suter,* no. 432, and *Cairo Survey,* no. C62.).
51. On al-Maridini see *Suter,* no. 421, and *Cairo Survey,* no. C47.
52. On Sibt al-Maridini see *Suter,* no. 445, and *Cairo Survey,* no. C97.
53. This treatise is published with translation and commentary in *King* 5, X.
54. Analyzed in *ibid.,* Appendix B, pp. 231–40.
55. On al-Khwarizmi see the article in *DSB* by G. Toomer, and *Sezgin,* V, pp. 228–41, and 6, pp. 140–43, where his treatise on the sundial is men-

tioned in 6, p. 143. On these sundial tables see the reference cited in note 7 above. I am currently collecting material for a survey of all available Islamic tables for marking sundials (see *King* 4, II, pp. 51–53 for a brief account).

56. On Ibn Sudun, Ibn al-Muhallabi, and Sibt al-Maridini (*Suter,* no. 445) see *Cairo Survey,* nos. C82, C67, and C97. The sample extracts from the tables of Sibt al-Maridini published in *Schoy* 2, pp. 352–60, are in fact by Ibn al-Muhallabi and are independent of Sibt al-Maridini's introduction translated by Schoy on pp. 344–51. Schoy was aware of a problem between text and tables (cf. *Schoy* 2, pp. 342–47), but had only one defective manuscript at his disposal. On Ibn al-Muhallabi's treatise on sundial theory see *King* 5, XVI, Section 4.
57. On al-Tizini see *Suter,* no. 450, and *Cairo Survey,* no. C95.
58. On al-Wafa'i see *Suter,* no. 437, and *Cairo Survey,* no. C61.
59. On the *da'irat al-mu'addil* see *Tekeli,* where al-Wafa'i's text is published with translation, and also *Brice, Imber & Lorch* and *Dizer.* See also *King* 5, XII, pp. 216–217, where this instrument is compared with the earlier instrument of Ibn al-Shatir.
60. *Cf. King* 5, XVI, Appendix, note 3.
61. al-Wafa'i's treatise is published with translation and commentary in *King* 3.
62. These tables are analyzed in *SATMI,* I, Section 9.9.
63. On al-Karadisi see *Suter,* no. 442, and *Cairo Survey,* no. C90.
64. These tables are analyzed in *SATMI,* II, Section 8.1.
65. On Ibn al-Mushrif see *Cairo Survey,* no. C43.
66. These tables are analyzed in *SATMI,* I, Section 9.8.
67. On Sudun see *Cairo Survey,* no. C81.
68. On al-Aqfahsi see *Suter,* no. 440, and *Cairo Survey,* no. C68.
69. On al-Sufi see *Suter,* no. 447, and *Cairo Survey,* no. C98.
70. *Cf. Kennedy* 1, nos. 12 and 37.
71. Analyzed in *SATMI,* I, Section 9.10
72. Analyzed in *SATMI,* I, Section 5.5.2.
73. Published with translation and commentary in *King* 5, XII.
74. On al-Husni see *Suter,* no. 511, and *Cairo Survey,* no. C124.
75. These tables are analyzed in *SATMI,* II, Section 8.3.
76. On 'Abd al-Qadir al-Minufi see *Suter,* no. 479, and *Cairo Survey,* no. C121.
77. On this see *King* 5, XI, pp. 373–76.
78. Some of this material is analyzed in *SATMI,* II.
79. *Cf. Price,* esp. pp. 399–400.
80. *Sédillot-fils,* p. 182 and plate 89 (cited in *Price,* p. 399, n. 2 !).
81. *Poulle,* esp. p. 130.
82. *Schmalzl,* pp. 71–99, and *King* 4, XIII, pp. 111–15.
83. *Kennedy* 2, pp. 189–90.

[182]

Bibliography and Bibliographical Abbreviations

Aaboe: A. Aaboe, "On the Tables of Planetary Visibility in the Almagest and the Handy Tables," *Kgl. Danske Vidensk. hist.-fil. Meddelelser* (1960), 37(8):3–20.

Aaboe Festschrift: Studies on the History of Science in Honor of Asger Aaboe, Acta Historica Scientiarum Naturalium et Medicinalium, Copenhagen, 1987.

Brice, Imber & Lorch: W. Brice, C. Imber, and R. Lorch, "The Da'ire-yi Muʿaddel of Seydi ʿAli Re'is," *Seminar on Early Islamic Science* (The University of Manchester), Monograph No. 1 (July 1986), reviewed in *King* 5, XIII.

Brockelmann: C. Brockelmann, *Geschichte der arabischen Litteratur,* 2 vols. (2d ed.), Leiden: E.J. Brill, 1943–49; Supplement, 3 vols., Leiden: E.J. Brill, 1937–42.

Cairo Survey: D. A. King, *A Survey of the Scientific Manuscripts in the Egyptian National Library,* Winona Lake, Ind.: Eisenbrauns, 1986.

Debarnot. M.-Th. Debarnot, *Al-Biruni: Kitab Maqalid 'Ilm al-Hay'a—La Trigonométrie Sphérique chez les Arabes de l'Est à la Fin du Xe siècle,* Damascus: Institut Français de Damas, 1985.

Dizer: M. Dizer, "The Daʾirat al-Muʿaddal in the Kandilli Observatory, and some Remarks on the Earliest Recorded Islamic Values of the Magnetic Declination," *Journal for the History of Arabic Science* (1977), 1: 257–262.

DSB: Dictionary of Scientific Biography, 14 vols. and 2 supp. vols. New York: Scribner's, 1970–80.

EI_2 The Encyclopædia of Islam, 2d. ed., 5 vols. to date, Leiden: E.J. Brill, 1960 to present.

Hartner: W. Hartner, *Oriens—Occidens: Ausgewählte Schriften zur Wissenschafts- und Kulturgeschichte,* Hildesheim: Georg Olms, 1968.

Irani: R.A.K. Irani, "The *Jadwal al-Taqwim* of Habash al-Hasib," Master's Dissertation, American University of Beirut, 1956.

Janin: L. A. Janin, "Le cadran solaire de la Mosquée Umayyade à Damas, " *Centaurus* (1972), 16:285–98, reprinted in *Kennedy & Ghanem,* pp. 107–21.

Jensen: C. Jensen, "Abu Nasr's Approach to Spherical Trigonometry as Developed in His Treatise *The Table of Minutes,*" *Centaurus* (1971), 16:1–19.

Kennedy 1: E. S. Kennedy, "A Survey of Islamic Astronomical Tables," *Transactions of the American Philosophical Society* (1956), 46(2):123–77.

Kennedy 2: E. S. Kennedy, *The Planetary Equatorium of Jamshid Ghiyath al-Din al-Kashi,* Princeton: Princeton University Press, 1960.

Kennedy 3: E. S. Kennedy, "Spherical Astronomy in Kashi's Khaqani Zij," *Zeitschrift für Geschichte der Arabisch-Islamischen Wissenschaften* (1985), 2: 1–46.

Kennedy & Ghanem: E. S. Kennedy and I. Ghanem, eds., *The Life and Work of*

Ibn al-Shatir: An Arab Astronomer of the Fourteenth Century, Aleppo: Institute for the History of Arabic Science, 1976.

*Kennedy **et al.***: E.S. Kennedy, Colleagues and Former Students, *Studies in the Islamic Exact Sciences*, Beirut: American University of Beirut Press, 1983.

King 1: D. A. King, "The Astronomy of the Mamluks," *Isis* (1983), 74:531–55, reprinted in *King* 4, III.

King 2: D. A. King, "al-Khwarizmi and New Trends in Mathematical Astronomy in the Ninth Century," *Occasional Papers on the Near East* (New York University, Hagop Kevorkian Center for Near East Studies), 2 (1983).

King 3: D. A. King, *The Astronomical Instruments of Ibn al-Sarraj*, to be published by the Benaki Museum, Athens.

King 4: D. A. King, *Islamic Mathematical Astronomy*, London: Variorum Reprints, 1986.

King 5: D. A. King, *Islamic Astronomical Instruments*, London: Variorum Reprints, 1987.

King 6: D. A. King, "Universal Solutions in Islamic Astronomy," in *Aaboe Festschrift*.

King 7: D. A. King, *Spherical Astronomy in Medieval Islam: The* Hakimi Zij *of Ibn Yunus*, Frankfurt: IGAIW (in press).

MAY: D. A. King, *Mathematical Astronomy in Medieval Yemen*, Malibu, Ca.: Undena Publications, 1983.

Mayer: L. A. Mayer, *Islamic Astrolabists and Their Works*, Geneva: Albert Kundig, 1956.

Michel & Ben Eli: H. Michel and A. Ben-Eli, "Un cadran solaire remarquable," *Ciel et Terre* (1965), 81.

Millás 1: J. Millás Vallicrosa, *Don Profeit Tibbon: Tractat de l'Assafea d'Azarquiel*, Biblioteca Hebraico-Catalana, vol. 4, Barcelona, 1933.

Millás 2: J. Millás Vallicrosa, *Estudios sobre Azarquiel*, Madrid & Granada: Consejo Superior de Investigaciones Científicas, Instituto Miguel Asín, 1943–50.

Poulle: E. Poulle, "Théorie des planètes et trigonométrie au XVe siècle, d'après un équatoire inédit, le sexegenarium," *Journal des Savants* (1966), 129–61.

Price: D. J. de Solla Price, "The Little Ship of Venice—A Middle English Instrument Tract," *Journal of the History of Medicine and Allied Sciences* (1960), 15:399–407.

Renaud: H. P. J. Renaud, "Les Ibn Baso," *Hespéris* (1937), 24: 1–12.

Samsó: J. Samsó Moya, "Una hipótesis sobre cálculo por approximación con el cuadrante šakkazi," *Al-Andalus* (1971), 36:117–26.

Samsó & Catalá: J. Samsó Moya and M.A. Catalá, "Un instrumento astronómico de raigambre Zarqali: el cuadrante šakkazi de Ibn Tibuga," *Memorias de la Real Academia de Buenas Letras de Barcelona* (1971), 13:5–31.

SATMI: D. A. King, *Studies in Astronomical Timekeeping in Medieval Islam*, I: *A Survey of Medieval Islamic Tables for Reckoning Time by the Sun and Stars*, and

II: *A Survey of Medieval Islamic Tables for Regulating the Times of Prayer*, to be published by Springer-Verlag.

Sayılı: Aydin Sayılı, *The Observatory in Islam*, Ankara, 1960 (Turkish Historical Society).

Schmalzl: P. Schmalzl, *Zur Geschichte des Quadranten bei den Arabern*, Munich: Salesianische Offizin, 1929.

Schoy 1: K. Schoy, *Gnomonik der Araber*, in E. von Basserman-Jordan, ed., *Die Geschichte der Zeitmessung und der Uhren*, Bd. 1F, Berlin & Leipzig: Vereinigung Wissenschaftlicher Verleger, 1923.

Schoy 2: K. Schoy, "Sonnenuhren der spätarabischen Astronomie," *Isis* (1924), 6:332–60.

Sédillot-fils: L. A. Sédillot, "Mémoire sur les instruments astronomiques des Arabes," *Mémoires de l'Académie Royale des Inscriptions et Belles-Lettres de l'Institut de France* (1844), 1:1–229, to be reprinted (Frankfurt: IGAIW).

Sédillot-père: J. J. Sédillot, *Traité des instruments astronomiques des Arabes composé au treizième siècle par Aboul Hhassan Ali de Maroc intitulé Jamiʿ al-mabadiʾ wa-'l-ghayat*, 2 vols., Paris: Imprimerie Royale, 1834–35, reprinted Frankfurt: IGAIW, 1984.

Seemann: H. Seemann, "Das kugelförmige Astrolab nach den Mitteilungen von Alfons X. von Kastilien und den vorhandenen arabischen Quellen," *Abhandlungen zur Geschichte der Naturwissenschaften und der Medizin* (Erlangen), Heft VIII (1925).

Sezgin: F. Sezgin, *Geschichte des arabischen Schrifttums*, 9 vols. to date, Leiden: E. J. Brill, 1967–present.

Suter: H. Suter, "Die Mathematiker und Astronomen der Araber und ihre Werke," *Abhandlungen zur Geschichte der mathematischen Wissenschaften*, (1900), 10, and "Nachträge und Berichtigungen," *ibid.* (1902), 14:157–85, reprinted Amsterdam: Oriental Press, 1982.

Tekeli: S. Tekeli, "(The) 'Equatorial Armilla' of Iz(z) al-Din b. Muhammad al-Wafai and (the) Torquetum," *Ankara Üniversitesi Dil ve Tarih—Coğrafya Fakültesi Dergesi* (1960), 18:227–59, (Ankara: Türk Tarih Kurumu Basimevi, 1962).

Waters et al.: The Planispheric Astrolabe (with a foreword by D. W. Waters), Greenwich: National Maritime Museum, 1976.

VIII

MIZWALA

MIZWALA (A.), sundial. This term and *sāʿa shamsiyya* are used in modern Arabic, but in mediaeval Islamic times horizontal sundials were referred to either as *rukhāma*, lit. "marble" or *basīṭa*, lit. "flat", and vertical sundials as *munḥarifa*, lit. "inclined". The gnomon was usually called *shakhṣ*, *shākhiṣ* or *miḳyās*. One expression of the Muslim concern with timekeeping and regulating the times of prayer [see MĪḲĀT] was an avid interest in gnomonics, the theory and practice of sundial construction. Muslim astronomers made substantial contributions to both aspects of the subject, and by the late mediaeval period there were sundials of one form or another in most of the major mosques in the Islamic world. Those which survived usually bear markings for the hours (seasonal or equinoctial) and for the midday (*ẓuhr*) and afternoon (*ʿaṣr*) prayers. Since the beginnings of the permitted intervals for these two prayers were defined in terms of shadow lengths, their regulation by means of sundials was singularly appropriate.

The earliest surviving Arabic treatise on sundials deals with their construction and is attributed to al-Khwārazmī [*q.v.*], active in Baghdād in the early 3rd/9th century. The work consists mainly of a set of tables of coordinates for constructing horizontal sundials serving 12 different latitudes. Each of al-Khwārazmī's sub-tables for a specific latitude displays for both of the solstices the solar altitude, the shadow of a standard gnomon, and the solar azimuth (see Pl. XVI, fig. 1). With these functions tabulated, construction of the sundial would have been almost routine. A 4th/10th-century treatise on the construction of vertical sundials has also survived. This is by one of the two Baghdād astronomers Ibn al-Adamī or Saʿīd ibn Khafīf al-Samarḳandī (the copyist of the unique manuscript was not sure). Two auxiliary functions are tabulated with which one can generate pairs of orthogonal coordinates useful for marking vertical sundials serving any terrestrial latitude and inclined at any angle to the local meridian. Also from the ᶜAbbāsid period, probably the 3rd/9th century, is the earliest text on the portable conical sundial.

Thābit b. Ḳurra (*fl.* Baghdād, *ca.* 300/900 [*q.v.*]) wrote a comprehensive work on sundial theory dealing with the transformation of coordinates between different orthogonal systems based on three planes: (1) the horizon, (2) the celestial equator, and (3) the variable plane of the sundial. As far as we know, Thābit's treatise, despite its merits, was not influential. Later Muslim astronomers were more interested in the practical side of gnomonics.

The major work on spherical astronomy and instrumentation in the later period of Islamic astronomy was the compendium by Abū ᶜAlī al-Marrākushī [*q.v.*], who worked in Cairo *ca.* 680/1280. There are lengthy sections on sundials, with numerous tables and diagrams. The discussion concentrates on descriptions of the mode of construction; there is little underlying theory. The text deals with horizontal, vertical, cylindrical and conical sundials as well as simple, portable vertical sundials, in plane, cylindrical or conical format. Al-Marrākushī's treatise was widely influential in later astronomical circles in Egypt, Syria and Turkey.

A contemporary of al-Marrākushī in Cairo, Shihāb al-Dīn al-Maḳsī, prepared a set of tables for marking vertical sundials for the latitude of that city. For each degree of inclination to the local meridian he tabulated the coordinates of the points of intersection of the lines for the seasonal hours and the *ᶜaṣr* curve with the shadow traces at the equinoxes and the solstices (see Pl. XVII, fig. 2). The astronomer Ibn al-Sarrādj, active in Aleppo some fifty years later, devised several ingenious sundials for all latitudes.

Only a few sundials survive from the mediaeval period. Hundreds or even thousands must have been

1. An extract from al-Khwārazmī's tables for sundial construction showing two pairs of sub-tables for each of latitudes 21°, 28°, 33°, 35° and 40°, based on obliquity 23;51° [see MAYL]. The final pair of tables is for latitude 29;30° but with obliquity 23;35°. These tables occur here in a treatise on astrolabes and sundials by al-Sidjzī (*fl.* Iran, *ca.* 975). Taken from ms. Istanbul Topkapı 3342, 8 + 9, with kind permission of the Director of the Topkapı Library.

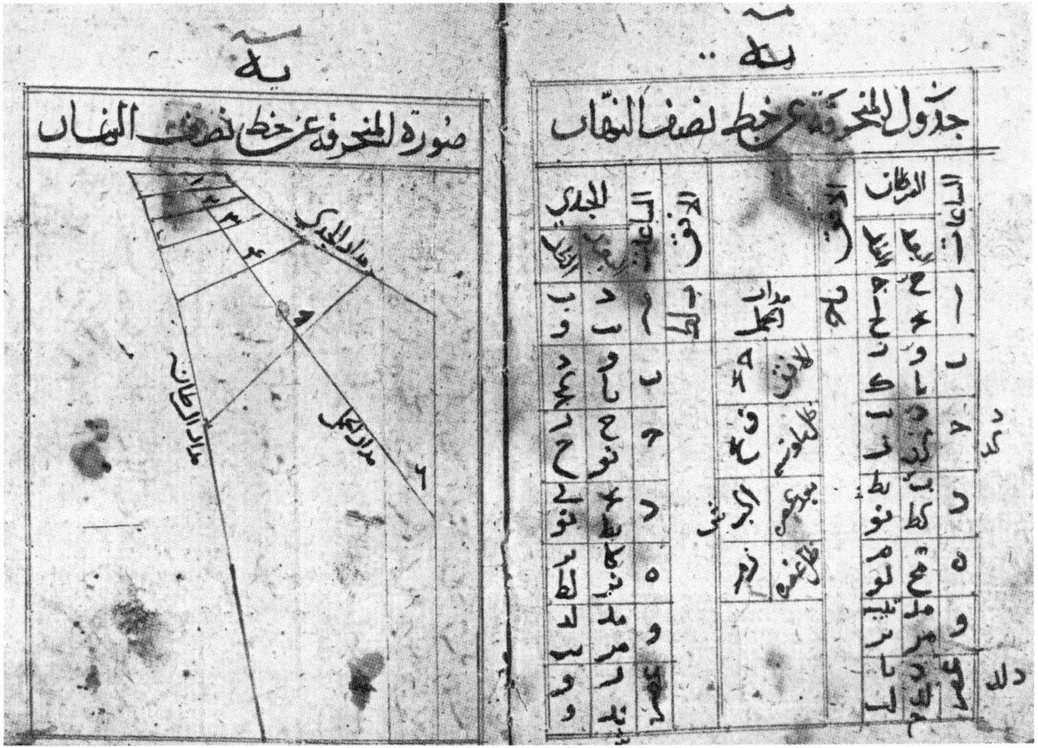

2. An extract from al-Maksī's tables for constructing vertical sundials for the latitude of Cairo. This particular sub-table serves an inclination of 15° to the meridian. Taken from ms. Cairo Dār al-Kutub *mīḳāt* 103, fols. 68b-69a, with kind permission of the Director of the Egyptian National Library.

constructed from the 3rd/9th century onwards, but the vast majority have disappeared without trace. The oldest surviving Islamic sundial (see Pl. XVIII, fig. 3) was made by Ibn al-Ṣaffār, an astronomer of some renown who worked in Cordova about the year 400/1000. Only one-half of the instrument survives, but the remains are adequate to establish that gnomonics was not the maker's *forte*. The sundial is of the horizontal variety and there are lines for each of the seasonal hours and the *ẓuhr* prayer; there would also have been markings for the *ᶜaṣr*. The gnomon is now missing, but its length is indicated as the radius of a circle engraved on the sundial. Several other, later Andalusī sundials which survive are singularly poor testimonials to their makers' abilities; yet proper sundials must have existed in mediaeval al-Andalus.

The Tunisian sundial shown in Pl. XVIII, fig. 4 is a much neater production. It was made in 746/1345-6 by Abu 'l-Ḳāsim b. Ḥasan al-S͟haddād, and it is of considerable historical interest because its markings display only the times of day with religious significance. For the afternoon (right-hand side) the curves for the *ẓuhr* and *ᶜaṣr* are marked according to the standard Andalusī/Mag͟hribī definitions. For the morning there is a curve for the *ḍuḥā*, symmetrical with the *ᶜaṣr* curve with respect to the meridian, and a line for the time of the *taʾhīb* one equinoctial hour before midday, this institution being associated with the communal worship on Friday [see DJUMᶜA]. It was the symmetry of the *ḍuḥā* and *ᶜaṣr* on this sundial that first led to an understanding of the definitions of the times of the daylight prayers in Islam.

The astronomer Ibn al-S͟hāṭir, chief *muwaḳḳit* of the Umayyad Mosque in Damascus, constructed in 773/1371-2 a horizontal sundial, some 2 m × 1 m in size, which is undoubtedly the most splendid sundial of the Middle Ages (see Pl. XIX, fig. 5). It was erected on a platform on the southern side of the main minaret of the Mosque. It could be used to measure time after sunrise in the morning and time before sunset in the afternoon, as well as time before and after midday. This means that it measured time relative to the *ẓuhr* and *maghrib* prayers, and the *ʿaṣr* curve enabled regulation of time relative to that prayer also.

No vertical sundials survive from the first few centuries of Islamic astronomy, but we know they were made because of the treatises on their use which were compiled from the 3rd/9th century onwards. The *munḥarifa*, meaning simply "vertical and inclined to the meridian", usually bore markings for each seasonal hour and the *ʿaṣr* prayer bounded by two hyperbolic shadow-traces for the solstices. Tables such as those of al-Maḳsī (see above) would be particularly useful for constructing such sundials on the walls of mosques.

Numerous sundials from the period after the 10th/16th century survive in mosques from Morocco to India. No inventory of these instruments has been made, beyond non-technical surveys of those in Istanbul. In any case, they constitute but a modest testimonial to a millennium of Muslim activity in gnomonics.

Bibliography: On Islamic sundial theory in general, see K. Schoy, *Gnomonik der Araber*, in E. von Bassermann-Jordan, ed., *Die Geschichte der Zeitmessung und der Uhren*, Band 1F, Berlin and Leipzig 1923, and idem, *Sonnenuhren der spätarabischen Astronomie*, in *Isis*, vi (1924), 332-60, both repr. in idem, *Beiträge zur arabisch-islamischen Mathematik und Astronomie*, Frankfurt 1988. A survey of Islamic tables for sundial construction is in preparation: see already D.A. King, *On the astronomical tables of the Middle Ages*, in *Studia Copernicana*, xiii (1975), 37-56, repr. in idem, *Islamic mathematical astronomy*, London, Variorum Reprints, 1986, II, esp. 51-3.

On a sundial used by Umayyad caliph ʿUmar b. ʿAbd al-ʿAzīz [*q.v.*] and other instruments which attest to the original association of the times of prayers with seasonal hours, see King, *On the origin of the prayers in Islam*, in *Oriens*, xxxiii (1990).

PLATE XVIII

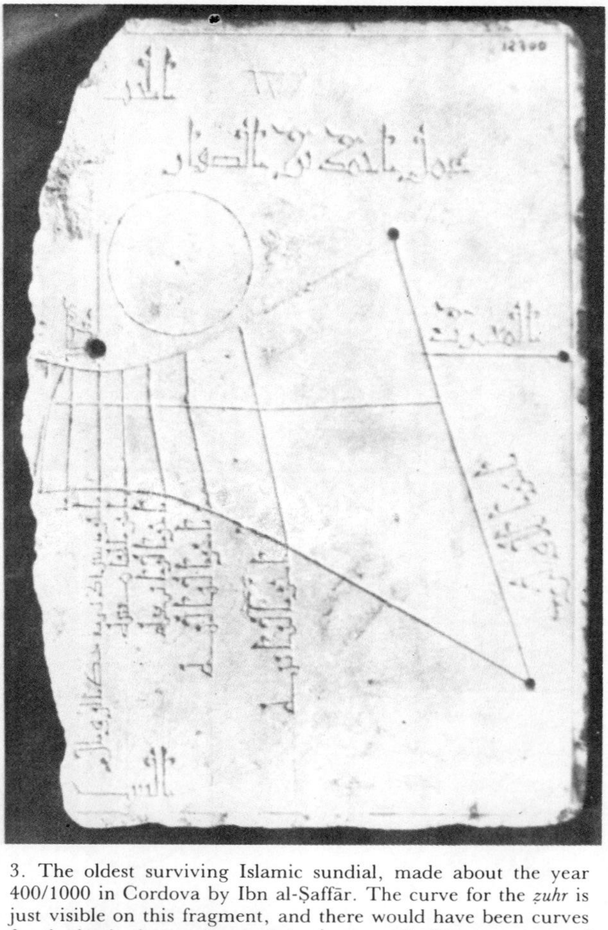

3. The oldest surviving Islamic sundial, made about the year 400/1000 in Cordova by Ibn al-Ṣaffār. The curve for the *ẓuhr* is just visible on this fragment, and there would have been curves for the beginning and end of the *ʿaṣr* as well. Photo courtesy of the Museo Arqueológico Provincial de Córdoba.

4. A 8th/14th-century Tunisian sundial indicating four times of day with religious significance: the *ḍuḥā* and the *taʾhīb* before midday, and the *ẓuhr* and the *ʿaṣr* after it. Property of the National Museum of Carthage, photo courtesy of the late M. Alain Brieux, Paris.

On al-Khwārazmī's tables for sundial construction, see B.A. Rosenfeld, ed., *Muhammad ibn Musa al-Khorezmi*, Moscow 1983, 221-34, and also King, *Al-Khwārizmī and new trends in mathematical astronomy in the ninth century*, in *Occasional Papers on the Near East* (New York University, Hagop Kevorkian Center for Near Eastern Studies), 2 (1983), esp. 17-22. On Thābit's treatise, see K. Garbers, *Ein Werk Ṯābit b. Qurra's über ebene Sonnenuhren*, in *Quellen und Studien zur Geschichte der Mathematik, Astronomie und Physik*, Abt. A, 4 (1936); and P. Luckey, *Thābit b. Qurra's Buch über die ebenen Sonnenuhren*, in *Quellen und Studien zur Geschichte der Mathematik, Astronomie und Physik*, Abt. B, Bd. 4 (1937-8), 95-148.

On al-Marrākushī's treatise, see J.-J. Sédillot, *Traité des instruments astronomiques des Arabes composé au treizième siècle par Aboul Hhassan Ali de Maroc...*, Paris 1834-5, repr. Frankfurt 1985, and L.A. Sédillot, *Mémoire sur les instruments astronomiques des Arabes*, in *Méms. de l'Acad. Royale des Inscriptions et Belles-Lettres de l'Institut de France*, i (1844), 1-229, repr. Frankfurt 1989.

On Andalusī sundials, see King, *Three sundials from Islamic Andalusia*, in *Jnal. for the Hist. of Arabic Science*, ii (1978), 358-93, repr. in idem, *Islamic astronomical instruments*, London, Variorum Reprints 1987, XV. Two more examples unearthed at al-Madīna al-Zahrā' are illustrated in [J. Vernet *et al.*, eds.], *Instrumentos astronómicos en la España medieval: su influencia en Europa*, Santa Cruz de la Palma 1985, 126-33. The Tunisian sundial is discussed in King, *A fourteenth-century Tunisian sundial for regulating the times of Muslim prayer*, in W. Saltzer and Y. Maeyama, eds., *Prismata. Festschrift für Willy Hartner*, Wiesbaden 1977, 187-202, repr. in King, *Instruments*, XVIII. On Ibn al-Shāṭir's sundial, see L. Janin, *Le cadran solaire de la Mosquée Umayyade à Damas*, in *Centaurus*, xvi (1971), 285-98, repr. in E.S. Kennedy and I. Ghanem, eds., *The life and work of Ibn al-Shāṭir: an Arab astronomer of the fourteenth century*, Aleppo 1976, 107-21.

Other mediaeval sundials are described in P. Casanova, *La montre du Sultan Noûr ad Dīn (554 de l'Hégire = 1159-1160)*, in *Syria*, iv (1923), 282-99; Janin and King, *Le cadran solaire de la Mosquée d'Ibn Ṭūlūn au Caire*, in *JHAS*, ii (1978), 331-57, repr. in King, *Instruments*, XVI; A. Bel, *Trouvailles archéologiques à Tlemcen. Un cadran solaire arabe*, in *R. Afr.*, xlix (1905), 228-31; Janin, *Quelques aspects récents de la gnomique tunisienne*, in *ROMM*, xxiv (1977), 202-21; H. Michel and A. Ben-Eli, *Un cadran solaire remarquable*, in *Ciel et Terre*, lxxxi (1965) (describing an Ottoman polar sundial in Acre); and J. Livingston, *The Mukḥula, an Islamic conical sundial*, in *Centaurus*, xvi (1972), 299-308. Several Ottoman and Ṣafawid sundials on *ḳibla* indicators are illustrated in King, *Maps and instruments for finding the direction of Mecca (9th-19th century)*, in *Johann Wolfgang Goethe-Universität (Frankfurt am Main), Institut für Geschichte der Naturwissenschaften, Preprint Series*, no. 13 (1989).

Non-technical surveys of the sundials are in A.S. Ünver, *Sur les cadrans solaires horizontaux et verticaux en Turquie*, in *Archives Internationales d'Histoire des Sciences*, xxviii-xxix (1954), 254-66, and W. Meyer, *Sundials of the Osmanic era in Istanbul*, in *Procs. of the International Symposium on the Observatory in Islam (Istanbul 1977)*, Istanbul 1980, 193-202, and *idem*, *Istanbul'daki güneş saatleri*, in *Sandoz Kültür Yayınları* (Istanbul), no.7 (1985).

PLATE XIX

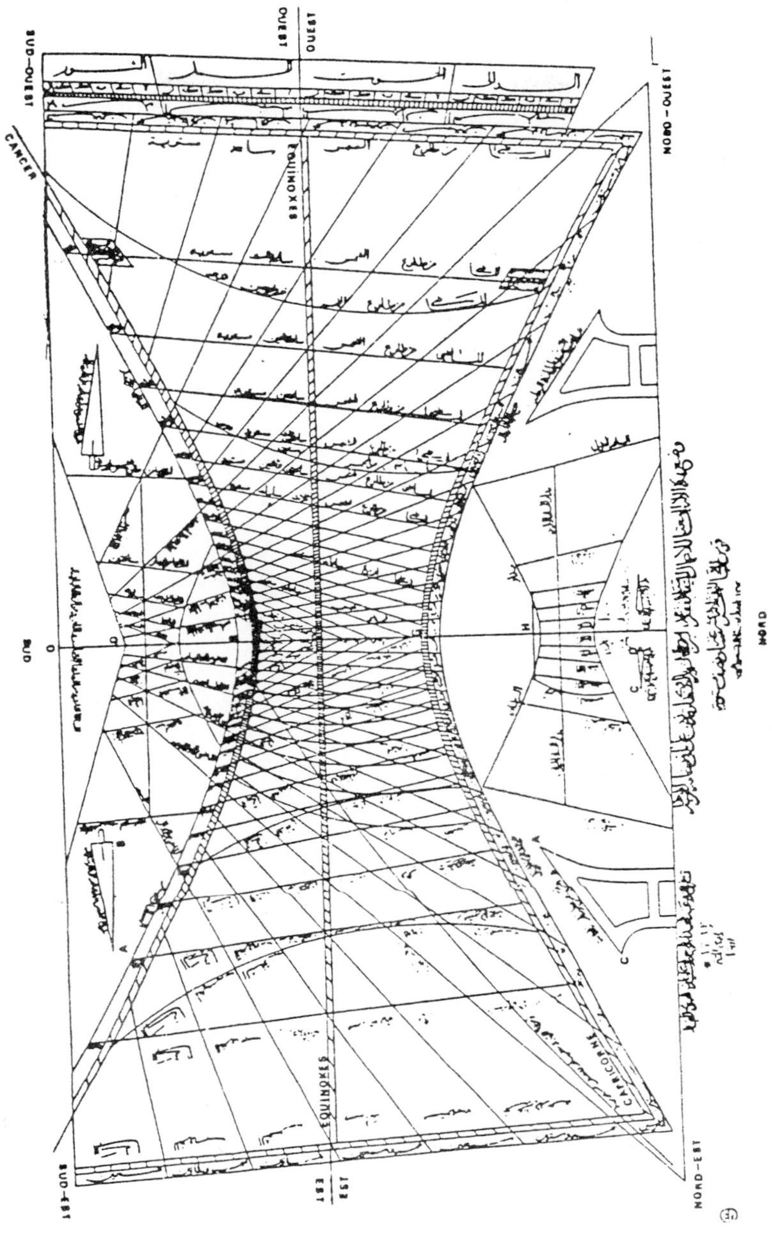

5. The markings of the splendid sundial of Ibn al-<u>Sh</u>āṭir which once graced the main minaret of the Umayyad Mosque in Damascus. The original sundial is in fragments, preserved in the garden of the nearby Archeological Museum. This copy is made from an exact replica made by the 13th/19th-century *muwakkit* al-Ṭanṭawī which is still *in situ* on the minaret. Courtesy of the Syrian Department of Antiquities and the late M. Alain Brieux, Paris.

IX

ḲIBLA: SACRED DIRECTION

ḲIBLA, the direction of Mecca, towards which the worshipper must direct himself for prayer.

Astronomical Aspects

The determination of the *ḳibla* was an important problem for the scientists of mediaeval Islam. Although essentially a problem of mathematical geography, the determination of the *ḳibla* can also be considered as a problem of spherical astronomy. Thus most Islamic astronomical handbooks or *zīdj*s, of which close on 200 were compiled during the millenium beginning in 750 A.D., contain a chapter on the determination of the *ḳibla*. In addition, several dozen mediaeval manuals for timekeeping deal with the topic. In contrast, the number of treatises dealing specifically with the *ḳibla* problem is quite few.

The *ḳibla* at a given locality is a trigonometric function of the local latitude, the latitude of Mecca, and the longitude difference from Mecca. The derivation of the *ḳibla* in terms of these three quantities was the most complicated of the standard problems of mediaeval Islamic spherical astronomy, and the solutions to the *ḳibla* problem proposed by the leading astronomers of mediaeval Islam bear witness to the development of mathematical methods from the 3rd/9th to the 8th/14th centuries and to the level of sophistication in trigonometry and computational techniques attained by these scholars. Already in the 3rd/9th century Muslim scholars had derived exact solutions using the construction of Greek mathematics known as the analemma (in which the various significant planes involved in a specific problem are either projected or folded into a single working plane, whereupon the geometrical solution can be derived graphically or the trigonometric solution can be derived by plane trigonometry) or using the classical Theorem of Menelaos for the complete spherical quadrilateral. Later *ḳibla* methods included trigonometric solutions based on projection methods or on the simpler corollaries of the Theorem of Menelaos. Certain Muslim astronomers contented themselves with approximate solutions, which were adequate for practical purposes. The final mathematical solution to the *ḳibla* problem was the table compiled by the 8th/14th-century astronomer al-Khalīlī, displaying the *ḳibla* for all latitudes and longitudes.

The mathematical problem

Fig. 1

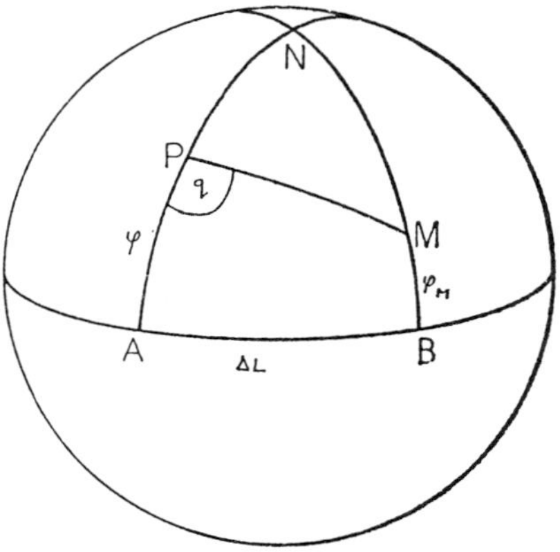

Fig. 1 shows a locality P and Mecca M on the terrestrial surface. The point N represents the north pole, and the meridians at P and M are shown as NPA and NMB, where A and B lie on the equator. In mathematical terms the *ḳibla* at P is defined by the direction of the great circle through P and M. In mediaeval Arabic the angle q between the arc PM and the local meridian NPS was called *inḥirāf al-ḳibla*, and the complementary angle between PM and the east-west line through P was called *samt al-ḳibla*. If φ and φ_M denote the latitudes of the locality and of Mecca (= PA nd MB), and ΔL denotes their longitude difference (= AB), then q is a function of φ, φ_M, and ΔL, and can be determined by spherical trigonometry. The modern formula, which can be derived from an application of the spherical co-tangent rule to ΔNPM, is:

$$q = \cot^{-1} \frac{\sin \varphi \cos \Delta L - \cos \varphi \tan \varphi_M}{\sin \Delta L}.$$

The exact solutions proposed by the mediaeval astronomers are less direct but ultimately equivalent to this.

Although the problem of determining the *ḳibla* is a problem of mathematical geography, it is mathematically equivalent to the astronomical problem of determining the azimuth or direction of a celestial body with given declination for a given hour-angle, and as such it was usually treated by the mediaeval astronomers. Indeed, the *ḳibla* problem may be transferred to the celestial sphere simply by considering the zenith of Mecca rather than Mecca.

Fig. 2

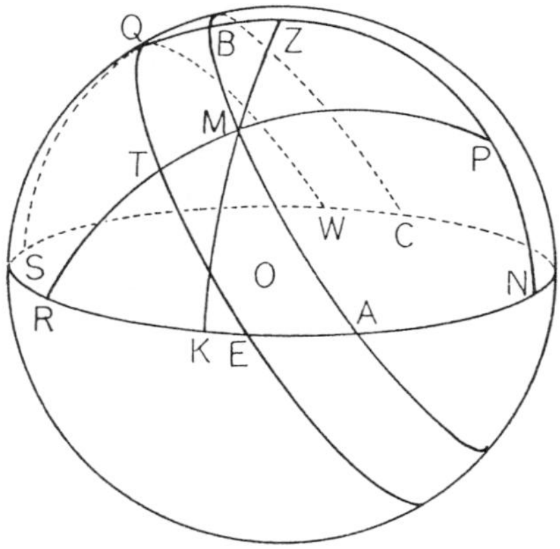

Fig. 2 shows the zenith of Mecca M on the celestial sphere for a locality O. The local horizon is NESW and the local meridian is NPZS. EQW is the celestial equator, P is the celestial pole, and ABC is the day-circle of M. PMTR is the meridian of Mecca. Now:

$$\text{PN} = \varphi, \text{MT} = \varphi_M, \text{and QT} = \Delta L,$$

and the problem is to determine the azimuth of M measured from the meridian by the arc SK $= q$.

Most mediaeval methods involve first finding the altitude of the zenith of Mecca above the local horizon, measured by the arc MK $= h$. This is equivalent to finding the complement of the distance between the two localities. Thus the problem of determining q (φ, φ_M, ΔL) is mathematically equivalent to determining the azimuth a (φ, δ, t) of a celestial body with declination δ (measured by MT) when the hour-angle is t (measured by QT) and the local latitude is φ (measured by PN). Indeed, several Islamic *ḳibla* methods state simply that if one faces the sun on the day when the solar declination is φ_M at the time when the hour-angle is ΔL (before or after midday, according as the locality is west or east of Mecca), then one is facing Mecca.

In the sequel a selection of methods is presented to illustrate the variety and sophistication of some of the few mediaeval *ḳibla* determinations that have been investigated in modern times. The notation has been modified in order to relate to that used in Fig. 2. For the details of the original constructions the reader is referred to the secondary literature listed in the bibliography. Those methods that are trigonometric in character are represented by means of trigonometric equations; in the original texts the relations are written out in words. The capital notation for trigonometric functions denotes that they are to a base other than unity: thus, for example, Sin $\theta = R \sin \theta$ where R is generally 60 and occasionally, in the case of works following the Indian-Sāsānid tradition, 150. Likewise Cos $\theta = R \cos \theta$, Vers $\theta = R$ vers $\theta = R\,(1 - \cos \theta)$, etc. The radius of the celestial sphere is taken to be R.

The trigonometric procedures outlined by the Muslim astronomers can also be performed geometrically using a grid of the kind which occurs on the circular instruments known in mediaeval Arabic as *al-dustūr* and *al-s̲h̲akkāziyya*, or the related quadrants known as *al-rubʿ al-mud̲j̲ayyab* and *rubʿ al-s̲h̲akkāziyya*. Most Islamic treatises on these instruments contain a chapter on the determination of the *ḳibla*.

Approximate solutions

A popular approximate method for determining the *ḳibla* which occurs in the *Zīdj* of the Syrian astronomer al-Battānī (*fl.* Raḳḳa, *ca.* 297/910) and in several unsophisticated Islamic astronomical works such as *al-Mulakhkhaṣ fi 'l-hay'a* by al-Djaghmīnī (*fl.* Khwārazm, ? *ca.* 725/1325) is the following.

Fig. 3

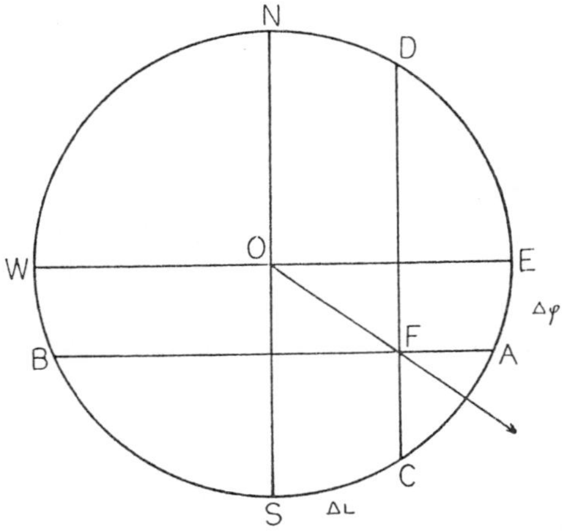

Fig. 3 shows the construction for a locality where Mecca is to the south east. Mark the cardinal directions NWSE on a horizontal circle centre O and radius R and measure arcs EA=WB=$\Delta\varphi=\varphi-\varphi_M$ southwards and SC=ND=ΔL eastwards. Next draw AB and CD, and denote their point of intersection by F. Then OF defines the *ḳibla*. This method is equivalent to an application of the formula

$$q = \mathrm{Sin}^{-1}\left\{\frac{R\,\mathrm{Sin}\,\Delta L}{\sqrt{\mathrm{Sin}^2\Delta\varphi + \mathrm{Sin}^2\Delta L}}\right\} = \mathrm{Tan}^{-1}\left\{\frac{R\,\mathrm{Sin}\,\Delta L}{\mathrm{Sin}\,\Delta\varphi}\right\}.$$

Certain Muslim astronomers also used tables based on this formula and displaying values of q ($\Delta\varphi$, ΔL) for each degree of both arguments from 1° to 20°. A feature of these tables is that the entries for $\Delta\varphi=\Delta L$ are all 45° 0′.

Another approximate solution to the *ḳibla* problem is outlined in a treatise related to al-Kh^w^ārazmī (*fl.* Baghdād, *ca.* 215/830). Here the formula

$$q = \text{Sin}^{-1}\left\{\frac{\left[\dfrac{\text{Sin}\,\Delta L\,\text{Cos}\,\varphi_M}{R}\right]R}{\sqrt{\text{Sin}^2\,\Delta\varphi + \left[\dfrac{\text{Sin}\,\Delta L\,\text{Cos}\,\varphi_M}{R}\right]^2}}\right\},$$

with the trigonometric functions to base $R = 150$ rather than $R = 60$, is outlined in words. The table displaying $q\ (\Delta\varphi, \Delta L)$ that accompanies this treatise was rather popular with later Muslim astronomers and exists in several manuscript copies, some of which are of Syrian, Yemeni, and Turkish provenance. Yet other approximate *ḳibla* tables based on non-trivial formulae are found in the *Ashrafī zīdj* of the Persian astronomer Sayf-i Munadjdjim (*fl. ca.* 710/1310), the *Zīdj* of the Persian astronomer Shams al-Munadjdjim al-Wābiknawī (*fl. ca.* 725/1325), and in a treatise on the quadrant by Ibn Ṭaybughā (*fl.* Aleppo, *ca.* 751/1350).

Exact solutions

Four exact solutions to the *ḳibla* problem are outlined below. The first and second illustrate the application and mathematical elegance of the analemma construction, and the way in which it can be used to derive complicated formulae of spherical trigonometry from a plane figure. The third and fourth illustrate the application and mathematical elegance of the Theorem of Menelaos and its corollaries. Al-Khalīlī's *ḳibla* table, which is perhaps the most sophisticated trigonometric table compiled in the mediaeval period, illustrates the competence of an 8th/14th century scholar in the algebra of functions and computational techniques.

A geometric *ḳibla* construction proposed by Ḥabash al-Ḥāsib (*fl.* Baghdād and Damascus, *ca.* 235/850) involves an analemma in which the working plane is considered consecutively as the meridian, equatorial, meridian, and horizon planes. Ḥaba<u>sh</u>'s method may be summarised as follows (see Fig. 4). On a circle centre O and radius R mark the cardinal directions NWSE, and then draw arc WQ $= \varphi$, arc QB $= \varphi_M$, and arc QT $= \Delta L$. Draw the diameter QOR and the parallel chord BC with midpoint G. Mark the point M_2 on OT such that $OM_2 = GC$ and draw the perpendicular M_2M_1 onto BC. Next draw M_1L parallel to WE and M_1IJ parallel to SN to cut WE in I and the circle in J. Finally, construct the point M_3 on M_1L such that $OM_3 = IJ$ and produce OM_3 to cut the circle at K. Then OK defines the *ḳibla*.

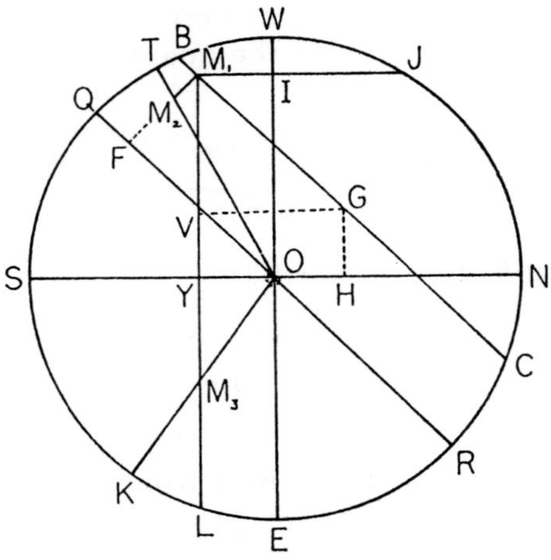

Fig. 4

This construction may be explained as follows. Firstly, QOR and BGC represent the projections of the celestial equator and the day circle of the zenith of Mecca in the meridian plane. Secondly, M_2 represents the projection of the zenith of Mecca in the equatorial plane. If we then imagine the equatorial plane to be folded into the meridian plane, M_2 moves to M_1, which is thus the projection of the zenith of Mecca in the meridian plane. Furthermore, M_1IJ is the projection in this plane of the almucantar through the zenith of Mecca, whose radius is thus IJ. Also M_1I and IJ measure the distances from the zenith of

Mecca to the prime vertical and to the line joining the local zenith to O, respectively. Finally, we consider the horizon plane the working plane; by virtue of the construction, M_3 is the projection of the zenith of the Mecca in this plane. Thus OK defines the *ḳibla*.

From such a geometric construction a trigonometric solution for the *ḳibla* problem can be derived with facility. Indeed, from the analemma construction for the *ḳibla* proposed by Ibn al-Hayt͟ham [*q.v.*] (*fl.* Cairo, d. 430/1039): a single formula for $q\ (\varphi, \varphi_M, L)$ equivalent to the modern one can be derived directly. Ibn Yūnus [*q.v.*] (*fl.* Cairo and Fuṣṭāṭ, d. 399/1009) proposed the following trigonometric solution to the *ḳibla* problem. Firstly from the quantity

$$h = \mathrm{Sin}^{-1} \left\{ \frac{\left\{\dfrac{\mathrm{Cos}\ \varphi\ \mathrm{Cos}\ \Delta L}{R}\right.\ \mathrm{Cos}\ \varphi_M + \mathrm{Sin}\ \varphi_M\ \mathrm{Sin}\ \varphi\left.\right\}}{R} \right\},$$

and then the *ḳibla* is defined by

$$q = \mathrm{Sin}^{-1} \left\{ \frac{\mathrm{Sin}\ \Delta L\ \mathrm{Cos}\ \varphi_M}{\mathrm{Cos}\ h} \right\}.$$

Ibn Yūnus offered no justification for this procedure, but his formulae can be derived from an analemma construction such as the one proposed by Ḥabas͟h. If in Fig. 4 we draw the perpendiculars GH and GV from G to SN and M_1Y, then we have

$$M_1V = \left\{ \frac{\mathrm{Cos}\ \varphi\ \mathrm{Cos}\ \Delta L\ \mathrm{Cos}\ \varphi_M}{R^2} \right\},$$

since

$$M_1V = M_1G\ \frac{\mathrm{Cos}\ \varphi}{R}$$

and

$$M_1G = OF = OM_2\ \frac{\mathrm{Cos}\ \Delta L}{R} = GC\ \frac{\mathrm{Cos}\ \Delta L}{R}$$

$$= \frac{\mathrm{Cos}\ \varphi_M\ \mathrm{Cos}\ \Delta L}{R}.$$

Also

$$\mathrm{GH} = \mathrm{OG}\,\frac{\mathrm{Sin}\,\varphi}{R} = \frac{\mathrm{Sin}\,\varphi_M\,\mathrm{Sin}\,\varphi}{R}\,.$$

Furthermore, since the arc JN measures h, we have $M_1Y = \mathrm{Sin}\,h$. But $M_1Y = MV + VY = MV + GH$. Ibn Yūnus' first formula follows immediately.

Next we observe that M_3Y and M_2F both measure the distance of the zenith of Mecca to the meridian and are hence equal. Thus

$$M_3Y = M_2F = OM_2\,\frac{\mathrm{Cos}\,\Delta L}{R} = \frac{\mathrm{Cos}\,\varphi_M\,\mathrm{Cos}\,\Delta L}{R}\,.$$

Also, since the arc JN measures h, and OM3 is by construction equal to IJ, we have OM3 = IJ = Cos h. Ibn Yūnus' second formula follows immediately, since q measures the arc SK.

Ibn Yūnus also compiled a table displaying the solar altitude in Cairo when the sun is in the azimuth of Mecca. His table gives values for each degree of solar longitude, corresponding roughly to each day of the year. Tables of this kind were contained in the main corpora of tables for timekeeping that were used in such centres as Cairo, Damascus, Jerusalem, and Istanbul, and an isolated table of this kind was compiled for the observatory at Marāg͟ha in north-western Persia. The solar altitude in the azimuth of Mecca is occasionally displayed graphically on the backs of astrolabes.

Al-Nayrīzī (*fl.* Bag͟hdād, *ca.* 287/900) solved the *ḳibla* problem by four successive applications of the Theorem of Menelaos. His solution involves finding successively the arcs TR, SR, MK and KS in Fig. 2, as follows. Firstly, find TR using

$$\frac{\mathrm{Sin\,PS}}{\mathrm{Sin\,SQ}} = \frac{\mathrm{Sin\,PR}}{\mathrm{Sin\,RT}}\,\frac{\mathrm{Sin\,TE}}{\mathrm{Sin\,EQ}}\,,$$

that is,

$$\frac{\mathrm{Sin}\,(180^\circ - \varphi)}{\mathrm{Sin}\,\varphi} = \frac{\mathrm{Sin}\,(90^\circ + \mathrm{TR})}{\mathrm{Sin\,TR}}\,\frac{\mathrm{Sin}\,(90^\circ - \Delta L)}{\mathrm{Sin}\,90^\circ}\,.$$

Secondly, find SR using

$$\frac{\text{Sin PQ}}{\text{Sin QS}} = \frac{\text{Sin PT}}{\text{Sin TR}} \frac{\text{Sin ER}}{\text{Sin ES}},$$

that is,

$$\frac{\text{Sin } 90^\circ}{\text{Sin } (90^\circ - \varphi)} = \frac{\text{Sin } 90^\circ}{\text{Sin TR}} \frac{\text{Sin ER}}{\text{Sin } 90^\circ},$$

whence ER and SR ($= \overline{\text{ER}} = 90^\circ - \text{ER}$). Thirdly, find MK ($= h$) using

$$\frac{\text{Sin SP}}{\text{Sin SZ}} = \frac{\text{Sin PR}}{\text{Sin RM}} \frac{\text{Sin MK}}{\text{Sin KZ}},$$

that is,

$$\frac{\text{Sin } (180^\circ - \varphi)}{\text{Sin } 90^\circ} = \frac{\text{Sin } (90^\circ + \text{TR})}{\text{Sin } (\text{TR} + \varphi_{\text{M}})} \frac{\text{Sin MK}}{\text{Sin } 90^\circ}.$$

Finally, find KS ($= q$) using

$$\frac{\text{Sin KS}}{\text{Sin SR}} = \frac{\text{Sin KZ}}{\text{Sin ZM}} \frac{\text{Sin MP}}{\text{Sin PR}},$$

that is

$$\frac{\text{Sin } q}{\text{Sin SR}} = \frac{\text{Sin } 90^\circ}{\text{Sin } (90^\circ - h)} \frac{\text{Sin } (90^\circ - \varphi_{\text{M}})}{\text{Sin } (90^\circ + \text{TR})}.$$

Al-Bīrūnī [*q.v.*] (*fl.* Khwarazm and Ghazna, d. after 442/1050) proposed several different methods for finding the *ḳibla*, based on a variety of different procedures. In his work on mathematical geography, the *Taḥdīd nihāyāt al-amākin*, al-Bīrūnī derived the longitude difference between Mecca and Ghazna mathematically using the distances between staging posts on the major caravan routes, and then derived the *ḳibla* at Ghazna using four different methods, including spherical trigonometry (using Menelaos' Theorem), solid geometry (using procedures equivalent to those standard in solving timekeeping problems), and the analemma. Al-Bīrūnī's solution to the *ḳibla* problem in his major astronomical work *al-Ḳānūn al-Masʿūdī*, compiled after the *Taḥdīd*, is more elegant than his solution by spherical trigonometry in the earlier work. It was also proposed about thirty years previously as an alternative solution by Ibn Yūnus. Al-Bīrūnī proved its correctness by spherical trigonometry. Ibn Yūnus presented it algebraically with no justification, but he appears to have derived most of his formulae for spherical astronomy by projection methods rather than by spherical trigonometry. Al-Bīrūnī's treatment of the problem illustrates the progress made by Muslim scholars in spherical trigonometry during the tenth century. Whereas his predecessor al-Nayrīzī had laboriously used Menelaos' Theorem, al-Bīrūnī used its simpler corollaries, the spherical Sine Rule and the "Rule of Four Quantities".

Al-Bīrūnī first outlined an algebraic procedure for finding q using four auxiliary arcs which we call θ_1, θ_2, θ_3, and θ_4. Since he used $R = 1$ rather than 60 his trigonometric functions are the same as the modern ones. First find θ_1, "the distance on the day circle", thus $\sin \theta_1 = \sin \Delta L \cos \theta_M$. Then find θ_2, "the local latitude adjusted for the horizon (of Mecca)", using $\sin \theta_2 = \frac{\sin \theta_M}{\cos \theta_1}$, and θ_3, "the correction to the latitude", using $\theta_3 = \varphi - \theta_2$. Then find θ_4, "the distance between the two localities", using $\cos \theta_4 = \cos \theta_3 \cos \theta_1$. Finally, q is given by

$$\sin q = \frac{\sin \theta_3 \cos \theta_1}{\sin \theta_4}.$$

Al-Bīrūnī's justification of this procedure is equivalent to the following. In Fig. 5, which is essentially the same as the diagrams in the manuscripts of the original text, the base circle represents the horizon, with N and S the north- and south-points. The local meridian is SZPN where Z is the local zenith and P is the celestial pole. M is the zenith of Mecca and GLJ and MPL are respectively the horizon and

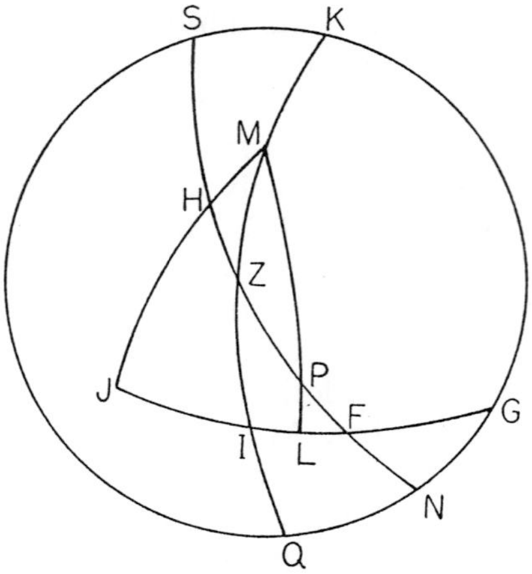

Fig. 5

meridian at Mecca. ZMK is the altitude circle of M and MHJ is a great circle with F as pole. Thus PN $= \varphi$, PL $= \varphi_M$, and MPZ $= \Delta L$, and it is required to find SK ($= q$). Al-Bīrūnī observed that (by the spherical Sine Rule):

$$\frac{\sin MP}{\sin MH} = \frac{\sin \angle MHP}{\sin \angle MPH}, \text{ i.e., } \frac{\cos \varphi_M}{\cos \angle F} = \frac{\sin 90^\circ}{\sin \Delta L},$$

whence $\angle$ F ($= \angle$ PFL) is known. Thus θ_1 is the complement of $\angle$ F. Similarly

$$\frac{\sin \angle F}{\sin \angle PLF} = \frac{\sin PL}{\sin PF}, \text{ i.e., } \frac{\cos \theta_1}{\sin 90} = \frac{\sin \varphi_M}{\sin PF},$$

whence PF is known.

Thus θ_2 is PF. Thirdly, since FN = PN — PF = φ-θ_2, θ_3 measures FN. Fourthly, al-Bīrūnī states that by the "Rule of Four Quantities"

$$\frac{\sin FZ}{\cos \angle G} = \frac{\sin FH}{\sin HJ}, \text{ i.e., } \frac{\cos \theta_4}{\cos \angle G} = \frac{\sin 90^\circ}{\cos \theta_1},$$

(note that $\cos \angle G = \cos \angle FGN = \cos IQ = \sin ZI$) so that θ_4 measures $\angle$ G.

Finally, he points out that (by the spherical Sine Rule)

$$\frac{\sin \angle G}{\sin \angle F} = \frac{\sin FN}{\sin GN}, \text{ i.e., } \frac{\sin \theta_4}{\cos \theta_1} = \frac{\sin \theta_3}{\sin GN},$$

whence q = SK = 90° — GN.

Al-Khalīlī (*fl.* Damascus, *ca.* 766/1365) compiled a *ḳibla* table based on an accurate formula and displaying q (φ, ΔL) for each degree of φ from 10° to 56° and each degree of ΔL from 1° to 60°. Al-Khalīlī's table thus contains a total of almost 3,000 entries, and the *ḳibla* is computed to degrees and minutes. The vast majority of the entries are either correct or in error by ± 1 or ± 2 minutes, a remarkable achievement. Table 1 shows a section of al-Khalīlī's table.

Al-Khalīlī does not describe the way in which he compiled his *ḳibla* table. However, in his introduction to the table he expresses his approval of the *ḳibla* method of al-Marrākushī (*fl.* Cairo, *ca.* 679/1280). This involved first finding h using

$$\text{Sin } h = \text{Sin } (\bar{\varphi} + \varphi_M) - \text{Vers } \Delta L \frac{\text{Cos } \varphi_M \text{ Cos } \varphi}{R^2},$$

and then applying the standard Islamic formula for finding the azimuth from the celestial altitude, namely

$$q = \text{arc Cos} \left\{ \frac{R \left[\frac{\text{Sin } h \text{ Tan } \varphi}{R} - \frac{R \text{ Sin } \delta}{\text{Cos } \varphi} \right]}{\text{Cos } h} \right\}.$$

Both of these formulae can be derived from Fig. 4. If al-Khalīlī did use precisely this method to compile his *ḳibla* table, it may be that he also used his universal auxiliary tables (*al-djadwal al-āfāḳī*) to facilitate the computation. These tables which were specifically designed for solving all of the standard problems of spherical astronomy for any latitude,

Table 1

Sample Entries from al-<u>Kh</u>alīlī's *Ḳibla* Table
(The errors in the minutes are shown after each entry)

$\Delta L\,/\,\varphi$	34°	35°	36°	37°
25°	66°55′ + 1	64°59′	63° 9′ + 1	61°20′ — 1
24	65 39 + 1	63 39 — 1	61 45 — 2	59 56 — 2
23	64 17	62 16 — 1	60 19 — 2	58 29 — 1
22	62 52	60 50	58 50 — 2	56 58 — 1
21	61 25 + 2	59 18	57 18	55 24
20	59 50 + 1	57 42 + 1	55 41 + 1	53 46 + 2
19	58 10	56 2 + 2	53 59 + 3	52 2 + 2
18	56 25 + 1	54 14 + 2	52 10 + 2	50 12 + 2
17	54 33	52 19	50 14 + 1	48 17 + 2
16	52 34	50 20 + 1	48 13	46 14 — 1
15	50 30 + 2	48 15 + 3	46 8 + 2	44 9 + 1
14	48 16 + 2	46 0 + 2	43 54 + 2	41 56
13	45 52	43 38 + 2	41 33 + 2	39 38 + 1
12	43 23 + 3	41 8 + 2	39 6 + 3	37 12 + 1
11	40 42 + 4	38 26 — 1	36 29 + 2	34 40 + 2
10	37 46	35 38	33 43	31 59
9	34 42 — 1	32 40 — 1	30 50 — 1	29 12
8	31 28 — 1	29 34	27 51	26 18 — 1
7	28 7 + 2	26 19 + 1	24 44 + 1	23 20 + 1
6	24 31 + 2	22 54 + 2	21 30 + 2	20 14 + 1
5	20 44 + 1	19 20 + 1	18 7 + 2	17 2 + 2
4	16 49 + 2	15 38 + 1	14 36	13 44 + 1
3	12 43	11 48 — 1	11 1 — 1	10 21
2	8 33	7 56	7 24	6 57 + 1
1	4 19 + 1	4 1 + 2	3 45 + 2	3 33 + 4

Table 2

Sample Entries from al-<u>Kh</u>alīlī's *Ḳibla* List
(The errors are derived using the mathematically correct formula and al-<u>Kh</u>alīlī's coordinates)

Locality	Longitude	Latitude	*Ḳibla*	Error
Gaza	37° 0′	32° 0′	42°46′	+ 1′
Hebron	56 30	31 35	45 21	— 6
Jerusalem	56 50	32 0	43 14	— 2
Damascus	60 0	33 30	29 4	+ 1
Beirut	59 15	33 20	31 59	— 4
Baʿlabakk	60 0	33 50	28 25	+ 1
Ḥamā	61 45	34 45	20 32	— 2
Aleppo	62 10	35 50	17 42	0
Mārdīn	64 0	37 55	9 48	0
Bag<u>h</u>dād	70 0	33 25	13 19	0
Kūfa	69 30	31 30	13 12	+ 1
Mecca	67 0	21 30	-	-

display the three functions

$$f_{\varphi}(\theta) = \frac{R \text{ Sin } \theta}{\text{Cos } \varphi}, \quad g_{\varphi}(\theta) = \frac{\text{Sin } \theta \text{ Tan } \varphi}{R},$$

$$\text{and } G(x, y) = \text{arc Cos} \left\{ \frac{Rx}{\text{Cos } y} \right\}.$$

The procedure for finding q $(\varphi, \Delta L)$ would be to first find h $(\varphi, \Delta L)$ using the simple formula of al-Marrākus͟hī and then to use the auxiliary tables to apply the formula

$$q(\varphi, \Delta L) = G \{x(\varphi, \Delta L), h\},$$

where

$$x(\varphi, \Delta L) = g_{\varphi}(h) - f_{\varphi}(\varphi_{M}).$$

This latter procedure is easily shown to be equivalent to the standard azimuth formula.

Al-K͟halīlī also computed the *ḳibla* for 44 localities in Palestine, Syria, and Iraq. These are likewise very carefully computed. Sample entries from this list are shown in Table 2. Several other *ḳibla* lists were compiled by mediaeval Muslim astronomers, and the geographical tables in late Islamic *zīd͟j*s often display the *ḳibla* alongside the latitudes and longitudes of important localities.

Alignment of Mosques

Now even though the mediaeval astronomer might have been aware of an exact formula for computing the *ḳibla*, the accuracy of his *ḳibla* determinations depended on the geographical data that he had at his disposal. Mediaeval longitude determinations, based either on simultaneous observations of lunar eclipses in different localities or on measuring distances between the localities, were generally not very accurate. Mediaeval latitude determinations, on the other hand, based on observations of the solar meridian altitude, were generally more accurate.

Even so, the most popular values used by Muslim astronomers for the latitude of Mecca were 21°, 21° 20′, 21° 30′, and 21° 40′, whereas the accurate value is 21° 26′. This explains why mediaeval mosques may be incorrectly oriented even though their *miḥrāb*s [*q.v.*] were erected in a *ḳibla* direction computed by competent mathematicians.

Another reason why mosques may be incorrectly aligned is that their *ḳibla*s were not computed from geographical data at all but were inspired by tradition. Thus, for example, mosques in the Maghrib and the Indian subcontinent generally face due east or due west, respectively. Likewise, in early Muslim Egypt, the *ḳibla* adopted was the azimuth of the rising sun at the winter solstice. Several mosques in Cairo face this direction, which was favoured as the *ḳiblat al-ṣaḥāba* but which is about 10° off the *ḳibla* computed mathematically using mediaeval geographical coordinates and about 20° off the true *ḳibla* for Cairo. No survey has yet been made of the orientation of mediaeval mosques. Such a survey would be of considerable interest for the history of Islamic architecture as well as the history of science.

Bibliography: Several of the following secondary sources contain descriptions and analyses of mediaeval *ḳibla* methods. There exist numerous Islamic astronomical works containing *ḳibla* methods that have not been investigated in modern times. On the *ḳibla* method of Ulugh Beg [*q.v.*] (*fl.* Samarḳand, d. 853/1449), which is none other than the method of Ibn Yūnus and al-Bīrūnī, see L. A. Sédillot, *Prolégomènes des tables astronomiques d'Oloug-Beg: traduction et commentaire*, (Paris 1853), 116-21. On the approximate methods of al-Battānī and al-Djaghmīnī, see C. A. Nallino, *al-Battānī sive Albatenii Opus Astronomicum* (Milan and Rome 1899-1907), i, 318-9, and ii, p. xxvii; and G. Rudloff and A. Hochheim, *Die Astronomie des Maḥmûd ibn Muḥammad ibn ʿOmar al-Ǵagmînî*, in *ZDMG*, xlvii (1893), 213-75 (esp. 271-2).

The first serious investigations of Islamic *ḳibla* methods were conducted by C. Schoy (see his article ḲIBLA in *EI*[1]). The methods of Ibn al-Haytham and al-Nayrīzī were discussed in his *Abhandlung des al-Ḥasan ibn al-Ḥasan ibn al-Haiṯam (Alhazen) über die Bestimmung der Richtung der Qibla*, in *ZDMG*, lxxv (1921), 242-53; and in his *Abhandlung von al-Faḍl b. Ḥâtim al-Nairîzî über die Richtung der Qibla*, in *SB Bayr. Akad.*, Math.-phys. Kl. (Munich 1922), 55-68 (also contains a list of *ḳibla* values for various cities, taken from an 8th/14th century Syrian source). Schoy's other studies on the *ḳibla* include *Die arabische Sonnenuhr im Dienste der islamischen Religionsübung*, in *Naturwissenschaftliche Wochenschrift*, N.F., xi (1912), 625-9; *Mittagslinie und Qibla*, in *Zeitschr. der Gesell. für Erdkunde zu Berlin* (1915), 551-76; *Die Mekka- oder Qiblakarte (Gegenazimuthale mittabstandstreue Projektion mit Mekka als Kartenmitte)*, in *Kartographische und schulgeographische Zeitschr.* (Vienna 1916), 184-5; and *Gnomonik der Araber*, in E. von Bassermann-Jordan, ed., *Die Geschichte der Zeitmessung und der Uhren*, Band 1F (Berlin-Leipzig 1923) (esp. 33-43 and 84-6 on the methods of al-Battānī, Ibn Yūnus, and Abu 'l-Wafāʾ).

On al-Bīrūnī's *ḳibla* methods, see his *al-Ḳānūn al-Masʿūdī*, ed. M. Krause, Hyderabad 1955, ii, 522-8; Schoy, *Die trigonometrischen Lehren des ... al-Biruni ...*, Hanover 1927, 70-1; and E. S. Kennedy, *A commentary upon Bīrūnī's Kitāb Taḥdīd al-Amākin*, Beirut 1973, esp. 198-215. Ḥabash's construction is discussed in E. S. Kennedy and Y. Id, *A letter of al-Bīrūnī: Ḥabash al-Ḥāsib's analemma for the Qibla*, in *Historia Mathematica*, i (1974), 3-11.

Al-Khalīlī's *ḳibla* table is analysed in D. A. King, *Al-Khalīlī's* Qibla *Table*, in *JNES* (1975), (which also contains references to other mediaeval *ḳibla* tables and a discussion of the determination of the *ḳibla* using a quadrant). Ibn Yūnus's table displaying the solar altitude in the azimuth of the *ḳibla* is discussed in idem, *Ibn Yūnus' Very useful tables for reckoning time by the sun*, in *Archive for History of Exact Sciences*, x (1973), 342-94 (esp. 368). Considerable additional information on *ḳibla* determinations is contained in the forthcoming publication by idem, *Studies in astronomical time-keeping in mediaeval Islam. ii. A survey of mediaeval Islamic tables for regulating the times of prayer.*

Several lists of geographical coordinates of cities and the corresponding *ḳibla* values, taken from Islamic astrolabes, are given in R. T. Gunther, *The astrolabes of the world*, i, Oxford 1932, see esp. 24-6. On mediaeval Islamic longitude determinations, see Schoy, *Längenbestimmung und Zentral-meridian bei den älteren Völkern*, in *Mitt. der Kaiserlich-Königlichen Geographischen Gesell.*, 58 (1915), 27-62; Kennedy's commentary on al-Bīrūnī's *Taḥdīd* (mentioned above); and F. Haddad and E. S. Kennedy, *Geographical tables of mediaeval Islam*, in *al-Abḥāth*, xxiv (1971), 87-102.

On the analemma in mediaeval Islamic astronomy, see the references cited in the study by E. S. Kennedy and Y. Id (mentioned above). On the development of spherical trigonometry and computational techniques in mediaeval Islam, see P. Luckey, *Zur Entstehung der Kugeldreieckrechnung*, in *Deutsche Mathematik*, v (1940), 405-46; Kennedy, *Al-Bīrūnī's Maqālīd ʿIlm al-Hayʾa*, in *JNES*, xxx (1971), 308-14, and the references there cited; and King, *Al-Khalīlī's auxiliary tables for solving problems of spherical astronomy*, in *Jnal. for the Hist. of Astronomy*, iv (1973), 99-110.

X

MAKKA: AS THE CENTRE OF THE WORLD

Introduction.

In Ḳur'ān, II, 144, Muslims are enjoined to face the sacred precincts in Mecca during their prayers. The Kaʿba was adopted by Muḥammad as a physical focus of the new Muslim community, and the direction of prayer, *ḳibla,* was to serve as the sacred direction in Islam until the present day.

Since Muslims over the centuries have faced the Kaʿba during prayer, mosques are oriented so that the prayer wall faces the Kaʿba. The *miḥrāb* [*q.v.*] or prayer-niche in the mosque indicates the *ḳibla*, or local direction of Mecca. Islamic tradition further prescribes that certain acts such as burial of the dead, recitation of the Ḳur'ān, announcing the call to prayer, and the ritual slaughter of animals for food, be performed in the *ḳibla,* whereas expectoration and bodily functions should be performed in the perpendicular direction. Thus for close to fourteen centuries, Muslims have been spiritually and physically oriented towards the Kaʿba and the holy city of Mecca in their daily lives, and the *ḳibla* or sacred direction is of fundamental importance in Islam [see KAʿBA and ḲIBLA, i. Ritual and legal aspects].

A statement attributed to the Prophet asserts that the Kaʿba is the *ḳibla* for people in the sacred mosque which surrounds the Kaʿba, the Mosque is the *ḳibla* for the people in the sacred precincts (*ḥaram*) of the city of Mecca and its environs, and the sacred precincts are the *ḳibla* for people in the whole world. To ʿĀ'isha and ʿAlī b. Abī Ṭālib, as well as to other early authorities, is attributed the assertion that Mecca is the centre of the world. The early Islamic traditions with Mecca as the centre and navel of the world constitute an integral part of Islamic cosmography over the centuries (see Wensinck, *Navel of the earth,* 36), although they do not feature in the most popular treatise on the subject from the late mediaeval period, namely, that of al-Suyūṭī [*q.v.*]; see Heinen, *Islamic cosmology*.

From the 3rd/9th century onwards, schemes were devised in which the world was divided into sectors (*dj̲iha* or *ḥadd*) about the Kaʿba. This sacred geography had several manifestations, but the different schemes proposed shared a common feature, described by al-Maḳrīzī, "The Kaʿba with respect to the inhabited parts of the world is like the centre of a circle with respect to the circle itself. All regions face the Kaʿba, surrounding it as a circle surrounds its centre, and each region faces a particular part of the Kaʿba" (*Kh̲iṭaṭ,* i, 257-8).

Islamic sacred geography was quite separate and distinct from the mainstream Islamic tradition of mathematical geography and cartography, which owed its inspiration to the *Geography* of Ptolemy [see DJUGHRĀFIYĀ and KHARĪṬA]. Indeed, it flourished mainly outside the domain of the scientists, so that a scholar such as al-Bīrūnī [*q.v.*] was apparently unaware of this tradition: see his introduction to astronomy and astrology, the *Tafhīm,* tr. R. R. Wright, London 1934, 141-2, where he discusses the Greek, Indian and Persian schemes for the division of the world, but makes no reference to any system centred on Mecca or the Kaʿba.

The orientation of the Kaʿba.

In the article KAʿBA, it is asserted that the corners of the Kaʿba face the cardinal directions. In fact, the Kaʿba has a rectangular base with sides in the ratio *ca.* 8:7 with its main axis at about 30° counter-clockwise from the meridian. When one is standing in front of any of the four walls of the Kaʿba, one is facing a significant astronomical direction; this fact was known to the first generations who had lived in or visited Mecca. In two traditions attributed to Ibn ᶜAbbās and al-Ḥasan al-Baṣrī [*q.vv.*], and in several later sources on folk astronomy, it is implied that the major axis of the rectangular base of the Kaᶜba points towards the rising of Canopus, the brightest star in the southern celestial hemisphere, and that the minor axis points towards summer sunrise in one direction and winter sunset in the other (Heinen, *Islamic cosmology,* 157-8).

For the latitude of Mecca, the two directions are indeed roughly perpendicular. (A modern plan of the Kaᶜba and its environs, based upon aerial photography, essentially confirms the information given in the texts, but reveals more: for epoch 0 AD, the major axis is aligned with the rising of Canopus over the mountains on the southern horizon to within 2°, and the minor axis is aligned with the southernmost setting point of the moon over the south-western horizon to within 1°. This last feature of the Kaᶜba is not known to be specifically mentioned in any mediaeval text, and its significance, if any, has not yet been established.) In early Islamic meteorological folklore, which appears to date back to pre-Islamic times, the Kaᶜba is also associated with the winds. In one of several traditions concerning the winds in pre-Islamic Arabia, the four cardinal winds were thought of as blowing from the directions defined by the axes of the Kaᶜba. This tradition is in some sources associated with Ibn ᶜAbbās (see MAṬLAᶜ and also Heinen, 157).

The term *ḳibla,* and the associated verb *istaḳbala* for standing in the *ḳibla,* appear to derive from the name of the east wind, the *ḳabūl.* These terms correspond to the situation where one is standing with the north wind (*al-shamāl*) on one's left (*shamāl*) and the Yemen on one's right (*yamīn*); see Chelhod, *Pre-eminence of the right,* 248-53; King, *Astronomical alignments,* 307-9. In other such traditions recorded in the Islamic sources, the limits of the directions from which the winds blow were defined in terms of the rising and setting of such stars and star-groups as Canopus, the Pleiades, and the stars of the handle of the Plough (which in tropical latitudes *do* rise and set), or in terms of cardinal directions or solstitial directions [see MAṬLAᶜ].

It appears that in the time of the Prophet, the four corners of the Kaᶜba were already named according to the geographical regions which they faced and which the Meccans knew from their trading ventures: namely, Syria, ᶜIrāḳ, Yemen, and "the West". As we shall see, a division of the world into four regions about the Kaᶜba is attested in one of the earliest sources for sacred geography. Since the Kaᶜba has four sides as well as four corners, a division of the world into eight sectors around it would also be natural, and, as we shall see, eight-sector schemes were indeed proposed. However, in some schemes, the sectors were associated with segments of the perimeter of the Kaᶜba, the walls being divided by such features as the waterspout (*mīzāb*) on the north-western wall and the door on the north-eastern wall (see Fig. 1).

The directions of sunrise and sunset at midsummer, midwinter and the equinoxes, together with the north and south points, define eight (unequal) sectors of the horizon, and, together with the directions perpendicular to the solstitial directions, define 12 (roughly equal) sectors. Each of these eight- and 12-sector schemes was used in the sacred geography of Islam.

The determination of the sacred direction

The article ḲIBLA, ii. Astronomical aspects, ignores the means which were used in popular practice for determining the sacred direction, since at the time when it was written, these had not yet been investigated. It is appropriate to consider them before turning to the topic of sacred geography *per se*.

From the 3rd/9th century onwards, Muslim astronomers working in the tradition of classical astronomy devised methods to compute the *ḳibla* for any locality from the available geographical data. For them, the *ḳibla* was the direction of the great circle joining the locality to Mecca, measured as an angle to the local meridian. The determination of the *ḳibla* according to this definition is a non-trivial problem of mathematical geography, whose solution involves the application of complicated trigonometric formulae or geometrical constructions. Lists of *ḳibla* values for different localities and tables displaying the *ḳibla* for each

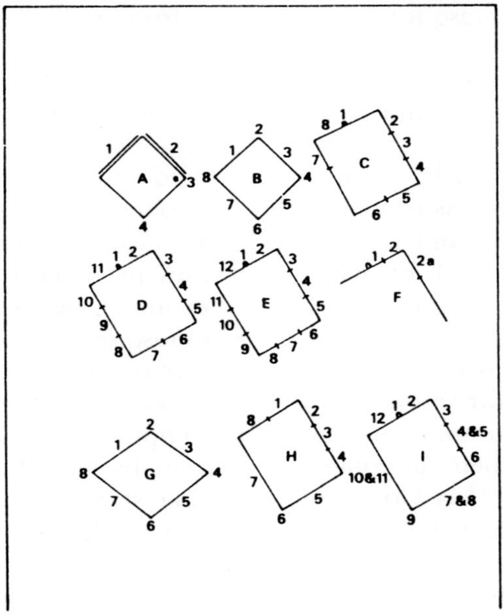

Fig. 1. Different schemes for dividing the perimeter of the Kaᶜba to correspond to different localities in the surrounding world. The Black Stone is in the south-eastern corner; the door is on the north-eastern wall; the blocked door is on the south-western wall; and the waterspout is on the north-western wall.

degree of longitude and latitude difference from Mecca were available. Details of this activity are given in ḲIBLA. ii. Astronomical aspects. However, mathematical methods were not available to the Muslims before the late 2nd/8th and early 3rd/9th centuries. And what is more important, even in later centuries, the *ḳibla* was not generally found by computation anyway.

In some circles, the practice of the Prophet in Medina was imitated: he had prayed southwards towards Mecca, and there were those who were content to follow his example and pray towards the south wherever they were, be it in Andalusia or Central Asia. Others followed the practice of the first generations of Muslims who laid out the first mosques in different parts of the new Islamic commonwealth. Some of these mosques were converted from earlier religious edifices, the orientation of which was considered acceptable for the *ḳibla*; such was the case, for example, in Jerusalem and Damascus, where the *ḳibla* adopted was roughly due south.

Other early mosques were laid out in directions defined by astronomical horizon phenomena, such as the risings and settings of the sun at the equinoxes or solstices and of various prominent stars or star-groups; such was the case, for example, in Egypt and Central Asia, where the earliest mosques were aligned towards winter sunrise and winter sunset, respectively. The directions known as *ḳiblat al-ṣaḥāba,* the "*ḳibla* of the Companions", remained popular over the centuries, their acceptability ensured by the Prophetic dictum: "My Companions are like stars to be guided by: whenever you follow their example you will be rightly guided".

Astronomical alignments were used for the *ḳibla* because the first generations of Muslims who were familiar with the Kaᶜba knew that when they stood in front of the edifice, they were facing a particular astronomical direction. In order to face the appropriate part of the Kaᶜba which was associated with their ultimate geographical location, they used the same astronomically-defined direction for the *ḳibla* as they would have been standing directly in front of that particular segment of the perimeter of the Kaᶜba. This notion of the *ḳibla* is, of course, quite different from that used by the astronomers. Such simple methods for finding the *ḳibla* by astronomical horizon phenomena (called *dalāʾil*) are outlined both in legal texts and in treatises dealing with folk astronomy. In the mediaeval sources, we also find *ḳibla* directions expressed in terms of wind directions: as noted above, several wind schemes, defined in terms of solar or stellar risings and settings, were part of the folk astronomy and meteorology of pre-Islamic Arabia.

The non-mathematical tradition of folk astronomy practiced by Muslims in the mediaeval period was based solely on observable phenomena, such as the risings and settings of celestial bodies and their passages across the sky, and also involved the association of meteorological phenomena, such as the winds, with phenomena in the sky [see ANWĀʾ, MANĀZIL, MAṬLAʿ and RĪḤ]. Adapted primarily from pre-Islamic Arabia, folk astronomy flourished alongside mathematical astronomy over the centuries, but was far more widely known and practised. Even the legal scholars accepted it because of Ḳurʾān, XVI, 16, "... and by the star[s] [men] shall be guided". There were four main applications of this traditional astronomical folklore: (1) the regulation of the Muslim lunar calendar; (2) the determination of the times of the five daily prayers, which are astronomically defined; (3) finding the *ḳibla* by non-mathematical procedures; and (4) the organisation of agricultural activities in the solar calendar (see King, *Ethnoastronomy*, and Varisco, *Agricultural almanac*).

Historical evidence of clashes between the two traditions is rare. Al-Bīrūnī made some disparaging remarks about those who sought to find the *ḳibla* by means of the winds and the lunar mansions (*Kitāb Taḥdīd nihāyāt al-amākin,* tr. J. Ali as *The determination of the coordinates of cities,* Beirut 1967, 12 (slightly modified): "When [some people] were asked to determine the direction of the *ḳibla,* they became perplexed because the solution of the problem was beyond their scientific powers. You see that they have been discussing completely irrelevant phenomena such as the directions from which the winds blow and the risings of the lunar mansions".

But the legal scholars made equally disparaging and far more historically significant remarks about the scientists. According to the 7th/13th-century Yemeni legal scholar al-Aṣbaḥī (ms. Cairo Dār al-Kutub, *mīḳāt* 984, 1, fol. 6a-b): "The astronomers have taken their knowledge from Euclid, [the authors of] the *Sindhind*, Aristotle and other philosophers, and all of them were infidels".

It is quite apparent from the orientations of mediaeval mosques that astronomers were seldom consulted in their construction. Indeed, from the available architectural and also textual evidence, it is clear that in mediaeval times several different and often widely-divergent *ḳibla*s were accepted in specific cities and regions. Among the legal scholars there were those who favoured facing the Kaʿba directly (*ʿayn al-Kaʿba*), usually with some traditionally acceptable astronomical alignment such as winter sunrise, and others who said that facing the general direction of the Kaʿba (*dj̲ihat al-Kaʿba*) was sufficient (see Pl. 1). Thus, for example, there were legal scholars in mediaeval Cordova who maintained that the entire south-eastern quadrant could serve as the *ḳibla* (see King, *Qibla in Cordova,* 372, 374).

Islamic sacred geography

The earliest known Kaʿba-centred geographical scheme is recorded in the *Kitāb al-Masālik wa 'l-mamālik,* ed. de Goeje, 5, of the 3rd/9th century scholar Ibn K̲h̲urradād̲h̲bih [*q.v.*]. Even if the scheme is not original to him, there is no reason to suppose that it is any later than his time. In this scheme, represented in Fig. 2, the region between North-West Africa and Northern Syria is associated with the north-west wall of the Kaʿba and has a *ḳibla* which varies from east to south. The region between Armenia and Kashmir is associated with the north-east wall of the Kaʿba and has a *ḳibla* which varies from south to west. A third region, India, Tibet and China, is associated with the Black Stone in the eastern corner of the Kaʿba, and, for this reason, is stated to have a *ḳibla* a little north of west. A fourth region, the Yemen, is associated with the southern

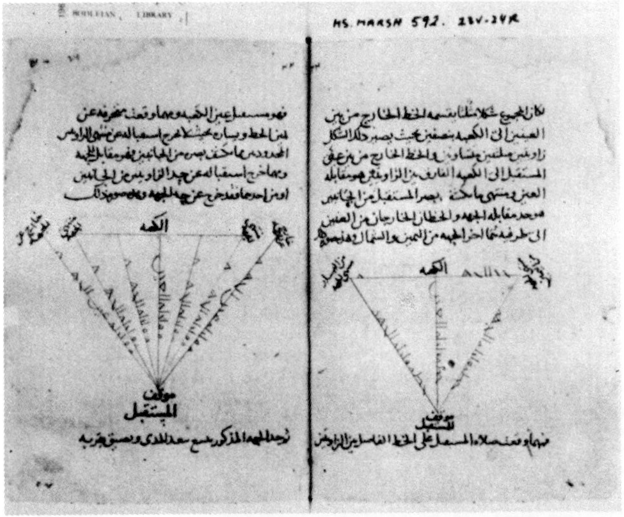

1. Diagrams in the unique manuscript of the *Tahdhīb* of al-Dimyāṭī displaying the notions of *ʿayn al-Kaʿba*, facing the Kaʿba head on, and *djihat al-Kaʿba*, facing the general direction of the Kaʿba, that is, anywhere within the field of vision (*ca.* 90°) of a person facing the Kaʿba head on. Taken from ms. Oxford Bodleian Marsh 592, fols. 23b-24a.

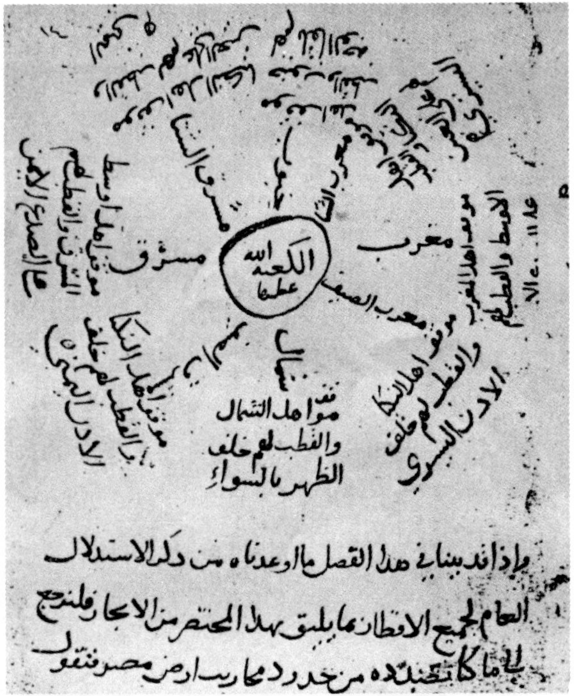

2. An illustration of al-Dimyāṭī's eight-sector *ḳibla* scheme in the unique copy of his shorter treatise on the *ḳibla*. The directions of the *ḳibla* are defined in each sector in terms of the Pole Star. Taken from ms. Damascus Ẓāhiriyya 5579, fol. 14a.

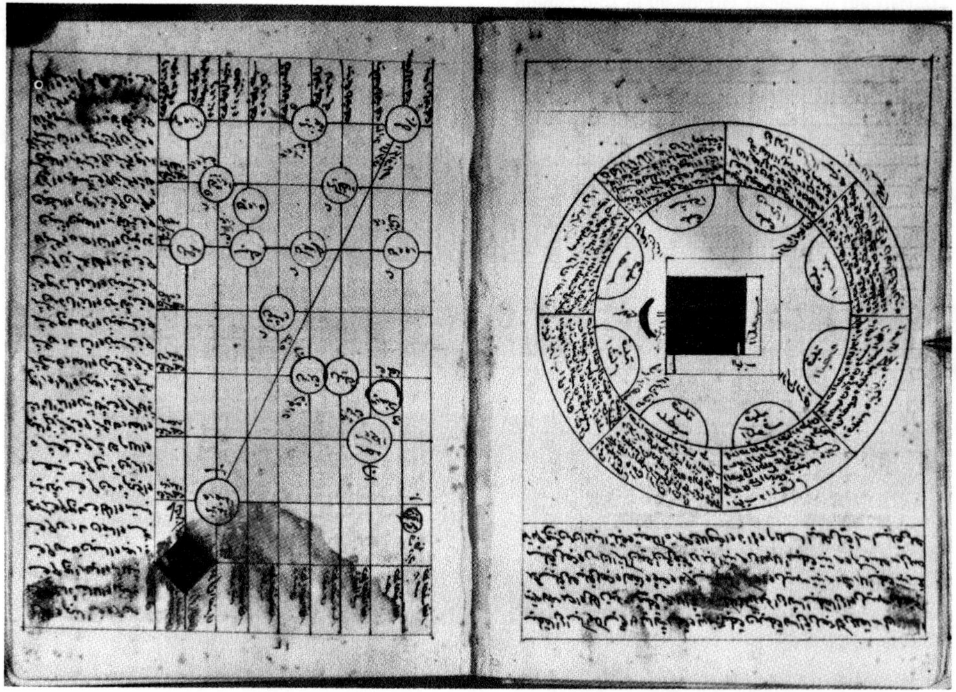

3. Two illustrations from an anonymous treatise on the *ḳibla* and the Kaᶜba of uncertain provenance. On the left is a latitude and longitude grid with various localities marked, as well as the Kaᶜba, shown in the upper left corner, inclined to the meridian. (No such diagram is contained in any scientific treatise from the mediaeval Islamic period.) On the right is an eight-sector *ḳibla* scheme not attested in any other known source. The main *ḳibla* indicators used in each sector are the risings or settings of prominent stars which should be *directly behind* the person facing the *ḳibla*: this suggests that they were determined by someone standing with his back to the appropriate part of the Kaᶜba looking towards those regions. Taken from ms. Cairo Ṭalᶜat *madjāmīᶜ* 811, fols. 59b-60a.

corner of the Kaᶜba and has a *ḳibla* of due north.

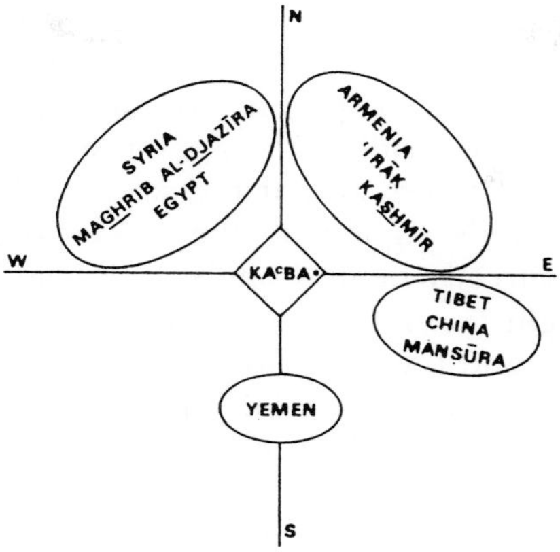

Fig. 2. A simple scheme of sacred geography in the published text of the *Kitāb al-Masālik* of Ibn Khurradādhbih.

The 4th/10th century legal scholar Ibn al-Ḳāṣṣ wrote a treatise entitled *Dalāʾil al-ḳibla* which is unfortunately not extant in its entirety. The Beirut ms. is lost, and the Istanbul and Cairo mss. (Veliyüddin 2453,2 and Dār al-Kutub, *mīḳāt* 1201) are quite different in content. In the Istanbul copy, Ibn al-Ḳāṣṣ states that the world is centred on the Kaᶜba and then presents a traditional Ptolemaic survey of the seven climates [see IḲLĪM]. In the Cairo copy, he surveys the different stars and star-groups used for finding the *ḳibla*.

The principal scholar involved in the development of sacred geography was Muḥammad b. Surāḳa al-ᶜĀmirī, a Yemeni *faḳīh* who studied in Basra and died in the Yemen in 410/1019. Little is known about this individual, and none of his works are known to survive in their original form. However, from quotations in later works, it appears that he devised a total of three distinct schemes, with eight, 11 and 12 sectors around the Kaᶜba. In each scheme, several prescriptions for finding the *ḳibla* in each region are outlined. Ibn Surāḳa explains in words and without recourse to any diagrams how one should stand with respect to the risings or settings of some four stars and the four winds; the actual direction which these prescriptions are intended to help one face is not specifically stated. Thus, for example, people in ᶜIrāḳ and Iran should face the north-east wall of the Kaᶜba, and to achieve this one should stand so that the stars of the Plough rise behind one's right ear, the lunar mansion al-Hanᶜa rises directly behind one's back, the Pole Star is at one's right shoulder, the East wind blows at one's left shoulder, and the West wind blows at one's right cheek, and so on (see Table 1). Ibn Surāḳa did not actually point out that the *ḳibla* in ᶜIrāḳ was toward winter sunset.

Ibn Surāḳa's eight-sector scheme is known from the writings of one Ibn Raḥīḳ, a legal scholar of Mecca in the 5th/11th century, who wrote a treatise on folk astronomy (extant in the unique Berlin ms. Ahlwardt 5664; see especially fols. 23a-25b). Several significant regions of the Muslim world were omitted from this scheme. A similar but more refined eight-sector scheme is proposed by the 7th/13th century Libyan philologist Ibn al-Adjdābī [*q.v.*] (*Kitāb al-Azmina*..., ed. I. Hassan, 120-35). Here eight sectors are neatly associated with the four walls and the four corners of the Kaᶜba, and the *ḳibla*s in each region are defined in terms of the cardinal directions and sunrise and sunset at the solstices. A cruder scheme based on the same notion is proposed by the 6th/12th century Egyptian legal scholar al-Dimyāṭī (ms. Damascus, Ẓāhiriyya 5579, fol. 14a). He represents the Kaᶜba by a circle and associates each of the eight regions around the Kaᶜba with a wind (see Pl. 2).

Yet another eight-sector scheme is presented in an anonymous treatise preserved in a 12th/18th century Ottoman Egyptian manuscript (Cairo, Ṭalᶜat, *madjāmīᶜ* 811,6, fols. 59a-61a (see Pl. 3). From internal evidence, it is clear that this scheme, in which the *ḳibla*s are actually defined in terms of the stars which rise or set behind one's back when one is standing in the *ḳibla* and in terms of the Pole Star, was already at

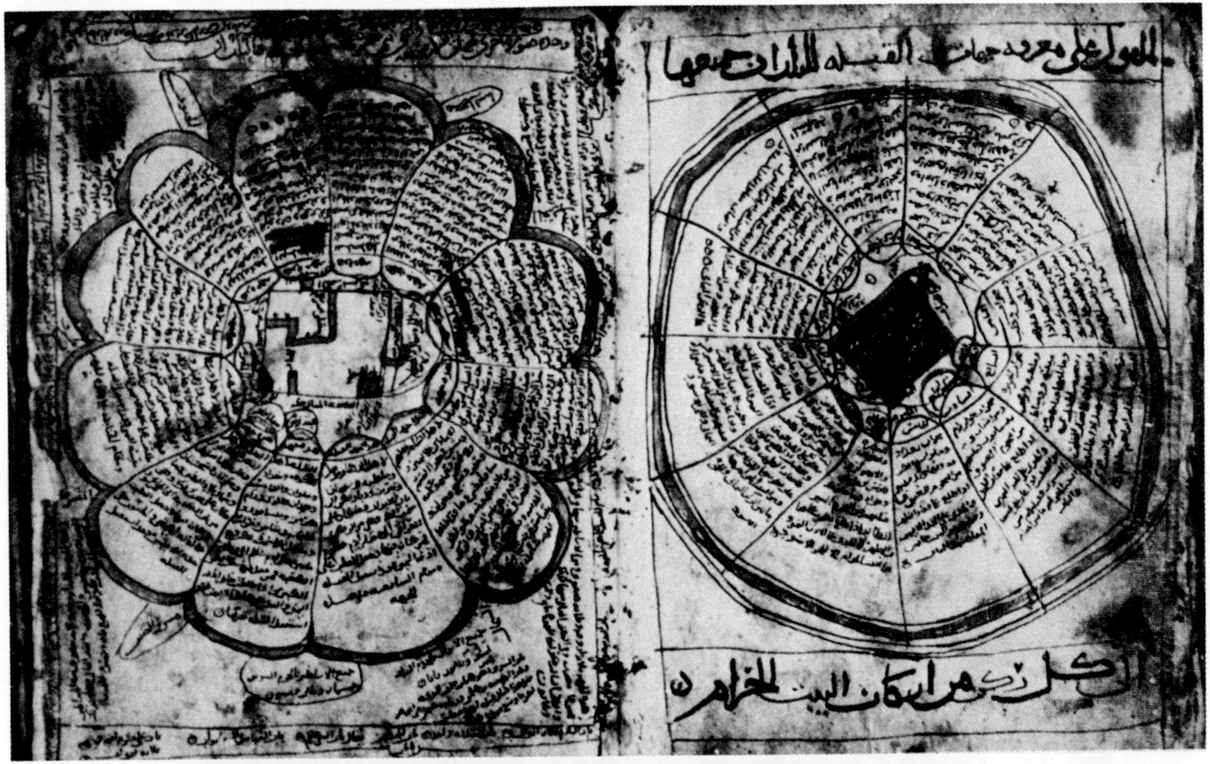

4. Two 12-sector diagrams occuring at the end of the unique complete copy of a treatise on folk astronomy by the 7th/13th century Yemeni scholar Muḥammad b. Abī Bakr al-Fārisī. The one on the left corresponds roughly to the scheme described in the text of al-Fārisī's treatise. Notice that the scheme is surrounded by a schematic representation of the Sacred Mosque. The one on the right is developed from the eight-sector scheme of Ibn Surāḳa. Taken from ms. Milan Ambrosiana Griffini 37, unfoliated.

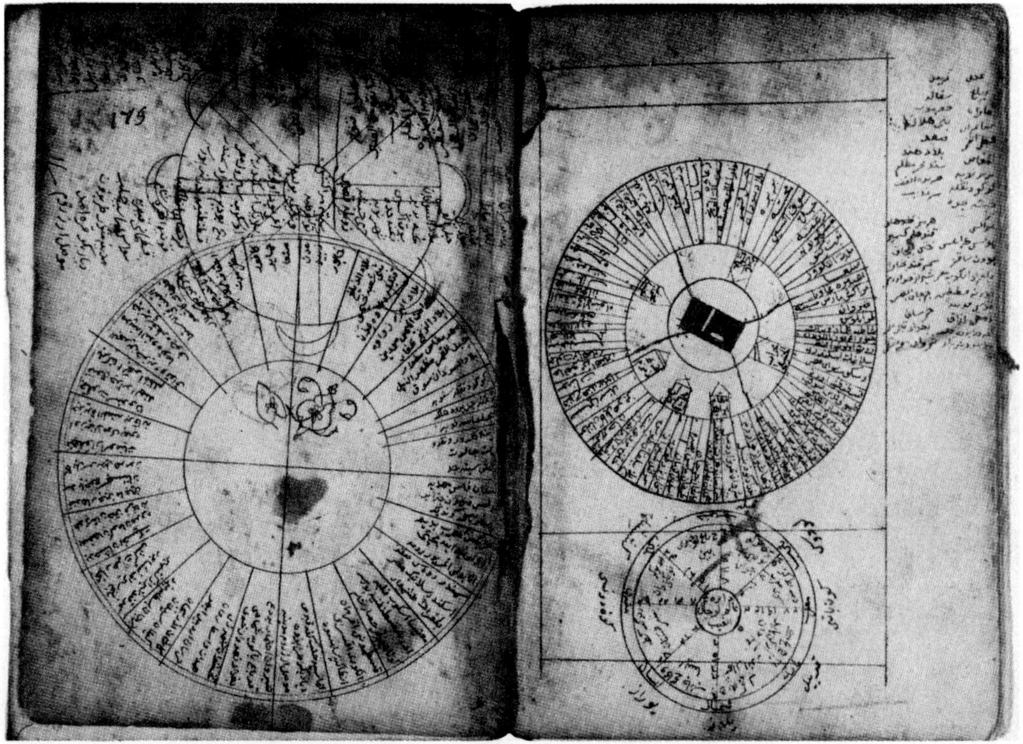

5. On these flyleaves of an Ottoman Turkish copy of a 10th/16th century Syrian *zīdj* are preserved four different schemes of sacred geography. Two are represented graphically, and two others by ordered lists of localities (upper left and upper right). In addition, there is (lower right) a diagram for locating the so-called *ridjāl al-ghayb*, intermediaries between God and man, belief in whom was widespread amongst Ṣūfīs in Ottoman times. Taken from ms. Paris B.N. ar. 2520.

least five centuries old when it was copied in this manuscript. For example, various 7th/13th century Yemeni astronomical sources contain 12-sector schemes based on precisely the same eight *ḳibla* directions. In the eight-sector scheme, Palestine had been omitted and two regions were associated with two entire walls of the Kaʿba. The individual who first devised this particular 12-sector scheme added a sector for Palestine and three more for segments of those two walls.

Ibn Surāḳa's 11-sector scheme is known from an 8th/14th (?) century Egyptian treatise (ms. Milan Ambrosiana, II.75 (A75), 20, fols. 174a-177b), and in it he has simply added three sectors to his eight-sector scheme and modified the prescriptions for finding the *ḳibla*. His 12-sector scheme is yet more refined. It was used by al-Dimyāṭī in his *Kitāb al-Tahdhīb fī maʿrifat dalāʾil al-ḳibla* (ms. Oxford, Bodleian Marsh 592, fols. 97b-101b, and 126a-28a), who complained that Ibn Surāḳa had placed Damascus and Medina in the same sector, and so he himself presented a 13-sector scheme, subdividing the sector for Syria and the Ḥidjāz. Ibn Surāḳa's 12-sector scheme was also used by the 7th/13th century Yemeni astronomer al-Fārisī in his treatise on folk astronomy. The unique copy of this work (ms. Milan Ambrosiana X73 sup.) includes diagrams of Ibn Surāḳa's 12-sector scheme and the different 12-sector scheme discussed above (see Pl. 4).

Several sources contain schemes in which the prescriptions for finding the *ḳibla* in each region of the world are based only on the Pole Star (*al-Djudayy* or *al-Ḳuṭb*). Although the earliest known scheme of this kind dates from the 6th/12th century, others must have been in circulation prior to this time, since al-Bīrūnī (*Taḥdīd*, tr. Ali, 13, modified) wrote: "Of the majority of people [who write about the *ḳibla* in non-mathematical terms], none are closer to the truth than those who use (*iʿtabarahu bi-*) the Pole Star known as *al-Djudayy*. By means of its fixed position, the direction of a person travelling can be specified approximately".

Table 1: *Ḳibla* indicators in the eight- and eleven-sector schemes of Ibn Surāḳa

BN = Banāt Naʿsh; EW = east wind; NW = north wind; PS = Pole Star; SW = south wind; WW = west wind

Ms. Berlin Ahlwardt 5664, fols. 23a-25b		Ms. Milan Ambrosiana A75,20, fols. 174a-177b	
1. Medina, Palestine (Waterspout)	BN setting behind Canopus rising in front (al-Dimyāṭī (ms. Oxford Marsh 592, fol. 100b) attributes: NW intermediate wind behind BN setting behind PS behind left shoulder)	1. As in 8-sector scheme ,,	BN setting behind Canopus rising in front Vega rising at left ear Vega setting behind right ear EW at left eye NW behind left ear WW behind right ear SW at right eyebrow
		2. Syria (Waterspout to Syrian Corner)	BN rising behind left ear PS at left shoulder, al-Haḳʿa rising at left EW at left cheek NW at joint of right shoulder WW at right ear towards nape SW in front
2. Djazīra, Armenia (Syrian Corner to *Muṣallā* of Adam)	Winter sunrise at bone behind left ear EW at left shoulder NW at right cheek WW in front SW at left eye	3. As in 8-sector scheme ,,	Capella rising between behind left ear and nape Capella setting at right side PS between right ear and behind nape Winter sunrise at bone behind left ear EW at left shoulder NW at right cheek WW at right side of neck SW at left eye
3. C. ʿIrāḳ, N. Iran, Transoxania (*Muṣallā* of Adam to Door)	BN rising behind right ear al-Hanʿa rising behind PS at right shoulder EW at left shoulder NW between right side of neck and nape WW at right cheek SW at left cheek	4. ,, ,,	BN rising at right ear al-Hanʿa rising between directly behind and behind left ear PS at right shoulder EW at left shoulder NW between right side of neck and nape WW at right cheek SW at left cheek
4. S. ʿIraḳ, S. Iran, China (Door to ʿIrāḳī Corner)	PS at right ear Vega rising behind al-Shawla setting in front Summer sunrise behind right shoulder EW at left shoulder NW at right ear WW in front SW at left side of neck	5. ,, ,,	PS at right ear Vega [rising] behind al-Shawla setting in front Summer sunrise behind right shoulder EW at left shoulder NW at right ear WW at right cheek SW at left eye

5. Sind, India, Afghanistan (ʿIrāḳī Corner to *Muṣallā* of the Prophet)	BN rising at left cheek (*sic*) PS behind (*sic*) EW behind left ear NW at right cheek SW at left shoulder	6.	BN rising at righ cheek PS at right eye EW behind right ear NW at right cheek WW at left cheek SW at left shoulder
6. Yemen, Hadramawt (*Muṣallā* of the Prophet to Yemeni Corner)	PS in front Canopus rising at right ear (*sic*) Canopus setting behind left ear Winter sunrise at right ear EW at right shoulder NW in front SW at right shoulder (*sic*)	7.	PS in front Canopus rising at right ear (*sic*) Canopus setting behind left ear Winter sunrise at right ear EW at right shoulder NW in front WW at left side SW at left shoulder
		8. Ethiopia (Yemeni Corner to Blocked Door)	Pleiades rising in front Sirius and Capella rising at right eye PS at left ear EW at right ear NW in front WW on left SW behind
		9. Sudan (Blocked Corner to 7 cubis short of Western Corner)	Capella rising in front Pleiades rising at right eye al-S̲h̲awla setting behind PS at left cheek Summer sunrise in front Winter sunset behind EW at right eye NW at left eyebrow WW at left ear SW at left shoulder
7. Andalusia, Mag̲h̲rib, Ifrīḳiya, Ethiopia (7 cubits from Western Corner to Corner itself)	Pleiades rising in front Sirius rising at right eye EW in front WW behind NW at left shoulder SW at right shoulder	10. Andalusia, Mag̲h̲rib, Ifrīḳiya As in 8-sector scheme	Pleiades rising in front Sirius rising at righ eye Capella setting behind nape EW in front WW behind NW at left shoulder SW at right shoulder
8. Egypt, coast of Mag̲h̲rib and Ifrīḳiya (**Western Corner at Waterspout)**	BN setting at right shoulder BN rising at left shoulder PS behind WW at right	11.	*al-Zubānā* (*al-Aḥmira*) rising in front BN setting at left shoulder BN rising at left ear NW behind left ear EW at left side

The most detailed scheme of this kind is recorded by the 7th/13th century Egyptian legal scholar Shihāb al-Dīn al-Ḳarāfī [*q.v.*] in his *Dhakhīra*, ed. Cairo, i, 489-508; in this, some nine regions of the world are identified and instructions for finding the *ḳibla* are given as follows: "[The inhabitants of] Sind and India stand with [the Pole star] at their [right] cheeks and they face due west, etc." See Fig. 3 for a simplified version of this kind of scheme.

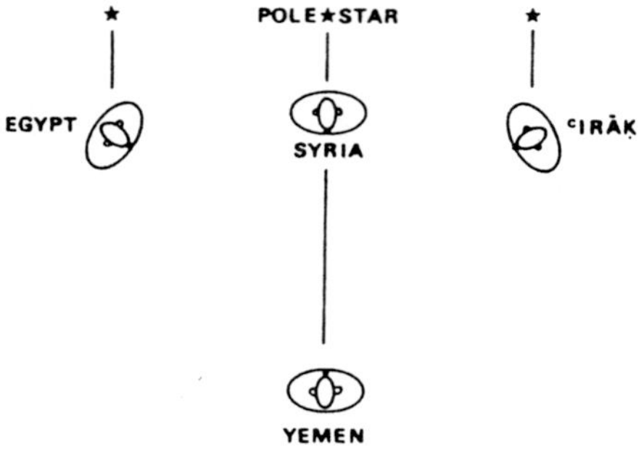

Fig. 3. A simple scheme for using the Pole Star to face Mecca recorded in a late Ottoman Egyptian text, typical of much earlier prescription for finding the *ḳibla*.

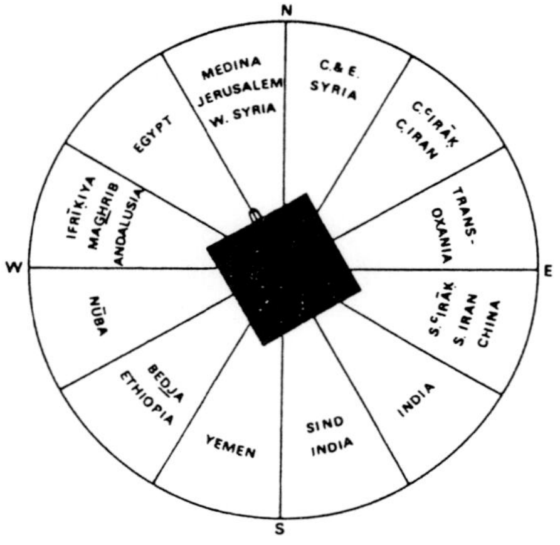

Fig. 4. A simplified version of the 12-sector scheme of sacred geography found in some manuscripts of the *Āthār al-bilād* of al-Ḳazwīnī.

At least one of the 12-sector schemes mentioned above must have been in circulation outside the Yemen before the 7th/13th century, because it was copied by the geographer Yāḳūt (*Buldān*, Eng. tr. Jwaideh, 51), who worked in Syria in *ca.* 600/1200. The instructions for finding the *ḳibla* are omitted from his diagram. A similar diagram is presented in al-Ḳazwīnī's *Āthār al-bilād,* 76, (see Fig. 4), and the same scheme is described in words in al-Ḳalḳashandī, *Ṣubḥ,* iv, 251-5. Another such simple 12-sector scheme occurs in the cosmography *Kharīdat al-ʿadjāʾib* of the 9th/15th century Syrian writer Ibn al-Wardī [*q.v.*], a work which was exceedingly popular in later centuries. In some copies of this, a diagram of an eight-sector scheme is presented. In others, diagrams of 18-, 34-, 35-, or 36-sectors schemes occur. In one manuscript of a Turkish translation of his treatise (ms. Istanbul Topkapı, Turkish 1340 = Bağdat 179), there is a diagram of a scheme with 72 sectors. In the published edition of the Arabic text (Cairo 1863, 70-1), extremely corrupt versions of both the 12- and the eight-sector schemes are included.

These simple diagrams were often much abused by ignorant copyists, and even in elegantly copied manuscripts we find the corners of the Kaʿba mislabelled and the localities around the Kaʿba confused. In some copies of the works of al-Ḳazwīnī and Ibn al-Wardī containing the 12-sector scheme, Medina occurs in more than one sector. In other copies, one of these two sectors has been suppressed and only 11 sectors appear around the Kaʿba (see Pl. 5).

Yet another scheme occurs in the navigational atlas of the 10th/16th century Tunisian scholar ʿAlī al-Sharafī al-Ṣafāḳusī (see Pl. 6). There are 40 *miḥrāb*s around the Kaʿba, represented by a square with its corners facing in the cardinal directions, and also by the fact that the scheme is superimposed upon a 32-division wind-rose, a device used by Arab sailors to find directions at sea by the risings and settings of the stars. Even though al-Ṣafāḳusī had compiled maps of the Mediterranean coast, the order and arrangement of localities about the Kaᶜba in his diagram in each of the available copies (mss. Paris, B.N. ar. 2273 and Oxford, Bodleian Marsh 294) are rather inaccurate. Again, no *ḳibla* indications are presented.

Mainly through the writings of al-Ḳazwīnī and Ibn al-Wardī, these simplified 12-sector schemes were copied right up to the 19th century. By then, their original compiler had long been forgotten, and Muslim scholars interested in the sciences were starting to use Western geographical concepts and coordinates anyway. In most regions of the Islamic world, traditional *ḳibla* directions which had been used over the centuries were abandoned for a new direction computed for the locality in question using modern geographical coordinates.

The orientation of Islamic religious architecture

A variety of different *ḳibla* values was used in each of the major centres of Islamic civilisation (see King, *Sacred direction*). In any one locality, there were *ḳibla*s advocated by religious tradition, including both cardinal directions and astronomical alignments advocated in texts on folk astronomy or legal texts, as well as the directions computed by the astronomers (by both accurate and approximate mathematical procedures). This situation explains the diversity of mosque orientations in any given region of the Islamic world. However, since very few mediaeval mosques have been surveyed properly for their orientations, it is not yet possible to classify them, and for the present we are forced to rely mainly on the information contained in the mediaeval written sources.

In Cordova, for example, as we know from a 6th/12th century treatise on the astrolabe, some mosques were laid out towards winter sunrise (roughly 30° S. of E.), because it was thought that this would make their *ḳibla* walls parallel to the north-west wall of the Kaᶜba. The Grand Mosque there faces a direction perpendicular to summer sunrise (roughly 30° E. of S.), for the very same reason: this explains why it

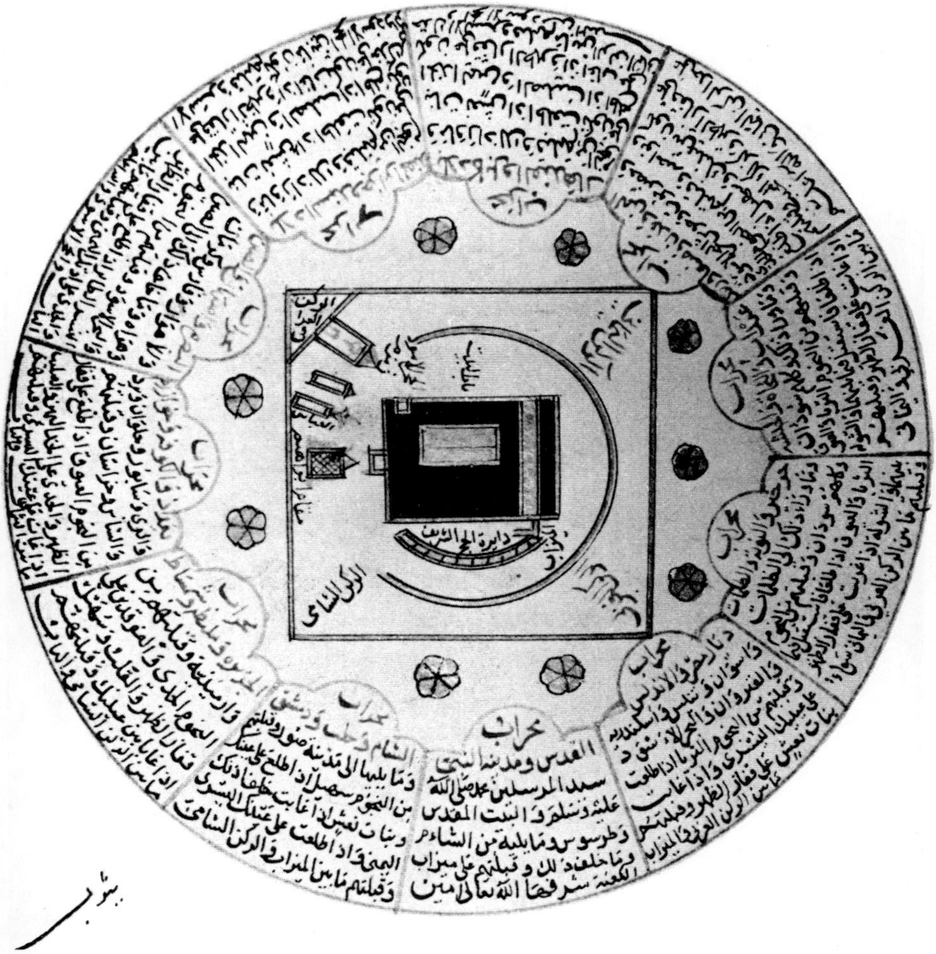

6. A defective diagram of a 12-sector scheme (simplified from that of Ibn Surāḳa) in an elegantly copied manuscript of Ibn al-Wardī's *Cosmography*. As in some diagrams of this kind in various copies of al-Ḳazwīnī's *Āthār al-bilād*, there are only 11 sectors: presumably at some stage in the transmission someone noticed that, because of copyists' errors in some copies of the text, Medina occurred in two sectors. The one omitted here is the one in which Medina had been entered by mistake. Taken from ms. Istanbul Topkapı Ahmet III 3020, fol. 52b.

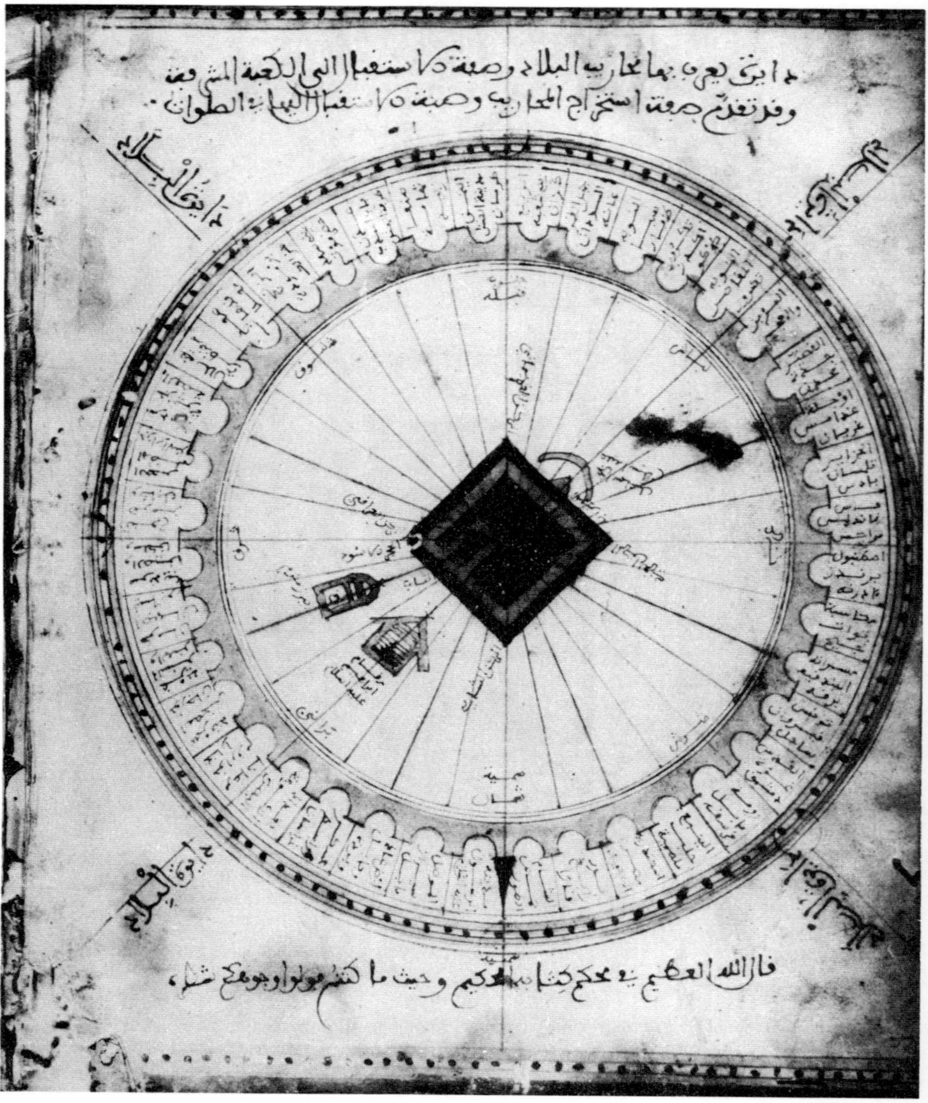

7. The 40-sector scheme of al-Ṣafāḳusī, superimposed on a 32-division windrose. The order of the localities around the Kaᶜba differs somewhat in the two extant copies of this chart. Taken from ms. Paris B.N. ar. 2278, fol. 2b.

faces the deserts of Algeria rather than the deserts of Arabia. In fact, the axis of the Mosque is "parallel" to the main axis of the Kaᶜba.

In Samarḳand, as we know from a 5th/11th century legal treatise, the main mosque was oriented towards winter sunset, in order that it should face the north-east wall of the Kaᶜba. Other mosques in Samarḳand were built facing due west because the road to Mecca left Samarḳand towards the west, and yet others were built facing due south because the Prophet, when he was in Medina, had said that the *ḳibla* was due south, and some religious authorities interpreted this as being universally valid.

Similar situations could be cited for other mediaeval cities. In some of these, the *ḳibla*, or rather, the various different directions accepted for the *ḳibla,* have played an important role in the development of the entire city in mediaeval times. Investigations of the orientations of Islamic cities are still in an early phase. However, the city of Cairo represents a particularly interesting case of a city oriented towards the Kaᶜba.

The first mosque in Egypt was built in Fusṭāṭ in the 1st/7th century facing due east, and then a few years later was altered to face winter sunrise (about 27° S. of E.). The first direction was probably chosen to ensure that the Mosque faced the Western Corner of the Kaᶜba, the second to ensure that it faced the north-western wall, but these reasons are not mentioned in the historical sources. When the new city of al-Ḳāhira was founded in the 4th/10th century, it was built with a roughly orthogonal street plan alongside the Pharaonic canal linking the Nile with the Red Sea. Now, quite fortuitously, it happened that the canal was perpendicular to the direction of winter sunrise. Thus the entire city was oriented in the "*ḳibla* of the Companions". The Fāṭimids who built al-Ḳāhira erected the first mosques in the new city (the Mosque of al-Ḥākim and the Azhar Mosque) in the *ḳibla* of the astronomers, which at 37° S. of E. was 10° south of the *ḳibla* of the Companions. Thus their mosques were skew to the street plan.

The Mamlūks built their mosques and *madrasas* in such a way that the exteriors were in line with the street plan and the interiors skew to the exteriors and in line with the *ḳibla* of the astronomers. When they laid out the "City of the Dead" outside Cairo, they aligned the street and the mausolea with the *ḳibla* of the astronomers. In the other main area of greater Cairo known as al-Ḳarāfa, both the streets and the mosques follow a southerly *ḳibla* orientation. Al-Maḳrīzī discussed the problem of the different orientations of mosques in Egypt, but without reference to the street plan of Cairo. Now that the methods used in mediaeval times for finding the *ḳibla* are understood, the orientation of mediaeval Islamic religious architecture in particular and cities in general is a subject which calls out for further investigation.

Concluding remarks

This purely Islamic development of a sacred geography featuring the world centred on the Kaʿba, provided a simple practical means for Muslims to face the Kaʿba in prayer. For the pious, to whom the "science of the ancients" was anathema, this tradition constituted an acceptable alternative to the mathematical *ḳibla* determinations of the astronomers. As noted above, it was actually approved of by the legal scholars, not least because of Ḳurʾān XVI, 16. The number and variety of the texts in which this sacred geography is attested indicate that it was widely known from the 4th/10th century onwards, if not among the scientific community. The broad spectra of *ḳibla* values accepted at different times in different places attest to the multiplicity of ways used by Muslims to face the Kaʿba over the centuries, and all of this activity was inspired by the belief that the Kaʿba, as the centre of the world and the focus of Muslim worship, was a physical pointer to the presence of God.

Bibliography: On the early Islamic traditions about Mecca as the centre of the world, see A. J. Wensinck, *The ideas of the Western Semites concerning the Navel of the Earth,* Amsterdam 1915, repr. in *Studies of A. J. Wensinck,* New York 1978. On early Islamic traditions about cosmology in general, see A. Heinen, *Islamic cosmology: a study of al-Suyūṭī's al-Hayʾa al-saniyya fī al-hayʾa al-sunniyya,* Beirut 1982.

On the Kaʿba, see in addition to the bibliography cited in KAʿBA, J. Chelhod, *A Contribution to the problem of the pre-eminence of the right, based upon Arabic evidence* (tr. from the French), in R. Needham, *Right and left,* Chicago 1973, 239-62; B. Finster, *Zu der Neuauflage von K. A. C. Creswell's Early Muslim Architecture,* in *Kunst des Orients,* ix (1972), 89-98, esp. 94; G. S. Hawkins and D. A. King, *On the orientation of the Kaʿba,* in *Jnal. for the Hist. of Astronomy,* xiii (1982), 102-9; King, *Astronomical alignments in medieval Islamic religious architecture,* in *Annals of the New York Academy of Sciences,* ccclxxxv (1982), 303-12; G. Lüling, *Der christliche Kult an der vorislamischen Kaaba ...,* Erlangen 1977, and other works by the same author; G. R. Hawting, *Aspects of Muslim political and religious history in the 1st/7th century, with especial reference to the development of the Muslim sanctuary,* University of London Ph. D. thesis, 1978, unpublished.

On Islamic folk astronomy, see in addition to the articles ANWĀʾ, MANĀZIL and MAṬLAʿ, King, *Ethnoastronomy and mathematical astronomy in the Medieval Near East,* and D. M. Varisco, *An agricultural almanac by the Yemeni Sultan al-Ashraf,* in *Procs. of the First International Symposium on Ethnoastronomy, Washington, D.C. 1983* (forthcoming).

All available sources on Islamic sacred geography (some 30 in number) are surveyed in King, *The sacred geography of Islam,* in *Islamic Art,* iii (1983) (forthcoming). For an overview of the *ḳibla* problem, see idem, *The world about the Kaʿba: a study of the sacred direction in Islam* (forthcoming), and its summary, *The sacred direction in Islam: a study of the interaction of science and religion in the Middle Ages,* in *Interdisciplinary Science Reviews,* x (1984), 315-28.

On the possibility of a *ḳibla* towards the east before the adoption of the *ḳibla* towards the Kaᶜba, see W. Barthold, *Die Orientierung der ersten muhammadanischen Moscheen,* in *Isl.*, xviii (1929), 245-50, and King, *Astronomical alignments,* 309.

On the orientation of Islamic religious architecture, see King, *op. cit.*, and on the situations in Cordova, Cairo and Samarḳand, see *Three sundials from Islamic Andalusia, Appx. A: Some medieval values of the Qibla at Cordova,* in *Jnal. for the Hist. of Arabic Science,* ii (1978), 370-87; *Architecture and astronomy: the ventilators of medieval Cairo and their secrets,* in *JAOS,* civ (1984), 97-133; *Al-Bazdawī on the Qibla in Transoxania,* in *JHAS,* vii (1983), 3-38. In 1983, a treatise on the problems associated with the *ḳibla* in early Islamic Iran by the 5th/11th century legal scholar and mathematician ᶜAbd al-Ḳāhir al-Bag̲h̲dādī was identified in ms. Tashkent, Oriental Institute 177; this awaits investigation. No doubt other treatises on the problems of the *ḳibla* in West and East Africa and in India were prepared, but these have not been located yet in the manuscript sources.

XI

MAṬLAᶜ: ASTRONOMICAL RISING-POINTS

AL-**MAṬLAᶜ** (A.), the rising point of a celestial body, usually a star, on the local horizon. This concept was important in Islamic folk astronomy [see ANWĀʾ and MANĀZIL on some aspects of this tradition], as distinct from mathematical astronomy [see ᶜILM AL-HAYʾA], because it was by the risings and settings of the sun and stars that the *ḳibla* [*q.v.*] or direction of Mecca was usually determined in popular practice. The terms used for the rising and setting points of the sun were usually *mas͟hriḳ* and *mag͟hrib*, *maṭlaᶜ* being generally reserved for stars. The directions of sunrise at the equinoxes and solstices were usually associated with the corresponding zodiacal signs [see MINṬAḲA] or seasons, thus e.g. *mas͟hriḳ al-d͟jady* and *mas͟hriḳ al-s͟hitāʾ* both refer to winter sunrise, since the sun enters the sign of Capricorn at midwinter.

In pre-Islamic Arabian folklore, the directions of the winds (see RĪḤ) were defined in terms of astronomical risings and settings (see Fig. 1) and one such wind-scheme is associated with the Kaᶜba itself (see Fig. 2). These wind schemes are recorded in later Arabic treatises on lexicography, folk astronomy, cosmography, as well as in encyclopaedias and various legal treatises on the *ḳibla*. The major axis of the rectangular base of the Kaᶜba points towards *maṭlaᶜ Suhayl*, the rising point of Canopus, and the minor axis roughly towards *mas͟hriḳ al-ṣayf*, the rising point of the sun at midsummer. The later Islamic attempts to define the *ḳibla* for different localities in terms of astronomical risings and settings stem from the fact that these localities were associated with specific segments of the perimeter of the Kaᶜba, and the *ḳibla*s adopted were the same as the astronomical directions which one would be facing when standing directly in front of the appropriate part of the Kaᶜba [see MAKKA iv].

The term *maṭlaᶜ* was also used to denote the "time of rising" in the expression *maṭlaᶜ al-fadjr*, daybreak or the beginning of morning twilight.

Bibliography: That given in the article ḲIBLA is to be supplemented with the information contained in D. A. King, *Astronomical alignments in medieval religious architecture*, in *Annals of the New York Academy of Sciences*, ccclxxxv (1982), 303-12, and *The sacred geography of Islam*, in *Islamic Art*, iii (to appear); G. S. Hawkins and King, *On the orientation of the Kaᶜba*, in *Jnal. for the History of Astronomy*, xiii (1982), 102-9. For a survey of the whole problem, see King, *The world about the Kaᶜba: a study of the sacred direction in Islam* (forthcoming); summaries are given in *Proceedings of the Second International Qurʾān Conference, New Delhi, 1982* and *Interdisciplinary Science Reviews*, ix (1984), pp. 315-328. See also MAṬĀLIᶜ.

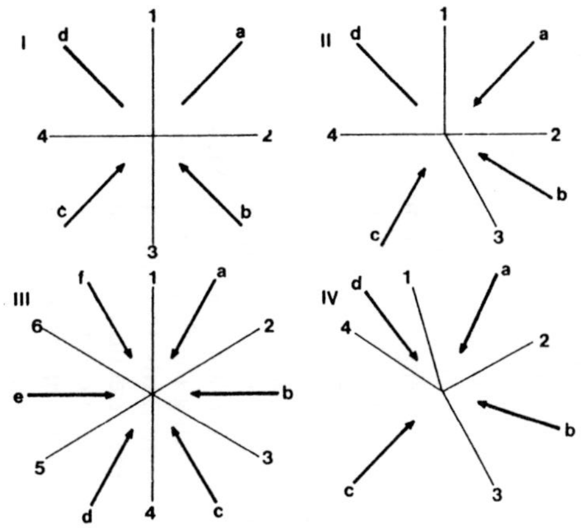

Fig. 1. Four early Arabian wind schemes defined in terms of astronomical risings and settings and attributed to early Muslim authorities.
I Khālid b. Ṣafwān. Limits of winds: 1-4 cardinal directions (defined in terms of the Pole Star and sunrise and sunset at the equinoxes); names of the winds: a *ṣabā*, b *djanūb*, c *dabūr*, d *shamāl*.
II ʿAlī b. Abī Ṭālib. Limits: 1, 2, 4 cardinal directions; names: as in I.
III Ibn Djandab. Limits: 1 north, 2 summer sunrise, 3 winter sunrise, 4 south, 5 winter sunset, 6 summer sunset; names: a *nakhbāʾ*, b *ṣabā* or *ḳabūl*, c *maḥwa*, d *djanūb*, e *dabūr*, f *shamāl*.
IV Ibn al-Aʿrābī. Limits: 1 setting (or rising ?) of the Banāt Naʿsh, 2 rising of the Pleiades, 3 rising of Canopus, 4 setting of Vega; names: as in I.

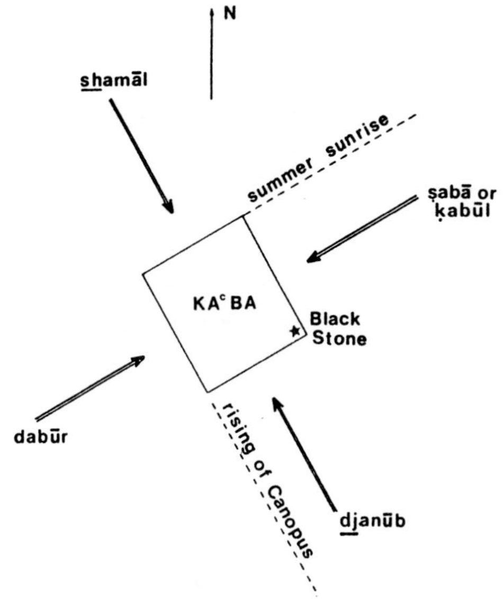

Fig. 2. The most popular early Arabian wind scheme, in which the four winds strike the walls of the Kaʿba head-on. The rectangular base of the Kaʿba points in astronomically significant directions, and so the limits of the four winds are likewise astronomically defined. The rising of Canopus and the solstitial risings and settings of the sun were widely used for finding the *ḳibla* in popular practice, in order to ensure that one would be "facing" a particular wall of the Kaʿba, that is, standing in a direction "parallel" to the appropriate axis of the Kaʿba.

XII

ON THE ORIENTATION OF THE KAᶜBA

GERALD S. HAWKINS,
DAVID A. KING,

1. *Introduction*

The Kaᶜba in Mecca is the most sacred sanctuary of the Muslim world. Newly-discovered textual evidence in the medieval Islamic sources points to the astronomical alignment of the base of the Kaᶜba. Only recently has a reliable plan of the Kaᶜba and its environs become available, and investigations of this plan essentially confirm the information provided in the medieval texts. In this paper we pool our respective disciplines of the history of Islamic science and archaeoastronomy in order to investigate the written and unwritten evidence.

2. *The Textual Evidence* (King)

The Kaᶜba was originally a pagan Arab sanctuary, and is of uncertain historical origin.[1] It has a roughly rectangular base and before the seventh century consisted of a walled enclosure with no roof and with walls just higher than the height of a man. The walls have been rebuilt several times but apparently always on the same foundations. A substantial corpus of medieval texts on folk astronomy, recently investigated for the first time, deals with the astronomical alignment of the Kaᶜba.[2] Such texts currently known to me date from the seventh to the seventeenth centuries. In brief, they imply that the major axis of the Kaᶜba points towards the rising of Canopus and that the minor axis points towards the rising point of the Sun at midsummer.[3] These alignments are also associated with the winds: each of the four winds blows so that it strikes one of the sides of the Kaᶜba.

Before considering the alignments of the Kaᶜba further, we should mention the sacred direction in Islam. Since the early seventh century the Kaᶜba has been the focus of Muslim prayer, and Islamic law requires that Muslims pray in the *qibla*, that is, facing the Kaᶜba and facing the local direction of Mecca.[4] Thus for centuries, Muslims have turned towards the Kaᶜba in prayer and have built their mosques facing Mecca, the direction being indicated by the *miḥrāb* or prayer-niche in the *qibla*-wall of the mosque.

It has recently become evident that astronomical alignments were widely used by Muslims over the centuries for finding the *qibla* and for orienting mosques towards the Kaᶜba.[5] The *qibla*-walls on some medieval mosques were intended to be "parallel" to one wall of the Kaᶜba, this "parallelism" being achieved by facing the mosque towards the same astronomical horizon phenomena as one would be facing when standing in front of the appropriate wall of the Kaᶜba. In the medieval Islamic world there existed a kind of sacred geography, in which the world was divided into sectors around the Kaᶜba, with each region associated with a particular segment of the perimeter of the Kaᶜba. The reason for the use of astronomical alignments, particularly cardinal and solstitial directions and the rising point of the star Canopus, can now be understood in the light of the astronomical alignment of the Kaᶜba itself. The

FIG. 1. An extract from MS Milan Biblioteca Ambrosiana 73 Sup. (unfoliated) of a thirteenth-century Yemeni treatise on folk astronomy. The extract is the first part of a chapter on determining the *qibla* (by non-scientific means), and begins with a statement about the winds and their astronomically-defined limits which correspond to the alignments of the sides of the Ka^{c}ba, and continues with a note about the luni-solar alignment of the Ka^{c}ba. The remainder of this chapter deals with twelve regions of the world about the Ka^{c}ba and their astronomically-defined *qibla*s.

use of astronomical alignments (and also directions associated with the winds) for the *qibla* was actually favoured by the religious authorities over the *qibla* directions which were computed by medieval Muslim astronomers using available geographic data and correct trigonometric formulae.[6]

As an example of a text dealing with the orientation of the Ka^{c}ba consider Figure 1, an extract from a thirteenth-century Yemeni treatise on folk astronomy. The author of this treatise was Muḥammad ibn Abī Bakr al-Fārisī, who worked in Aden *c.* 1290.[7] al-Fārisī was a competent astronomer, author of a substantial astronomical handbook with tables (*zīj*) computed especially for the Yemen, yet in his treatise on folk astronomy he gives no hints that he was also competent in mathematical astronomy. The extract is the first part of a chapter on the determination of the *qibla* by means of the stars and the winds. (In his other works al-Fārisī describes how to find the *qibla* using geometrical procedures.) The chapter begins with a statement about the winds and their astronomically-defined limits, which correspond to the alignments of the sides of the Ka^{c}ba. (This association of the Ka^{c}ba and the winds is found in several other medieval texts, and has been investigated by me elsewhere.) The relevant section translates as follows:

On the knowledge of the places from which the winds blow.

[The winds] are four [in number], and each one faces a corner of the Ka^{c}ba. Other winds in between these are called *nakbā'* [intermediary

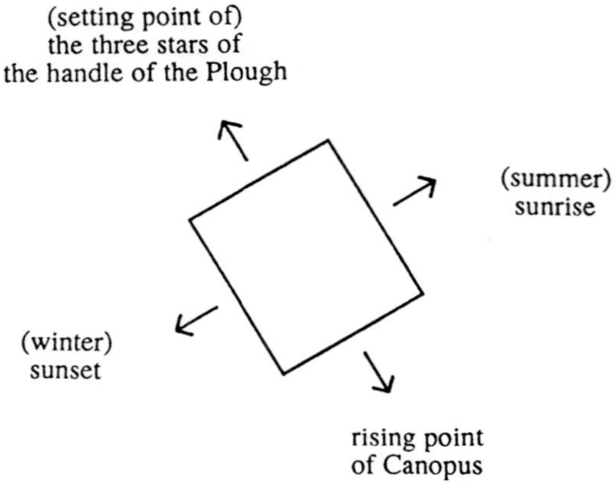

FIG. 2. The alignments of the Kaᶜba implied by the first section of the text illustrated in Figure 1.

> winds]. The first of the winds is the *saba* which is easterly of the Kaᶜba, facing [the wall between] the Black Stone and the southern corner, and it blows from between [summer] sunrise and the rising point of Canopus. The second is the *janūb* which is southerly, and it blows from the rising point of Canopus to the [winter] sunset. The third one is the *dabūr* which is westerly, and it blows from between the [winter] sunset to [the setting point of] the three stars of the handle of the Plough [Banāt Naᶜsh]. The fourth one is the *shamāl* which is northerly, and it blows from between [the setting point of] those stars and the [summer] sunrise.

The situation implied by this text is illustrated in Figure 2.

The text displayed in Figure 1 goes on to describe a luni-solar alignment for the Kaᶜba, which is inconsistent with the first set of alignments and also with reality (Figures 3 and 4). Nevertheless, it is of some interest because it is the only known text on this subject which mentions the Moon. It translates as follows:

> Useful note: Know that the Sun rises on Kānūn I [= December] 19th, which is the shortest day of the year and the beginning of winter [*add*: between the south corner and the east corner of the Kaᶜba, and sets] between the south corner and the west corner, which is where the lunar crescent first appears in that month. The Sun rises on Ḥazīrān [= June] 19th, which is the longest day of the year and the beginning of summer, facing the front [north-east] side of the Kaᶜba and sets opposite the water-spout [on the north-west side], which is where the lunar crescent first appears in that month. It rises on Ādhār and Aylūl [= March and September] 19th, which are the days of the equinoxes, over the mountain of Abū Qubays, facing the corner of the Black Stone [*lit.*: the black corner], and sets in these two months opposite the western corner, which is where the lunar crescent first appears in those [two] month[s].

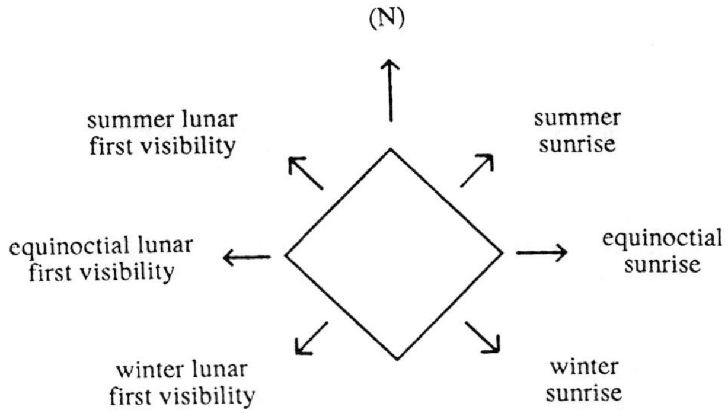

FIG. 3. The alignments of the Kaᶜba implied by the second section of the text illustrated in Figure 1.

An obscure marginal note in a later hand states that all this was correct at the time of the Hijra, but "now" the Sun rises on Kānūn II (= January) 9th.

I am of the opinion, after having read this kind of statement about the orientation of the Kaᶜba and its association with the winds in numerous Islamic sources, that we have here not records of observations of medieval visitors to the Kaᶜba but written versions of pre-Islamic meteorological and astronomical folklore. Whether this folklore was shared by the polytheist Arabs who first built the sanctuary we cannot know. The early Muslims developed their own highly complicated legends about the early history of the Kaᶜba, and the information on its orientation occurs not in the traditions associated with the Prophet or in later histories of Mecca, but in treatises on cosmography, on folk astronomy, and on the determination of the *qibla* by non-scientific means.[8]

None of the texts refers explicitly to the extreme lunar alignment for the Kaᶜba found by Dr Hawkins, although the winter Moon is indeed mentioned by al-Fārisī. Likewise, none of the texts points out that one diagonal of the Kaᶜba is roughly north–south, but this is implied in the names for these corners, which are associated with Syria and the Yemen, respectively. The

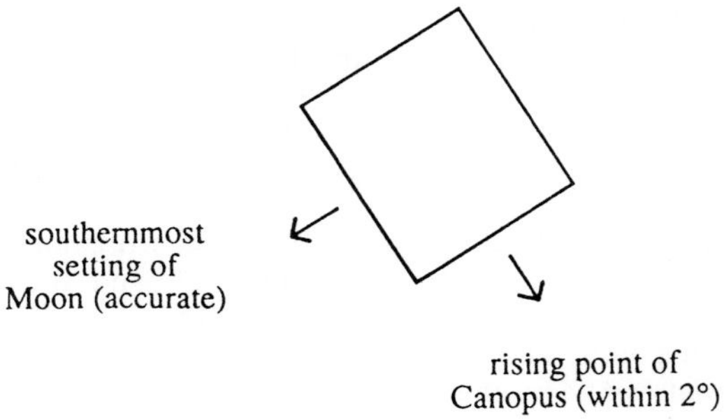

FIG. 4. Actual alignments of the Kaᶜba.

other diagonal is closer to a solstitial alignment than to the east–west line, but for this we have some textual information. The ninth-century legal scholar of Cordova Ibn Ḥabīb stated that the *qibla* at Cordova was towards the rising point of α Sco because that star rises at (the corner of the Kaᶜba containing) the Black Stone.[9] Now various medieval mosques in Cordova were aligned at 30° S of E facing winter sunrise, which is the same direction. On the other hand the Great Mosque in Cordova, built in the eighth century, is at 60° S of E, apparently "facing" the north-west wall of the Kaᶜba so that the *qibla* wall of the mosque is "parallel" to that wall of the Kaᶜba.[10] What concerns us here is the remark that the eastern corner of the Kaᶜba "faces" α Sco. As we shall see, there is some truth in this, but in this instance I suspect that we are dealing with a "tourist report".

Whatever the original significance—if any—of these alignments that are mentioned in the medieval texts, now confirmed by investigations of a modern plan of the Kaᶜba and its environs, we know at least that the solstitial and Canopic alignments were later imitated in certain mosque orientations in order to have the mosques "facing" a particular wall of the Kaᶜba. If the lunar alignment was in some way related to the luni-solar calendar of the pagan Arabs[11] there is no mention of this association in any medieval Islamic texts that are currently known; however, in the decades preceding the abolition of intercalation by the Prophet Muḥammad, pronouncements about intercalation were apparently made by professional intercalators in front of the Kaᶜba.[12] Future investigations of the material relating to the calendar in the medieval Islamic treatises on folk astronomy may cast some light on pre-Islamic practice. Certainly, the significance of the alignments of the Kaᶜba will have to be interpreted in the light of other sites which are possibly astronomically-aligned, in both North Arabia (especially Nabatean sites) and South Arabia, and not least the various megalithic sites in Central Arabia.[13]

3. *Astronomical Considerations* (Hawkins)

From the point of view of an astronomer, the medieval text references are not exact or specific, though there are general references to the rising and setting of the Sun, Moon, and stars. The stars mentioned are Canopus, known astronomically as α Car, and the stars in the handle of the Plough—ϵ, ζ and η UMa. There are inconsistencies wherein the medieval writer has apparently inverted rising and setting, and there are ambiguities, such as what does "opposite the western corner" mean? Nor are the astronomical dates clearly established—the one text cited was written *c.* 1290 A.D., but the original Kaᶜba was constructed over a thousand years earlier. However, for the information of historical scholars, measurements have been obtained and astronomical calculations made.

The altitude of the skyline surrounding the Kaᶜba was obtained from the map published by the Municipality of Mecca in 1967, and the walls and diagonals of the Kaᶜba were also read off the map. The grid of the map is aligned to the cardinal points, and the overall error in calculated declination is estimated to be about 0°·5. The equations used are given by me elsewhere,[14] and the results are given in Table 1. It will be noted that the base of the Kaᶜba is not square, but is elongated in the northwest–southeast direction. This is

TABLE 1

Feature	Azimuth east of north °	Skyline altitude °	Declination on skyline °
Perpendicular to wall . . .			
Northeast	57·2	7·7	+33·3
Southeast	145·9	3·2	−48·7
Southwest	238·5	4·4	−27·3 (for Moon*)
Northwest	326·9	6·7	+54·8
Axes . . .			
Major	147·8	3·2	−50·2
Major	327·8	6·7	+55·6
Minor	56·4	7·7	+34·0
Minor	236·4	4·4	−28·7 (for Moon*)
Line of wall . . .			
Northeast	147·2	3·2	−49·6
,,	327·2	6·7	+55·0
Southeast	235·9	4·4	−29·1 (for Moon*)
,,	55·9	7·7	+34·4
Southwest	328·5	6·7	+56·1
,,	148·5	3·2	−50·7
Northwest	56·9	7·7	+33·6
,,	236·9	4·4	−28·2 (for Moon*)
Diagonal . . .			
1	5†	4†	+72†
2	106	4	−13
3	185	4	−64
4	286	4	+16

*The declination for the Moon is calculated for the Moon's centre when the disk or crescent is standing on the skyline.
†Approximate.

called the "major axis". Also the corners are not right angles, deviating about a degree therefrom. Astronomical objects are listed in Table 2.

Any significance to be attached to the accord between the results of the calculations and medieval texts can only be established by a critical examination of the textual evidence, but a few comments might be made *ab initio*. The alignments to the Sun as referred to in the texts are accurate to no better than 5° or 10°. This amounts to twenty solar diameters and would not have been accepted using the criteria of archaeoastronomy, such as with the pattern of alignments first found at Stonehenge.[15] The alignment to the stars of the Plough seems to be more relevant to the medieval period than to the time of the laying of the foundations of the Kaᶜba. The alignment to Canopus is accurate to about 2° in A.D. 0. This line is marked by the major axis and the southwest wall.

The most accurate alignment is to the Moon at its extreme southern swing at declination −28°·9. It shows up with an accuracy of better than 1° in the minor axis and the foundation line of the southeast wall. This particular moonset is also within 2° of the normal to the face of the southwest wall. In a previous publication[16] reference was made to this possible Moon-line before the textual evidence was discovered. Along this line the crescent Moon would set over the nearby hills at the winter solstice once every eighteen or nineteen years. Although this may seem unlikely because of the long interval of waiting, it should be

TABLE 2

	Declination, A.D. 0 °	Declination, A.D. 1290 °
Moon (midwinter)	−28·9	−28·7
Sun (midsummer)	+23·8	+23·6
Canopus	−52·6	−52·4
ϵ UMa	+67·1	+59·8
ζ UMa	+65·9	+58·7
η UMa	+60·0	+52·9

noted that the hitting of an extreme is more significant than other non-specific phases which are intermediate in the long cycle. Moreover, the lowest Moon at the winter solstice might have had calendric significance. In that regard one should note that by standing by the stone of Abraham and looking over the front (northeast) wall of the holy shrine one was looking at this winter crescent Moon.

In regard to the minor axis aligning with the crescent at the extreme, and the major axis with the rising of Canopus, one should note the incorporation of astronomical right angles into other structures. For example, Stonehenge incorporated the right angle between the mid-summer sunrise and extreme winter moonrise at that latitude.[15] In Egypt, at the temple of Khonsu at Thebes, the minor axis pointed to the extreme summer New Moon crescent, and the major axis to the star Canopus.[17] In the latter example the Moon is aligned at its northern setting point, and Canopus also at its setting, these directions at the latitude of Thebes and the date of construction, being perpendicular.

Acknowledgements

The research of the second author is currently supported by the National Science Foundation under Grant no. SES-8007145; this support is gratefully acknowledged. Both authors thank the Biblioteca Ambrosiana in Milan for permission to publish a photograph of a manuscript in their collection.

REFERENCES

1. On the various Islamic traditions concerning the Kaᶜba see A. J. Wensinck, article "*Kaᶜba*" in *The encyclopaedia of Islam* (1st and 2nd eds, the latter with deletions and additions by J. Jomier).
2. This material is discussed in D. A. King, *The world about the Kaᶜba: A study of the sacred direction in medieval Islam* (in preparation; to be published by Islamic Art Publications, S.p.A.). See especially Section 3.
3. Previous writings on the orientation of the Kaᶜba include J. Chelhod, "A contribution to the problem of the pre-eminence of the right, based upon Arabic evidence", in R. Needham (ed.), *Right and left* (Chicago and London, 1973), 239–62, especially pp. 248–53; B. Finster, "Zu der Neuauflage von K. A. C. Creswells *Early Muslim architecture*", *Kunst des Orients*, ix (1972), 89–98, especially p. 94; and G. Lüling, *Der christliche Kult an der vorislamischen Kaaba* . . . (Erlangen, 1977), especially pp. 43–52. None of these authors was aware of the actual alignment of the Kaᶜba or of any medieval accounts of its astronomical alignment.
4. See further A. J. Wensinck, article "*Kibla* (ritual and legal aspects)" in *The encyclopaedia of Islam* (1st and 2nd eds).
5. See already D. A. King, "Astronomical alignments in medieval Islamic religious architecture", *Annals of the New York Academy of Sciences*, 1982. A more detailed discussion is in the work cited in ref. 2.
6. See D. A. King, article "*Kibla* (astronomical aspects)" in *The encyclopaedia of Islam* (2nd ed.), for details of the impressive achievements of the Muslim astronomers in solving the *qibla* problem.

7. For more information on the author and his works see D. A. King, *Mathematical astronomy in the medieval Yemen* (Publications of the American Research Center in Egypt; Malibu, California, 1982).
8. For a bio-bibliographical survey of Arabic literature dealing with folk astronomy and meteorology, see F. Sezgin, *Geschichte des arabischen Schrifttums*, vii: Astrologie, Meteorologie und Verwandtes (Leiden, 1980), 203 ff.
9. *Cf.* H. P. J. Renaud, "Astronomie et astrologie marocaines", *Hespéris*, xxix (1942), 41–63, especially p. 58.
10. *Cf.* D. A. King, "Some medieval values for the *qibla* at Cordova", *Journal for the history of Arabic science*, ii (1978), 370–87, where the remark that the Grand Mosque in Cordova faces due south (based on Creswell) is to be suppressed.
11. On this see the article "*Zamān* (time)" by W. Hartner in *The encyclopaedia of Islam* (1st ed.), and the bibliography there cited, as well as the article "*Nasī*' (intercalation)" by A. Moberg in the same work. Further information is provided in various articles such as those in *Islamic culture*, xvii (1943), 327–30; xxi (1947), 135–53; and xxii (1948), 174–80.
12. *Cf.* M. Hamidullah, "Intercalation in the Qur'ān and the Ḥadīth", *Islamic culture*, xvii (1943), 327–30, especially p. 328, note 1.
13. See A. Negev, "The Nabateans and the Provincia Arabia", in H. Temporini and W. Haase (eds), *Aufstieg und Niedergang der römischen Welt*, viii (Berlin and New York, 1977), 520–686, for references to Nabatean temples; W. Dostal, "Zur Megalithfrage in Südarabien", in E. Graf (ed.), *Festschrift Werner Caskel* (Leiden, 1968), 54–61, for a survey of stone formations in Southern Arabia; and on the stone circles of Central Arabia see A. H. Masri, "The historical legacy of Saudi Arabia", *Atlal*, i (1977), 9–19, especially p. 13, and various archaeological survey reports in the same journal, especially i (1977), 34–36 and ii (1978), 37–40.
14. G. S. Hawkins, *Beyond Stonehenge* (New York, 1973), 291–94. See also *idem*, *Archaeoastronomy in the Americas*, special report, Center for Archaeoastronomy, University of Maryland (1979), 17.
15. See G. S. Hawkins, "Stonehenge decoded", *Nature*, cc (1963), 306–8.
16. G. S. Hawkins, "The sky when Islam began", *Archaeoastronomy bulletin*, iii (1978), 6–7, especially p. 7a.
17. *Cf.* G. S. Hawkins, "Astroarchaeology: The unwritten evidence", in A. F. Aveni (ed.), *Archaeoastronomy* (Austin, Texas, 1975), 131–62. The alignment with Canopus is discussed in Hawkins, *Mind steps and the cosmos* (New York, in press).

XIII

Astronomical Alignments in Medieval Islamic Religious Architecture

SINCE THE EARLY days of Islam, Muslims have faced the sacred Ka'ba in Mecca when praying. With the advent of Islam, Mecca became the navel of the earth in Islamic tradition, and the Ka'ba, a pre-Islamic pagan shrine of uncertain origin and date, came to be a focus, if not an object, of veneration.[1] Thus, for centuries, mosques have been built so that they are aligned towards Mecca, the *mihrāb*, or prayer-niche, indicating the *qibla*, or local direction of Mecca.[2]

From the eighth century onwards, Muslim astronomers devoted much attention to the problem of determining the *qibla* of any locality from the geographical coordinates of Mecca and of that locality.[3] They derived geometric and trigonometric solutions of considerable sophistication, and even compiled tables displaying the *qibla* for each degree of latitude and longitude. They also devised approximate methods with which the *qibla* for any locality could be found easily and accurately enough for all practical purposes. Such approximate methods for finding the direction of Mecca were widely known in the medieval Islamic world from the ninth century onwards, and most medieval Islamic astronomical handbooks (*zījes*) contained chapters on the determination of the *qibla* by mathematical means, as well as geographical tables displaying latitudes, longitudes, and *qiblas* of important cities.

The legal aspects of the religious duty of Muslims to face Mecca in prayer, and the mathematical techniques that were available for finding the *qibla*, are now rather well documented in the modern scholarly literature. There are, however, other pertinent questions relating to the *qibla*, the answers to which lead to rather exciting new areas of research.[4] First, for example, how could the earliest Muslims, innocent of any knowledge of geography, let alone the exact sciences, have determined the *qibla*? Second, why are so many medieval mosques, according to modern historians of Islamic architecture, not properly aligned towards Mecca? And third, how can one account for a substantial corpus of medieval Islamic literature on "folk astronomy," in which instructions are presented for finding the *qibla* by means of the sun, the stars, and even the winds?

The earliest *qibla* determinations were, in fact, associated with the risings and settings of the sun and fixed stars, and mosque orientations in the seventh and eighth centuries, and even thereafter, were made by astronomical alignments. Thus, for example, some of the earliest mosques in Egypt and Andalusia faced the rising sun at midwinter, and some of the earliest mosques in Iraq, Iran, and Transoxania (Central Asia) faced the setting sun at midwinter. Directions perpendicular to the solstitial directions were also used. Cardinal orientations were popular as well, even where they were really quite inappropriate. Occasionally, of course, mosques were built on the sites of churches and pagan temples, without modification of the orientation of the earlier edifices.

The astronomical orientations of these early mosques were often used for mosques built later in the medieval period, even though the "correct" mathematically determined *qibla* was then available. Many Islamic cities had several accepted *qiblas*; this we know from medieval Arabic historical, legal, and astronomical texts, as well as from the surviving monuments. Also, entire cities with more-or-less orthogonal street plans were sometimes laid out facing either an accepted astronomical orientation for the *qibla* or a mathematically computed *qibla*. Religious architecture in these cities could thus be aligned with the street patterns. More commonly, mosques would be oriented to the *qibla* regardless of the street patterns, or would be aligned externally with the street patterns and internally with the *qibla*.

The old city of Cairo presents a particularly interesting case study. The city was founded in the tenth century alongside the Pharaonic Red Sea Canal. The canal, the direction of which was dictated by local topography, is fortuitously perpendicular to the *qibla* used by the first Muslims in Egypt in the seventh century, which was towards the winter sunrise. The more-or-less orthogonal street plan of the old city was thus aligned towards the winter sunrise. The religious architecture erected in the old city in later centuries is, in the main, externally aligned with the

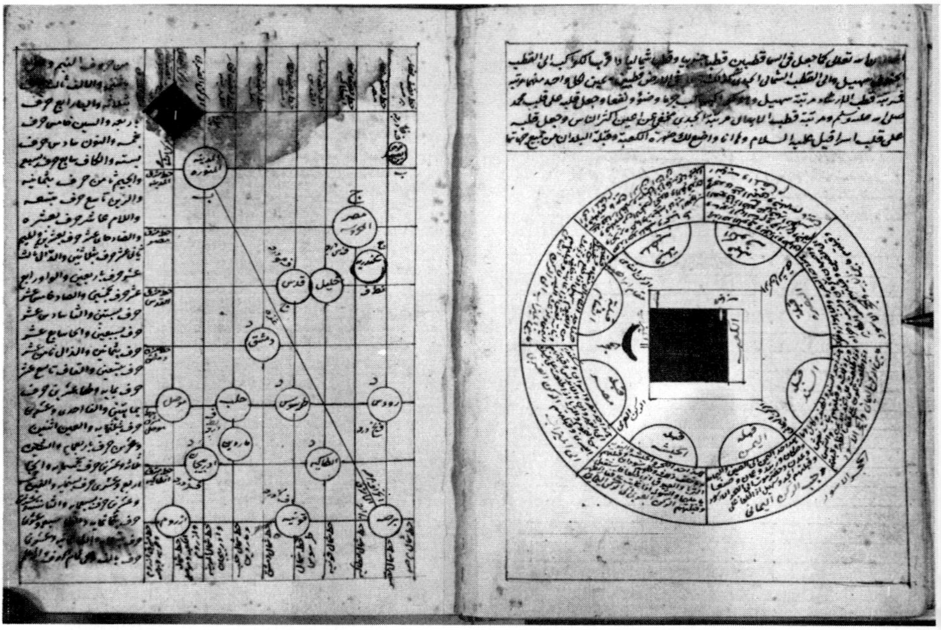

FIGURE 1. Diagrams in a treatise of uncertain provenance on the *qibla* and the Ka'ba contained in a seventeenth-century Egyptian manuscript. The diagram on the right identifies eight regions of the earth about the Ka'ba, and states which astronomical directions one should face in each in order to be standing in the *qibla*, as well as which part of the circumference of the Ka'ba is associated with each region. The diagram on the left shows various localities on a latitude-longitude grid and their positions relative to the Ka'ba. Notice that the Ka'ba is shown inclined to the meridian.

street plan, but the interiors of these mosques and madrasas are twisted to face the *qibla* determined by the astronomers, which differs from the direction of winter sunrise by 10°. The choice between the various available *qibla* directions in a city sometimes varied according to the particular legal school. Thus, for example, in Samarqand, one school, the Shafiites, favored due south whilst another, the Hanafites, favored due west; meanwhile, the main mosque in the city was aligned towards the winter sunset and the astronomers advocated yet another direction for the *qibla*.

A corpus of medieval Arabic texts dealing with folk astronomy, recently investigated for the first time, explains these solstitial, perpendicular to solstitial, and cardinal orientations in terms of the notion of a world divided into eight or twelve sectors about the Ka'ba (FIGURE 1).[5]

The same sources associate astronomical alignments, both solar and stellar, with the limits of the winds, that is, the boundaries of the directions from which they blow. These texts point even to the astronomical alignment of the Ka'ba itself; they tell us that the minor axis of the rectangular base of the Ka'ba is solstitially aligned towards summer sunrise and winter sunset, and that the major axis is aligned towards the local rising point of the star Canopus.[6] (These directions are very roughly perpendicular for the latitude of Mecca). One particular medieval text associates the alignments of the axes and the east-west diagonal of the Ka'ba with the directions of the first appearance of the lunar crescent after sunset about the times of the solstices and the equinoxes. The solstitial and Canopic alignments of the Ka'ba mentioned in the texts are only roughly confirmed by the available plans. The most accurate available map of Mecca, based on aerial photography, reveals that the minor axis of the Ka'ba is aligned (to within 0.5°) towards the southernmost setting of the moon at the winter solstice over the local horizon, and that the major axis is aligned (to within 2°) towards the local rising point of Canopus (epoch A.D. 0).[7]

The astronomical alignments of the sides of the Ka'ba are associated in certain medieval texts with the limits of the four winds (FIGURE 2). I suspect that this association reflects a pre-Islamic meterological tradition, and do not find surprising the fact that such a tradition is recorded only in books on folk astronomy and the determination of the *qibla* rather than books on the history of Mecca. Yet other medieval texts point to the physical associations of the Ka'ba (FIGURE 3), but these are relatively late and should not be taken to represent original conceptions about the sanctuary. Nevertheless, I suspect that the close correspondence between the actual orientation of the Ka'ba and the astronomical orientations mentioned in the texts is not fortuitous, and I am currently inclined to think that the Ka'ba may have been intentionally laid out in accordance with a wind theory. There is no mention of the specific lunar alignment of the Ka'ba (which was first noted by Dr. G. S. Hawkins) in any known medieval texts, and I have no evidence beyond the accurate lunar alignment of the minor axis of the Ka'ba to lead me to anticipate any association with the regulation of the lunisolar calendar of the pre-Islamic Arabs.[8] The alignment of the Ka'ba should, of course, not be considered in isolation but, rather, in the context of other sites with possible astronomical orientations, namely those in Northern Arabia (especially Nabatean sites),[9] Central Arabia (especially some hitherto unexplained stone circles),[10] and Southern Arabia (especially an-

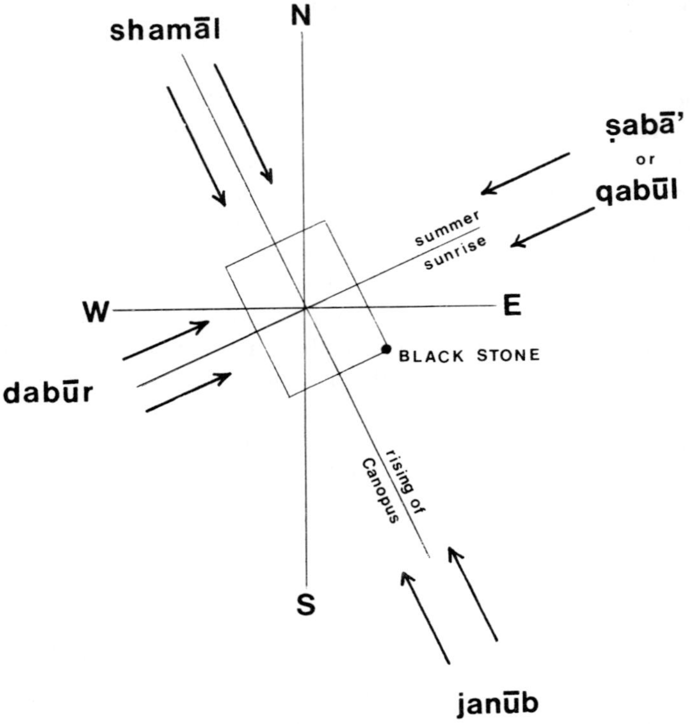

FIGURE 2. The astronomical alignments of the axes of the Ka'ba as described in the medieval Arabic sources, showing their relationship to the four winds. This relationship is recorded, for example, in al-Bīrūnī's treatise *al-Tafhīm*. The astronomically defined limits for the winds are recorded, for example, in E.W. LANE's *Arabic-English Lexicon*.

cient temples associated with astral worship, as well as more primitive rings and lines of stones).[11] None of these have been properly investigated yet.

The possibility that the Ka'ba is astronomically aligned in accordance with a wind theory calls to mind the more sophisticated octagonal "Tower of the Winds" in Athens, which Vitruvius tells us was built in the first century B.C. as an architectural representation of an eight-wind system. As we now know from the recent investigations of Prof. Derek de Solla Price of Yale University, the Athenian Tower was far more than that. In the light of recent archaeological investigations and new textual evidence, Price sees the tower as an architectural representation of "an interlocking set of theories covering virtually all creation and comprehend-

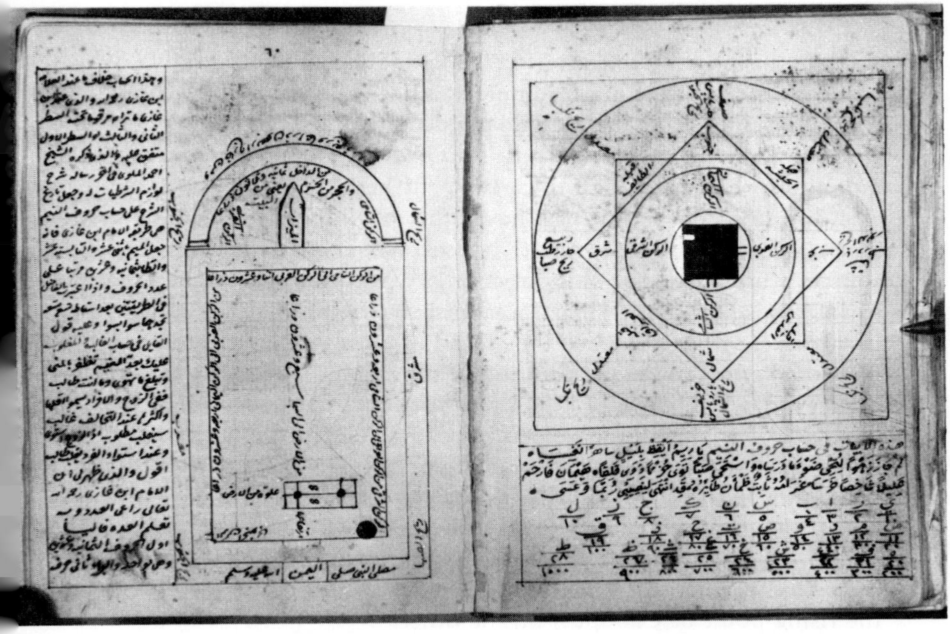

FIGURE 3. Two more diagrams in the same treatise mentioned in the caption to FIGURE 1. The one on the right shows the Ka'ba at the center of two superimposed squares usually associated with the representation of the four elements and their qualities. The seasons and their associated weather conditions, the winds, and the *qiblas* of four main geographical areas are featured in this diagram. The diagram on the left shows, in an exaggerated fashion, the rectangular base of the Ka'ba, with the Black Stone set in its southeast corner.

ing cosmology, chemistry and physics, meteorology, and medicine."[12] There is also evidence that public clocks in the Graeco-Roman world, incorporating either anaphoric dials as contained in the Tower of the Winds or sundials, were intended to represent the universe as well as to display the time of day. Further research on the Nabatean sites of Northern Arabia and Jordan may provide evidence of some connection between the Tower of the Winds as a microcosm of the Greek universe and the Ka'ba as a microcosm of the Arab universe. A first-century Graeco-Roman sundial is already known from one Nabatean site in Northern Arabia.

The relationship between the Ka'ba and the north and unfortunate left-hand side and the south and fortunate right-hand side has already been discussed by J. Chelhod.[13] Implicit in these associations is that one should be facing the rising sun, or the southwest wall of the Ka'ba. Indeed, in this position one is facing the wind called in Arabic *qabūl*, which blows from the direction of summer sunrise. The term *qibla* (based on a root q-b-l) appears to have originally denoted the direction in which one stood to face (the Arabic verb is *istaqbala*, based on the same root) the *qabūl* (also based on the root q-b-l). Furthermore, when one stands in this direction, one is facing one of the two favored directions of the pagan Meccans, whose round houses we know from the medieval sources were opened to this direction. W. Barthold has argued that the earliest mosques in the Hijaz faced east, but attributed this to Christian influence;[14] rather, if indeed they faced east, this is perhaps to be seen as a residual effect of the predilection of the pagan Arabs for the east. The *qibla* was soon changed to the south for these mosques in the Hijaz, in order that they should face Mecca. It appears that the alignment of the major axis of the Ka'ba was also important: the pagan Arabs favored the south because the south wind brings rain in the Hijaz. The Black Stone, the most sacred of several stones that were once venerated at the Ka'ba, is set in the southeast corner of the edifice, facing the east (determined by summer sunrise) on the one side, and the south (determined by the rising point of Canopus) on the other.

Certainly the Ka'ba also follows the tradition of the Semitic baetyls or sanctuaries of the ancient Near East.[15] It is built by the sacred spring of Zemzem in a valley flanked by sacred hills. It houses a sacred stone, and originally housed more of the same. In the decades before the ascendance of Muhammad, the Ka'ba is said to have contained some 360 idols of various gods, and it seems to have been associated with the worship of the sun and moon at some stage in its history: images of both the sun and moon were part of the paraphernalia associated with the Ka'ba before the advent of Islam.[16] We now know that the rectangular base of the Ka'ba is astronomically oriented. This orientation may have been deliberate, either to face the building to the winds or to the southern limit of the setting moon. On the other hand, the astronomical alignments of the sides of the Ka'ba may also be quite fortuitous. They may have been noticed by the early Muslims and used to facilitate *qibla* determinations.

The first Muslims – who built mosques as far apart as Andalusia and Central Asia – could not have known the actual direction of Mecca, but they were aware, I think, that the Ka'ba, which they wanted to face, was oriented in a certain way. Thus, they knew that, when facing a particular wall or corner of the Ka'ba in Mecca, one was facing a particular solar or stellar rising or setting point; they assumed that, away from Mecca, if one faced in that same astronomical direction one would still be facing the same wall or corner of the Ka'ba. Each wall or corner of the Ka'ba was associated with a specific region of the world, and so the *qiblas* in these regions were astronomically defined. Such associations are mentioned in one tenth-century text, but I suspect that they originated somewhat earlier. Two independent medieval texts, one from the eleventh century and the other from the thirteenth, tell of a religious scholar from Samarqand who went to Mecca to verify that the *qibla* in his home city was indeed towards the winter sunset, and of an Egyptian who went to Mecca to verify that the *qibla* in Cairo was indeed towards the winter sunrise.

Only a preliminary study has been made of mosque orientations, this based on all available plans in the modern scholarly literature.[17] Most of these are unreliable as far as orientations are concerned, and information is altogether lacking on local horizon conditions. However, some of the patterns that emerge can be explained in the light of the new material gleaned from medieval texts dealing with *qibla* determinations by non-mathematical procedures. Of particular interest are historical and legal texts that describe the various *qiblas* that were used in such localities as Central Asia, Egypt, the Maghrib, and Andalusia.

It is clear that the study of alignments in medieval Islamic religious architecture, as well, perhaps, as the various pre-Islamic Arabian sites mentioned above, constitutes an important new chapter in archaeoastronomy. Not a single Islamic or pre-Islamic Arabian monument has been surveyed yet with the rigor demanded by this relatively new discipline. Furthermore, only a very few of the relevant medieval texts are published, and none is yet available in any Western language. Nevertheless, there are ample indications that further research in this area will be worthwhile.

ACKNOWLEDGMENTS

This material is based upon work supported by the National Science Foundation under grant no. SES-8007145. This support is gratefully acknowledged. It is also a pleasure to thank the Egyptian National Library in Cairo for permission to publish photographs of a manuscript (no. ṬJ 811) in their collection.

NOTES AND REFERENCES

1. For a survey of modern knowledge about the Ka'ba, including the Muslim legends about its foundation, see A.J. Wensinck, "Ka'ba," *Encyclopaedia of Islam*, 1st and 2nd

eds. (the latter with additions and deletions by J. Jomier). On the notion of Mecca as the navel of the earth, see A.J. Wensinck, "The Ideas of the Western Semites Concerning the Navel of the Earth" (Amsterdam; 1916), reprinted in *Studies of A.J. Wensinck* (New York: Arno Press, 1978).

2. On the religious duty to face Mecca in prayer, see A.J. Wensinck, "*Ḳibla* (ritual and legal aspects)," *Encyclopaedia of Islam*, 2nd ed.

3. A survey of medieval Islamic *qibla* determinations by mathematical procedures is contained in D.A. King, "*Ḳibla* (astronomical aspects)," *Encyclopaedia of Islam*, 2nd ed., which contains references to all the available published literature.

4. For more details on the topic of this paper see D.A. King, *The World about the Ka'ba: A Study of the Sacred Direction in Medieval Islam* (Islamic Art Publications, S.p.A., in preparation).

On orientations in Andalusia, see D.A. King, "Some Medieval Values of the Qibla at Cordova," *Journal for the History of Arabic Science*, vol. 2 (1978), pp. 370–87. On orientations in Cairo, see D.A. King, "Architecture and Astronomy: On the Ventilators of Medieval Cairo and Their Secrets," in a volume of studies in honor of Franz Rosenthal, ed. J. Lassner (American Oriental Society, in press). On orientations in Islamic Central Asia see D.A. King, "al-Bazdawī on the Qibla in Early Medieval Transoxania," *Journal for the History of Arabic Science* (in press).

5. On this tradition of sacred geography in medieval Islam see D.A. King,[4] *World*. Such diagrams are also found in certain medieval Arabic cosmographies already available in published form.

6. There is, as yet, no published material in any Western language on the folklore of the winds in pre-Islamic Arabia. See D.A. King,[4] *World* for some material relevant to the *qibla* and the Ka'ba. For a bio-bibliographical survey of early Islamic writings on meteorology, see F. Sezgin, *Geschichte des arabischen Schrifttums*, vol. 7 (Astrologie-Meteorologie) (Leiden: E.J. Brill, 1979).

The relationship between the Ka'ba and the winds is recorded, for example, in al-Bīrūnī's treatise *al-Tafhīm* [photo-offset ed. with trans. by R.R. Wright entitled *The Book of Instruction in the Elements of the Art of Astrology by . . . al-Bīrūnī*, (London: Luzac & Co., 1934), p. 19, Section 130]. The astronomically-defined limits for the winds are also recorded by medieval Arab philologians and are included, for example, in E.W. Lane, *Arabic-English Lexicon*, (London, 1863–93), articles *shamāl*, *qabūl*, *ṣabā*, *janūb*, and *dabūr*. The earliest attestations of these associations are in remarks attributed to the celebrated scholars Ibn 'Abbās, a Companion of the Prophet (fl. ca. 650), and Ḥasan al-Baṣrī (fl. ca. 700).

7. For the written and unwritten evidence for the astronomical alignment of the Ka'ba, see G.S. Hawkins and D.A. King, "On the Orientation of the Ka'ba," *Journal for the History of Astronomy* (in press).

8. On the lunisolar calendar in pre-Islamic Arabia, see W. Hartner, "*Zamān* (time)," *Encyclopaedia of Islam*, 1st ed., and the bibliography there cited, as well as A. Moberg, "*Nasī'* (intercalation)," *Encyclopaedia of Islam*, 1st ed. Further information is provided in various articles, such as those in *Islamic Culture*, vol. 17 (1943), pp. 327–30, vol. 21 (1947), pp. 135–53, and vol. 22 (1948), pp. 174–80. See also J. Nebez and W. Schlosser, "Ein kurdisches Mond-Observatorium aus neuer Zeit," *Zeitschrift der Deutschen Morgenländischen Gesellschaft*, vol. 122 (1972), pp. 140–44.

9. For references to Nabatean temples, see A. Negev, "The Nabateans and the Provincia Arabia," in *Aufstieg und Niedergang der Römischen Welt*, eds. H. Temporini and W. Haase (Berlin: 1977), vol. 8, pp. 520–686.

10. On the most prominent of the stone circles in Arabia, likened to a modest Stonehenge, see W.G. PALGRAVE, *Narrative of a Year's Journey through Central and Eastern Arabia*, 2 vols. (London: Macmillan, 1865), vol. 1, pp. 250–52, and M. ALLAN, *Palgrave of Arabia* (London: Macmillan, 1972), pp. 208–9. (Both of these references were kindly brought to my attention by Ms. Ellen Schultz of New York University.) See further, A.H. MASRI, "The Historical Legacy of Saudi Arabia," *Atlal: The Journal of Saudi Arabian Archaeology*, vol. 1 (1977), pp. 9–19, especially p. 13; R.C. ADAMS *et al.*, "Saudi Arabian Archaeological Reconnaissance 1976," *Atlal: The Journal of Saudi Arabian Archaeology*, vol. 1 (1977), pp. 21–40, especially pp. 34–36; and P.J. PARR *et al.*, "Preliminary Report on the Second Phase of the Northern Province Survey 1397/1977," *Atlal: The Journal of Saudi Arabian Archaeology*, vol. 2 (1978), pp. 29–50, especially pp. 37–40.

11. For a survey of the stone formations in Southern Arabia, see W. DOSTAL, "Zur Megalithfrage in Südarabien," in *Festschrift Werner Caskel*, ed. E. Graf (Leiden: E.J. Brill, 1968), pp. 54–61. For an introduction to the archaeology of the area, see B. DOE, *Southern Arabia* (London: Thames and Hudson, 1971).

12. On the "Tower of the Winds" in Athens, see D.J. DE SOLLA PRICE, *Science Since Babylon*, 2nd ed. (New Haven: Yale University Press, 1976), pp. 78–80, and the references there cited.

13. See J. CHELHOD, "A Contribution to the Problem of the Pre-eminence of the Right, based upon Arabic Evidence," (translated from the French), in *Right & Left*, ed. R. Needham (Chicago: University of Chicago Press, 1973), pp. 239–62, especially pp. 248–53. Although Chelhod assumed that the corners of the Ka'ba are cardinally aligned, and was not familiar with any of the sources used in the present study, he was able to recognize the function of the Ka'ba as a "microcosm of the universe." On the orientation of the Ka'ba see also B. FINSTER, "Zu der Neuauflage von K.A.C. Creswells *Early Muslim Architecture*," *Kunst des Orients*, vol. 9 (1972), pp. 89–98, especially p. 94. The notion that the pagan Arab Ka'ba was converted into a church with an apse facing Jerusalem is propounded in G. LÜLING, *Der christliche Kult an der vorislamischen Kaaba* . . . (Erlangen: H. Lüling, 1977), pp. 43–52. See also F. LANDSBERGER, "The Sacred Direction in Synagogue and Church," *Hebrew Union College Annual*, vol. 28 (1957) pp. 181–203. (Landsberger's plans showing synagogues and churches facing Jerusalem remind me of plans of mosques that show the *qibla* wall facing Mecca.)

14. On the orientation of the earliest mosques in the Hijaz towards the east, see W. BARTHOLD, "Die Orientierung der ersten muhammedanischen Moscheen," *Der Islam*, vol. 18 (1929), pp. 245–50.

15. See P.H. LAMMENS, "Les Sanctuaires préislamites dans l'Arabie occidentale," *Mélanges de l'Université Saint-Joseph, Beyrouth*, vol. 11:2 (1926), pp. 37–173 (pp. 1–137 in the separatum).

16. On pre-Islamic Arabian beliefs concerning the heavens, see J. HENNINGER, "Über Sternkunde und Sternkult in Nord- und Zentralarabien," *Zeitschrift für Ethnologie*, vol. 79 (1954), pp. 82–117. See also T. FAHD, *Le Panthéon de l'Arabie Centrale à la vielle de l'Hégire* (Paris: Paul Geuthner, 1968), on pre-Islamic religion in general.

17. See D.A. KING,[4] *World*, Section 4 and Appendix A.

XIV

THE EARLIEST ISLAMIC MATHEMATICAL METHODS AND TABLES FOR FINDING THE DIRECTION OF MECCA

Contents

1. Introduction

The Kaʿba in Mecca is the physical focus of Muslim worship, a pointer to the presence of God. The qibla, or local direction of Mecca, is the sacred direction of Islam. In the first decades of Islam, before the Muslims came into contact with the mathematical sciences, the qibla was found by simple procedures either inspired by Prophetic tradition or involving astronomical risings and settings.[1] The Kaʿba itself is astronomically aligned, a fact known by the first generations of Muslims and recorded by such authorities as Ibn ʿAbbās and al-Ḥasan al-Baṣrī; in order to "face" one of its four sides from a distant locality, it was considered appropriate to stand in the same astronomically-defined direction as one would be facing if one were standing directly in front of that wall. In the tenth century, these procedures were formalized and developed into a sacred geography. Throughout the centuries they were advocated by the legal scholars.

From the eighth century onwards, Muslim astronomers concerned themselves with the determination of the qibla by scientific methods. The qibla can be found from the geographical coordinates of Mecca and of the locality in question, and, as is now well-known, by the end of the ninth century Muslim astronomers had derived exact solutions by trigonometry and by geometric construction.[2] Several later Muslim scholars considered all possible cases of the problem, and others even compiled tables displaying the qibla as an angle to the meridian for each degree of terrestrial latitude and longitude difference from Mecca. These theoretical and practical solutions constitute just one example, albeit an important one, of what I call "universal solutions" in Islamic mathematical astronomy.[3] In practice the astronomers

[1] On the traditional methods for finding the qibla, see *King* 3 and the article "Makka (4. As centre of the world)" in *EI*², and the literature there cited.

[2] For an introduction to medieval computational solutions to the problem of determining the qibla, see my article "*Ḳibla* (*astronomical aspects*)" in *EI*² (already outdated). For studies of individual methods, see also *Schoy* 2 and 3; *Kennedy & Id*; *Kennedy* 3; *King* 1, Part III, Ch. 28; and, more recently: *Debarnot* 1, pp. 260–264; *Lorch* 2; *Berggren* 1, 2 and 3; *Carandell*; and *Kennedy* 7, pp. 38–42. On Islamic qibla tables, see *King* 4, XIII, especially pp. 120–122; *Lorch* 1; and also Sections 7, 10 and 12 of the present study.

[3] See *King* 8.

were rarely consulted on the matter of the qibla; the legal scholars had far more influence in Muslim society. This explains the wide variation in orientation of medieval mosques. But our present concern is with *scientific* methods, as indicated by the title of this paper.

The purpose of this study is to describe some newly discovered material including the earliest known mathematical methods and tables for finding the qibla. I propose to investigate these for the light that they cast on the mathematical geography[4] and the computational techniques[5] of the formative period of Islamic science. The methods display a complete range of solutions from the crudest conceivable cartographic method to an ingenious, if clumsy, exact method, and the tables are based on a simple approximate formula and a highly sophisticated one, respectively.

Two manuscripts, one now preserved in Tashkent and the other in Istanbul, and both recently investigated for the first time, constitute the major sources for this study. The first is MS Tashkent Oriental Institute 177,3, fols. 148r–151v, copied in the thirteenth (?) century probably in Iran;[6] the second is MS Istanbul Aya Sofya 4830, fols. 187r–188r, copied *ca.* 1230 in Damas-

[4] For early Islamic geography and cartography, see the articles "*Djughrāfiyā* (= geography) and "*Kharīṭa* [= map(s)]" by S. Maqbul Ahmad in *EI*²; Schoy 1 and 5; and *Kennedy* 8 and 9. A forthcoming volume of *Sezgin* will deal with geography. The major Islamic treatise on mathematical geography is *al-Bīrūnī* 1.

[5] The best surveys of astronomy in this period are *Pingree* 1 and 2, and the same author's article "*'Ilm al-hay'a* (= astronomy)" in *EI*².

For early Islamic mathematics, the survey listed as *Youschkevitch* (corresponding to *Juschkewitsch*, pp. 175–325) is unsurpassed, but to this must eventually be added the kind of material on mathematical methods used in astronomical writings, some of which is documented in *Kennedy et al.*, *Kennedy Festschrift*, and *King* 4. For an overview of recent research on Islamic mathematics, see *Berggren* 5.

[6] This manuscript is not listed in *Matvievskaya & Tllashev*; it was located by using the index cards for the Oriental Institute collection. The treatise on the qibla follows a copy of al-Fāsī's *Ta'rīkh Makka* (fols. 1r–96r), and a treatise on the qibla in early Islamic Iran (fols. 97r–144v) by 'Abd al-Qādir ibn Ṭāhir Shāpūrī, also known as al-Baghdādī (on whom, see *Brockelmann*, I, p. 482 and SI, pp. 666–667, and *Sezgin*, V, p. 357). The remaining works in this *majmū'a* are not related to our subject, with the exception of some geographical tables – see Section 10(d) below.

cus.[7] Together the two sources appear to provide almost the complete text of an early Arabic treatise on the qibla. No compiler is mentioned in either manuscript. The heavy style of the Arabic and the nature of the mathematical methods identify the material as early Abbasid, and Baghdad is specifically mentioned in the text. Where the two versions overlap they are identical in meaning if not in vocabulary and style.

I have previously argued that the author of most of the material in the treatise, if not all of it, might be the celebrated astronomer al-Khwārizmī,[8] who was one of the scientists associated with the Abbasid Caliph al-Ma'mūn in the early ninth century. However, the approximate methods mentioned in the text perhaps predate even al-Khwārizmī, although it is unlikely that they go back to the mid-eighth century, when such astronomers as al-Fazārī and Ya'qūb ibn Ṭāriq were active, for their surviving writings on mathematical geography, scant as they are, are in a different style.[9] The exact method *could* have been proposed by al-Khwārizmī, and it must predate the mid-ninth century when less complicated equivalent solutions had been formulated in Baghdad. On the other hand, the table contained in the treatise is highly sophisticated, and as far as we know, al-Khwārizmī did not have the necessary skills to compile it. I now suspect that the table was compiled by another ninth- or possibly even tenth-century astronomer. The "treatise" is thus an anonymous collection of material originally compiled over a period of two centuries, if not longer.

The other main source for the present study is the astronomical handbook (*zīj*) of the tenth-century Syrian astronomer al-Bat-

[7] This manuscript contains a variety of mathematical and astronomical tracts but has never been properly catalogued. Some, but not all, of the contents are listed in *Krause* and *Sezgin*,V and VI. A detailed description of this manuscript would be a useful contribution. On the anonymous geographical tables in it, see Section 10(d) below.

[8] On al-Khwārizmī, see the article by G. Toomer in *DSB*. For lists of manuscripts of his works, see *Sezgin*, V, pp. 228–241, and VI, pp. 140–143 and 293. On some other recently-discovered minor astronomical works by him, including his lunar crescent visibility table, some tables for timekeeping and for construction of sundials and astrolabes, see *King* 2 (now substantially revised).

[9] On the surviving fragments of their works, see *Pingree* 3 and 4, and *Sezgin*, V, pp. 216–218, VI, pp. 122–127, and VII, pp. 101–102.

tānī.[10] He proposed another approximate method for finding the qibla which was widely used amongst astronomers in later centuries. I refer to this as the "standard" approximate method.

In this study, the material is arranged mainly according to increasing complexity. This approach has the advantage of separating the approximate from the exact methods.

In Section 2 I discuss the mathematics underlying the medieval qibla methods analyzed in this paper. The four approximate methods are essentially cartographic, in that they relate to attempts to solve the qibla problem by representing it in two dimensions. The two exact methods which I discuss belong more to the realm of spherical astronomy.

In Section 3 I discuss a cartographic construction for finding the qibla, found only in the Tashkent manuscript. The method casts some light on the early development of Islamic cartography, and its significance is discussed in more detail in a parallel study in which various "maps" based on such constructions are analyzed for the first time.[11]

The Istanbul manuscript contains an approximate mathematical method which was actually used by the astronomers commissioned by the Caliph al-Ma'mūn to find the qibla at Baghdad, as we now know from newly-discovered textual evidence on the earliest Muslim longitude observations.[12] This method is analyzed in Section 4.

The simple geometric construction advocated by al-Battānī is considered in Section 5. This differs from the other approximate methods in that it was widely used over the centuries.

Both the Tashkent and Istanbul manuscripts contain a slightly more sophisticated approximate method: this is analyzed in Section 6.

Some medieval tables displaying the qibla as a function of latitude and longitude difference from Mecca and based on the

[10] On al-Battānī, see the article by W. Hartner in *DSB* and also *Sezgin*, V, pp. 287–288, VI, pp. 182–187, and VII, pp. 158–160. On his qibla method, see *Nallino* 1, III, pp. 206–207 (text), and I, pp. 318–319 (commentary).

[11] See *King* 7 (forthcoming).

[12] A treatise by Ḥabash on these observations has recently been published in *Langermann*. A treatise by Yaḥyā ibn Aktham, the judge appointed by al-Ma'mūn to oversee the observations, is compared with Ḥabash's account in *King* 6.

standard approximate method are discussed in Section 7. (One of these is actually attributed to al-Khwārizmī.)

The Istanbul manuscript also includes an exact method for finding the qibla, which I discuss in Section 8. It constitutes a veritable *tour de force*, given the complexity of the problem when tackled by solid geometry. In Section 9 I discuss the relationship of this accurate method to the so-called "method of the *zījes*," which appears to date from the mid-ninth century and which was used by several later Muslim astronomers.

In both the Tashkent and Istanbul manuscripts there is a table for finding the qibla. The table is based on a complicated approximate formula. However, closer inspection reveals that the values are accurately computed for arguments different from those specified in the instructions – see Section 10.

Three medieval references to one or other of these two qibla tables are discussed in Section 11. These two tables appear to be the earliest of several compiled by Muslim astronomers over the centuries: a survey of the others is presented in Section 12. I have not been able to crack the formulae underlying some of these tables, so the topic is by no means exhausted. Also, I draw attention to the fact that the celebrated eleventh-century scientist Ibn al-Haytham[13] compiled a qibla table, as yet not positively identified.

In the translations of the relevant Arabic texts I have attempted to maintain the heavy flavour of the original Arabic and have distinguished between numbers written out in words and those expressed as numerals (either in the standard alphanumerical *abjad* system or in Indian numerals). In the original tables, numbers are expressed sexagesimally (*i. e.* to base 60) and are written in the *abjad* notation.[14]

2. The Basic Geographical and Mathematical Concepts

Fig. 1 displays the qibla problem. Point X denotes a general locality and point M, Mecca. The meridians at M and X are

[13] On Ibn al-Haytham, see the article by A. I. Sabra in *DSB*, and *Sezgin*, V, pp. 358–374, and VI, pp. 251–261, to which add his autobiography, now published with annotated translation in *Heinen*.

[14] On this convention, see *Irani*.

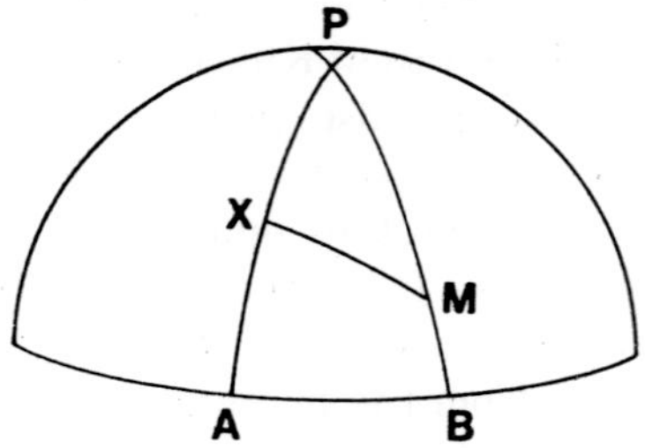

Figure 1 :
The diagram shows a general locality X and Mecca M
relative to the equator AB and the North Pole P.
The qibla at X is measured by the angle q between the meridian of X, PXB,
and the great circle XM joining X to M.
It can be found from the latitude of X (φ = XB), the latitude of M (φ_M = MA),
and their longitude difference (ΔL = AB).

PMB and PXA, where P is the north pole and AB is the equator. Given the latitudes of M and X, that is, the arcs MA and XB, as well as the longitude difference, that is, the arc AB (also measured by the angle APB), it is required to find the angle MXA where XM is the arc of a great-circle joining X and M. This is the angle most commonly used to specify the qibla in the medieval sources.

In what follows, I denote the latitude and longitude of a general locality by (φ, L), the latitude and longitude of Mecca by (φ_M, L_M), and the qibla, measured as an angle from the meridian, by q. (al-Battānī perversely uses the complement, which we label q'.) I further denote:

$$\Delta L = L \sim L_M \text{ and } \Delta\varphi = \varphi \sim \varphi_M.$$

The qibla at a given locality q is a trigonometric function of φ, φ_M, and ΔL, as is the distance between the locality and Mecca, represented by arc XM in Fig. 1 and denoted by D.

Tables of geographical coordinates were available to the Muslims in Ptolemy's *Geography*, where some eight thousand localities are listed.[15] This work was translated into Arabic several times during the Abbasid period. Geographical tables in Arabic

[15] On Ptolemy's *Geography*, see G. Toomer in the article "Ptolemy" in *DSB*, XI, esp. pp. 198–200 and 205; *Neugebauer* 2, pp. 220–224, and 3, pp. 880–888; and S. Maqbul Ahmed in the article "*Djughrāfiyā* (= geography)" in *EI*², II, esp. pp. 577–578.

were prepared by al-Khwārizmī (545 cities) and later by al-Battānī (273).[16] Most medieval *zīj*es or astronomical handbooks contain tables of geographical coordinates, and many known examples have been analyzed by E. S. Kennedy and his former associates at the American University of Beirut using a computer.[17]

To interpret in symbols the mathematical methods which are described in words in the medieval texts, we use the standard notation:

$$\text{Sin}_R\ \alpha = R \sin \alpha,\ \text{Cos}_R\ \alpha = R \cos \alpha, \text{ and}$$
$$\text{Vers}_R\ \alpha = R - \text{Cos}_R\ \alpha = R\,(1 - \cos \alpha),$$

for medieval trigonometric functions which are to a base R other than unity.[18] In most Islamic sources we find $R = 60$ for Sin, Cos, and Vers, following the Greek tradition for the Ptolemaic Chord function. In a few early Islamic sources, including the Abbasid material discussed in this paper, we also find $R = 150$, following the Indian-Sasanian tradition. The reader will observe that the medieval sources examined in this paper do not exploit the tangent or cotangent functions, preferring more complicated procedures with sines and cosines. The solution of a plane right-angled triangle for an angle q in terms of the ooposite and adjacent sides x and y would thus be:

$$q = \text{arc Sin}\,(x \cdot R\ /\ \sqrt{x^2 + y^2}).$$

The tangent and cotangent functions were available but were generally used only when the argument was the solar altitude, that is, when the functions measured shadows.

The evidence presented in this paper reveals that the earliest Muslim qibla determinations were cartographic, in that they can all be derived by considering plane representations of the part of the terrestrial globe between the locality and Mecca. Although the highly sophisticated cartographic techniques advocated in the *Geography* of Ptolemy may have been available in Arabic they were apparently not used systematically in Muslim cartography, perhaps for the very reason that they do not lend them-

[16] See *von Mžik* and *Nallino* 1, pp. 458–532, on the former and *Nallino* 2, pp. 458–532, on the latter.

[17] See *Kennedy & Haddad*. I have used Prof. Kennedy's current (1986) copy of the output in my investigations of geographical coordinates for this paper. (The output is now published by the IGAIW, Frankfurt.)

[18] On these, see *Kennedy* 1, pp. 139–140, and *Kennedy* 5.

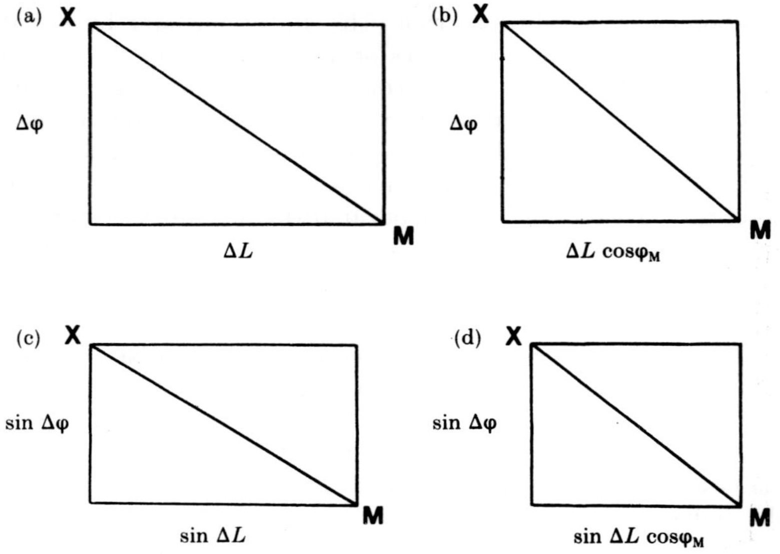

Figures 2a–d:
The two-dimensional representations of the latitude and longitude differences between a locality X̣ and Mecca M which underly the four approximate methods presented in Sections 3–6.

selves to a simple determination of the qibla. Rather, Muslim cartographers were mainly content to use a crude cylindrical projection, in which meridians and parallels of latitude are mapped into an orthogonal grid of squares – see Fig. 2a. Such, apparently, was the celebrated map of the earth prepared for the Caliph al-Ma'mūn, as we know from the tenth-century geographer Suhrāb.[19] In this projection, distances along the meridians are preserved, as well as distances along the equator but not along parallels of latitude. As we shall see, the method outlined in Section 3 involves such a projection.

Distances reckoned in degrees of longitude on a parallel of latitude φ are of lengths $\cos\ \varphi$ times the corresponding distances on the equator. In the projection of Marinos (*fl.* Tyre *ca.* 110 A. D.), distances along the parallel of latitude of Rhodes (36°) are preserved, the ratio of units of longitude to units of latitude being

[19] See *Kennedy* 8.

4/5 (≈ cos 36°).[20] No Islamic maps with this distinctive feature are known, but the qibla method described in Section 4 is best explained in terms of a Marinos-type projection preserving distances on the parallel of Mecca – see Fig. 2b.

al-Battānī (or his source), however, considered the determination of the qibla from maps based on such projections to be too crude. Rather he/they preferred to represent longitude and latitude differences between two localities in two dimensions by perpendicular segments whose length were the corresponding sines – see Section 5.[21] No explanation is provided in our sources, and we are forced to speculate about what might have been in the back of the mind of the originator. Fig. 3 displays a somewhat unorthodox representation of the situation on the globe. Through locality X are drawn the meridian at X (ABX) and the east-west line (shown as the perpendicular *great circle* CXD). ADBC is the great circle of which X is the pole. Through Mecca M are drawn the meridian (drawn as a small circle parallel to the meridian at

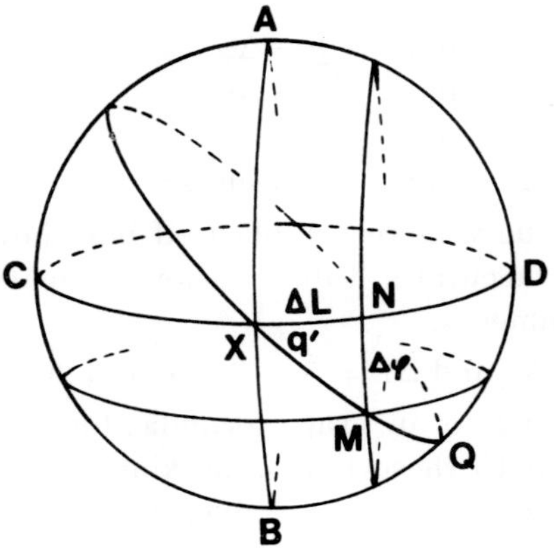

Figure 3:
An unorthodox representation of the coordinates of
a locality X and Mecca M
whose sole virtue is that it enables one to derive al-Battānī's qibla method.

[20] On his projection, see *Neugebauer* 2, p. 220, and 3, pp. 735, and 879–880, and *Kennedy* 9.

[21] An example of the solution of a right spherical triangle by considering a right plane triangle whose sides are the sines of the corresponding arcs is discussed in *Hogendijk*, commenting on the table of lunar brightness attributed to Ya'qūb ibn Ṭāriq (*Kennedy* 2, p. 127).

X) and the east-west line (drawn as a small circle parallel to the east-west line through X). The arc XM is produced to meet ADBC at Q and its direction is supposed to define the qibla at X. Note that, approximately:

$$\mathrm{XN} = \Delta L \quad \text{and} \quad \mathrm{MN} = \Delta\varphi,$$

and also that:

$$\mathrm{XD} = \mathrm{XQ} = 90°.$$

An orthogonal projection of this unhappy scene in the plane ADBC yields the desired two-dimensional configuration shown in Fig. 2c. The author of the method presented in Section 6 preferred to apply a Marinos-type correction factor to the quantity sin ΔL, yielding the situation displayed in Fig. 2d.

The four approximate methods described in Sections 3–6 and represented schematically in Figs. 2a–d may be represented algebraically in modern notation as follows:

$$q = \arctan (x/y),$$

where:

$x = \Delta L$	and	$y = \Delta\varphi$	(Formula 1)
$x = \Delta L \cos \varphi_M$	and	$y = \Delta\varphi$	(Formula 2)
$x = \sin \Delta L$	and	$y = \sin \Delta\varphi$	(Formula 3)
$x = \sin \Delta L \cos \varphi_M$	and	$y = \sin \Delta\varphi$	(Formula 4).

They are increasingly more complicated but they are not correspondingly more accurate. Table 1 shows a matrix of values of the qibla for arguments:

$$\Delta\varphi \text{ and } \Delta L = 5°, 10, 15, \text{ and } 20,$$

computed according to an exact formula (for a specific value of φ_M) and according to these four approximate formulae, together with the errors according to the convention:

$$\text{error} = \text{computed value} - \text{exact value}.$$

(Note that because of the nature of the function $q(\Delta\varphi, \Delta L)$ the values are similar to those for $\Delta\varphi$ and $\Delta L = 1°, 2, 3$ and 4 – see Table 2.) It is clear from Table 1 that the "best" approximation is in fact the simplest, namely Formula 1. The introduction of the factor cos φ_M or of the sines or both, presumably to produce a more accurate procedure, cannot be regarded as entirely successful. But then we should bear in mind that the relatively small errors (at least for $\Delta\varphi$ and $\Delta L < 20°$) introduced by using any of these approximate methods are far less than the "errors" caused by using astronomical alignments for the qibla.

Table 1

Sample values of the qibla computed using an exact formula and four approximate methods
(The underlying value of φ_M is 21;30°)

Exact formula

ΔL \ $\Delta\varphi$	5°	10	15	20
5°	43;28°	25;16	17;32	13;26
10	63;25	44;11	32;50	25;54
15	73; 8	56;53	45; 6	36;53
20	78;57	65;36	54;41	46;14

Values computed using approximate methods, and corresponding errors

Formula 1 (Section 3 / Fig. 2a):

45; 0	26;34	18;26	14; 2	1;32	1;18	0;54	0;36
63;26	45; 0	33;41	26;34	0; 1	0;49	0;51	0;40
71;34	56;19	45; 0	36;52	−1;34	−0;34	−0; 6	−0; 1
75;58	63;26	53; 8	45; 0	−2;59	−2;10	−1;33	−1;14

Formula 2 (Section 4 / Fig. 2b):

42;56	24;57	17;14	13; 6	−0;32	−0;19	−0;18	−0;20
61;45	42;56	31;49	24;57	−1;40	−1;15	−1; 1	−0;57
70;17	54;23	42;56	34;54	−2;51	−2;30	−2;10	−1;59
74;58	61;45	51; 8	42;56	−3;59	−3;51	−3;33	−3;18

Formula 3 (Section 5 / Fig. 2c):

45; 0	26;39	18;37	14;18	1;32	1;23	1; 5	0;52
63;21	45; 0	33;52	26;55	−0; 4	+0;49	1; 2	1; 1
71;23	56; 8	45; 0	37; 7	−1;45	−0;45	−0; 6	+0;14
75;42	63; 5	52;53	45; 0	−3;15	−2;31	−1;48	−1;14

Formula 4 (Section 6 / Fig. 2d):

42;56	25; 2	17;24	13;20	−0;32	−0;14	−0; 8	−0; 6
61;39	42;56	31;58	25;17	−1;46	−1;15	−0;52	−0;37
70; 6	54;12	42;56	35; 9	−3; 2	−2;41	−2;10	−1;44
74;41	61;23	50;53	42;56	−4;16	−4;13	−3;48	−3;18

Now we turn to the exact solutions. In the eighth and early ninth century, the methods of spherical trigonometry were not yet available to Muslim astronomers.[22] Likewise, the powerful mathematical tool of the analemma, essentially a procedure for reducing problems of the sphere to a plane, that is from three dimensions to two, was apparently not yet known to them.[23] Thus problems of spherical astronomy – of which the determination of the qibla is one since the problem may be reformulated as determining the direction on the local horizon of the zenith of Mecca – were solved by direct means such as solid geometry. These methods are typical of Indian astronomy, whose procedures characterize the earliest period of Islamic astronomy.

The new exact method presented in Section 8 can be explained by reference to Fig. 4. After providing a translation of the text, I shall represent the methods expounded by means of an algebraic notation, introducing symbols a, b, c, *etc.* to denote various quantities used in this qibla method which can be represented as lengths in Fig. 4, where the radius of the terrestrial sphere is R. I use the same notation in my commentary but for simplicity put $R = 1$.

In order to explain the basic quantities used, we first draw the perpendicular AC from A to OB. Then since arc AB = ΔL, we have:

$$AC = \sin \Delta L = a.$$

Now we draw the parallel of latitude (which is not a great circle) through M to meet PB in N and mark on the axis PO the point Q, centre of the circle containing arc MN. Since MA = NB = φ_M, we have:

$$QO = \sin \varphi_M = b \quad \text{and} \quad QM = QN = \cos \varphi_M = c.$$

We now draw ML perpendicular to QN. Since in sector QMN

$$\text{arc MN} = \Delta L \quad \text{and} \quad QM = \cos \varphi_M,$$

we have:

$$ML = \sin \Delta L \cos \varphi_M = d.$$

[22] On Islamic spherical astronomy, see *King* 1 and *Kennedy* 7. The best overview of early Islamic spherical astronomy is *Debarnot* 1.

[23] On the analemma, see *Luckey* 1 and *Neugebauer* 3, II, pp. 839–857. On various applications in Islamic astronomy, see, for example, *Schoy* 2; *Kennedy et al.*, pp. 495–498; *Kennedy & Id*; *King* 1, *passim*; *Berggren* 2; and *Carandell.*

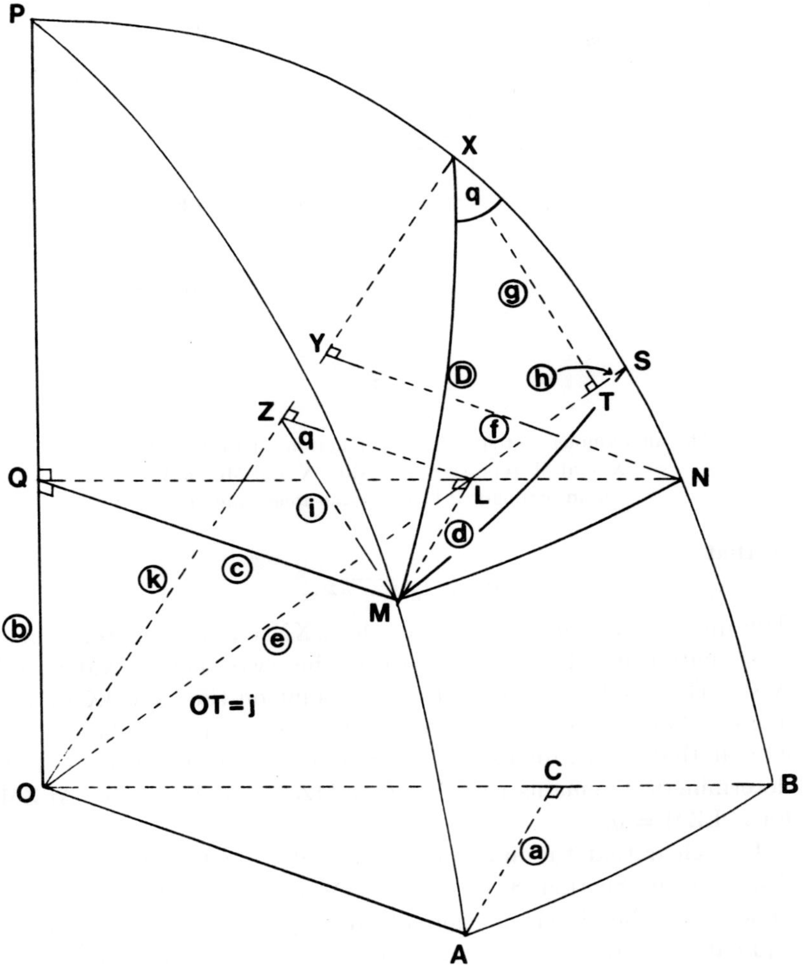

Figure 4:
The exact method of Section 8 can be explained
with reference to this diagram.

Now we draw NY perpendicular to OX. Since arc NX = $\Delta\varphi$ we have:

$$NY = \sin \Delta\varphi = f.$$

Notice that in ΔMLO:

$$ML = d,\ MO = 1,\ \angle L = 90^\circ,$$

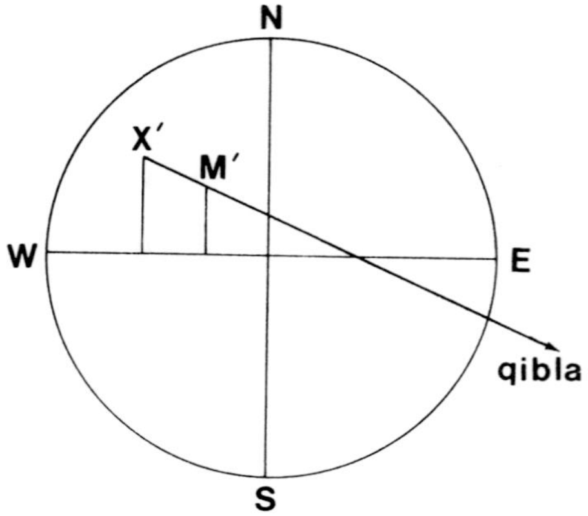

Figure 5:
The anonymous cartographic construction for finding the qibla.
X′ and M′ represent a locality X and Mecca M
on an orthogonal latitude and longitude grid.

so that:

$$\mathrm{OL} = \sqrt{1 - d^2} = e.$$

The qibla q, measured by the angle MXN on the sphere, is also represented by the angle between the two planes XMO and XNO. This angle is measured at any point on the axis OX by the angle between lines in each plane that are perpendicular to the axis at that point, for example, LZ and MZ. Since ML is already determined, it remains to find LZ or MZ in order to solve ΔLZM for $\angle$ LZM $= q$.

It is clear that the individual responsible for the exact method discussed in Section 8 was aware of the geometrical problem underlying the determination of the qibla in terms ultimately equivalent to those I have just outlined. It seems that he considered the qibla problem as one of mathematical geography, working from the mathematics of the terrestrial sphere, rather than one of sperical astronomy, as formulated by various later scientists. (The equivalent astronomical problem is to determine the azimuth of the zenith of Mecca on the local horizon. In Fig. 4, P becomes the celestial pole, AB the celestial equator, X the local zenith, and M the zenith of Mecca. Then the problem is to determine the azimuth (q) on the horizon of given latitude (φ) of a celestial body with given declination (φ_M) when the hour-angle (ΔL) is known. See further Section 9.)

Before one could use the mathematically-computed qiblas for determining the direction of prayer, laying out mosques, *etc.*, it would be necessary to establish the cardinal directions. The most commonly advocated method is the geometrical construction known as the "Indian circle" (*al-dā'ira al-hindīya*), and this is actually mentioned in the text presented in Section 6. The procedure is simply to bisect the angle between the two observed directions of sunrise and sunset on a particular day, or between the azimuths of morning and evening observations when the sun has the same altitude.[24]

3. An Approximate Cartographic Construction for the Qibla

The Tashkent manuscript but not the Istanbul manuscript contains a description of a simple cartographic construction for finding the qibla. The method is approximate, although this is not stated in the text (fols. 148r–149r):

بسم الله الرحمن الرحيم

معرفة تقويم سمت القبلة لأى بلد شئت إذا أردت أن تقوم سمت القبلة لأى بلد أردت فاعمل الدائرة التى تستخرج بها أنصاف النهار وعظمها واقسمها بأربعة أقسام متساوية بخطين يقطع أحدهما الآخر على مركز الدائرة واكتب على كل ناحية منها اسم ناحيته ثم اقسم كل ربع من أقسام هذه الدائرة بتسعين جزءًا قسمة مستوية ثم اقسم الخط الذى يكون من ناحية الشمال من حرف الدائرة إلى المركز على خط نصف النهار بتسعين جزءًا قسمة مستوية ثم انظر كم طول المدينة التى تريد أن تعرف سمت قبلتها من المغرب فما كان من الأجزاء فعد مثله من العلامة التى فى حرف[1] الدائرة من نقطة المشرق إلى مركز الدائرة التى هى من خط الجنوب إلى المغرب وهو على خط الاستواء من المشرق إلى المغرب فحيث انتهى العدد فتعلم عليه كان ذلك المو[ضع] (؟) أنصاف نهار تلك الأرض فربع الدائرة بخطين من تلك العلامة مثل الخـ[ـطين الأ] ولين ودع الأولين فإنك لا تحتاج

[24] On this construction, see *al-Biruni* 3, pp. 49–50; *Wiedemann*; and *Kennedy* 4, II, pp. 80–90.

98

إليهما واكتب عـ[ـلى] أطراف الخطوط جهاتها المشرق والمغرب والجنوب والشمال واجعل هذه الخطوط بينة[2] خارجة أطرافها[3] على الدائرة ثم اعرف طول مكة وعرضها وطول مدينتك التى أنت فيها وعرضها ثم خذ من خط المشرق على حرف الدائرة بقدر طول مكة إن أخذت طولها من المشرق أو أخذت طولها من المغرب فقدر بقدر أجزاء طول مكة من الخطوط التى على طرف الدائرة فإذا استكملت عدد طولها فخط من ذلك الموضع خطاً من حرف الدائرة على الخط الذى يلي الجنوب من حرف الدائرة إلى وسطها فإذا استكملت ذلك فخذ البركار فضع أحد طرفيه على موضع عرض مكة والطرف الآخر في وسط الدائرة ثم أدرها دائرة تكون في جوف الدائرة العظمى فحيث أصابت هذه الدائرة من خط مكة فتعلم عليه فإنه موضع مكة من الأرض فاحفظه ثم انظر طول المدينة التى أنت فيها من المغرب أردت أو من المشرق فخذ مثلها واعمل بطولها وعرضها فما عملت بطول مكة وعرضها حتى تعرف صوضعها من الأرض واحفظه ثم ضع أحد طرفى البركار فى موضع مكة وضع طرفه الآخر فى موضع المدينة ثم أدرها دائرة تكون مكة فى وسطها والمدينة فى حرفها[4] ثم ربع الدائرة بخطين أحدهما من المدينة التى تحسب لها إلى مكة إلى ما يحاذى به من جانب الدائرة الأخرى فهو جهة القبلة والخط الآخر بتربيع القبلة[5] حيث وقع إن شاء الله

[1] جوف   [2] بينه   [3] مكرر فى الاصل   [4] جوفها   [5] الدايرة (ولعل القبلة اصح)

Translation:

"In the Name of God, the Merciful and Compassionate. To determine (*taqwīm*) the azimuth of the qibla for any locality. If you want to do this, draw the circle for finding the meridian[a] and make it large. Then divide it into four equal divisions by two (perpendicular) lines which intersect at the centre of the circle. Write at the end of each the name of (the cardinal) directions, and divide each quadrant of the circle into ninety equal parts. Then divide the line which joins the north end of the circumference of the circle to the centre along the meridian into ninety equal parts.

Next look at the value of the longitude, measured from the

west, of the city whose qibla you want to find, and count that number of degrees from the mark which is on the circumference of the circle, (counting) from the east (??) point towards the centre of the circle, which is from the line of the south towards the west (??), which is on the equator from the east to the west. Make a mark where you stop counting. This point will be the meridian[a] of that locality [*sc.*: it will be the point where the meridian cuts the east-west line representing the equator], so draw two perpendicular lines inside the circle (*fa-rubbiʿi ʾl-dāʾira*) like the first two lines. Leave the first two because you no (longer) need them, and write at the ends of the (two new) lines the cardinal directions: east, west, south and north. Make these lines clear with their ends protruding outside the circle.

Next find the longitude and latitude both of Mecca and of the city where you are, and then count from the east point[b] on the circumference of the circle by the amount of the longitude of Mecca. Whether you reckon longitude from the east or from the west, measure the degrees of longitude of Mecca from the (east or west) points[c] which are on the circumference of the circle. When you have done this, draw a line from that point from the circumference of the circle towards the direction of south, (drawing it) from the circumference of the circle to (the east-west line which bisects the circle)[d]. When you have done this, take a compass and place one of its ends on the place of the latitude of Mecca and the other one at the (point where the meridian intersects the east-west line)[e], and draw a circle inside the large circle. At the point where this circle intersects the line (representing the meridian) of Mecca, make a mark: it will be the position of Mecca on the earth, so keep it in mind. Then look at the longitude of the city where you are, measured either from the west or from the east, and do the same with its longitude and latitude as you did for Mecca, so that you can find its position on the earth, which you keep in mind.

Next put one of the ends of the compass on the position of Mecca and the other one on the position of the city, and draw a circle with Mecca at the centre and the city on its circumference. Then draw two perpendicular lines inside (this) circle (*rubbiʿi ʾl-*

[a] text: meridians [b] text: line [c] text: lines [d] text: the middle (of the circle) [e] text: the middle of the circle

dā'ira bi-khaṭṭayn), one from the city for which you are calculating to Mecca and beyond to the point on the side of the other circle. Wherever (this line) falls, it will (represent) the direction of the qibla, and the other line (resulting from) the construction will be [the line perpendicular to the qibla ??], if God wills."

Commentary:

The text is not always easy to follow, but the procedure is clear. The author wishes to mark the positions of a general locality and of Mecca inside a circular frame in which the east-west line represents linearly 180° of the equator and the perpendicular diameter represents latitudes likewise: he then simply joins them to find the qibla. Thus in Fig. 5, we mark points X′ and M′ to represent any locality X and Mecca M on such a grid (here longitudes are measured from the west). Then the segment X′M′ indicates the direction of the qibla on the horizon of X. (Clearly some remote localities in the oekumene would lie outside this circular frame, but this possibility is not considered.)

Except for the fact that the frame is circular (in order to represent the horizon), the projection is a simple cylindrical type such as was described by Suhrāb – see Fig. 2a in Section 2. The qibla should not, of course, be read on a scale on the horizon circle, although on various later qibla "maps" in which Mecca is not at the centre, the qibla is determined in this way. Only in one later Islamic source known to me is a "map" presented displaying positions of various localities relative to Mecca specifically for finding the qibla. This source, an early-thirteenth-century Egyptian treatise on folk astronomy, has been analyzed elsewhere.[25] Latitudes and longitudes (the latter measured from the west) are measured linearly on an orthogonal grid centred on a point of the equator 90° east of "west", and the "map" is bounded by a circle representing the horizon.

The underlying formula for the qibla as it can be determined from such "maps" would be equivalent to:

$$q = \text{arc tan } (x/y),$$

where:

$$x = \Delta L \text{ and } y = \Delta\varphi.$$

[25] See *King* 7.

No such formula is proposed in the Abbasid text (though see Section 4) or in any later sources known to me.

It is of interest that the author suggests the possibility that longitudes be measured from the east, and he implies that the base meridian would be 180° east of the point taken as "west". It is known that al-Fazārī, as well as Ḥabash (*fl.* Baghdad and Damascus *ca.* 850) and al-Hamdānī (*fl.* Yemen and Baghdad *ca.* 950) measured longitudes from the east.[26]

The text in the Tashkent manuscript continues with a description of a practical method for finding the qibla by means of the sun:

فإذا أردت أن تضع مسجداً فانظر موضع مكة فاثبت فيه عوداً وانظر موضع المسجد
فضع فيه خشبة قائمة معتدلة قدر ثلثة وسامير أو أكثر ثم ارصد ظل العود الذى فى الدائرة
حتى إذا وقع ظله على خط ما بين مدينتك إلى مكة فذلك سمت القبلة وانظر ظل الخشبة
ومد خيطاً على الذى تعلمت فإنه موضع حائط المسجد من جانب يمين المسجد أو يساره
فربع المسجد من ذلك الخط حتى تستوى قبلته وهذه صورته

"If you want to lay out a mosque, fix a gnomon/stick (*'ūd*) at the position of Mecca (as determined above), and erect vertically a straight piece of wood at the position of the mosque. The wood can be three *wasāmīr* (??) long or more. Next observe the shadow of the gnomon/stick which is in the circle until it falls on the line between your city and Mecca: that will be the azimuth of the qibla. Look at the shadow of the piece of wood and extend a thread on that which you marked (?): it will be the position of the right-hand or left-hand wall of the mosque. Construct the mosque parallel and perpendicular to this line (? thread) so that its qibla will be correct. This is a picture:"

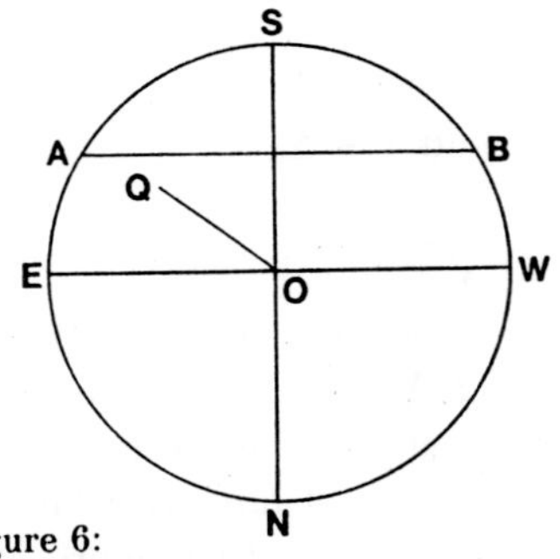

Figure 6:
A simple diagram
in the Tashkent manuscript.

A: "point of entry of the sun"
B: "point of exit of the sun"
NS: "meridian"
OQ: "stick"

[26] See *Pingree* 4, pp. 115–116, and *Kennedy* 9.

The diagram is shown as Fig. 6. It adds little to the exposition.

4. An Approximate Computational Method

This method occurs only in the Istanbul manuscript (fol. 187v):

باب أيضاً فى عمل القبلة مختصر تأخذ فضل ما بين الطولين فتضربه فى جيب تمام عرض مكة فما بلغ فاقسمه على جملة الجيب واحفظه ثم خذ جيب فضل ما بين العرضين فتضربه فى مثله وتضرب الذى حفظت من المقسوم على جملة الجيب فى مثله واجمعهما وخذ جذر ما اجتمع فاحفظه ثم خذ الذى قسمت على جملة الجيب فاضربه فى جملة الجيب واقسمه على الجذر الذى حفظت فما خرج فقوسه فما كان فهو سمت القبلة

"Another method for finding the qibla (but) abbreviated. You take the difference between the two longitudes and multiply it by the Sine of the complement of the latitude of Mecca. Divide the product by the total Sine and keep it in mind. Then take the difference (text: Sine of the difference!) between the two latitudes and multiply it by the same and multiply the quantity divided by the total Sine that you kept in mind by the same. Add the two (products) and take the root of the sum and keep it in mind. Then take (the quantity) that you divided by the total Sine and multiply it by the total Sine and divide (the product) by the root that you kept in mind. Take the arc corresponding to the result and it will be the azimuth of the qibla."

Commentary:

In this method we form the two quantities

$$x = \Delta L \text{ Cos } \varphi_M \ / \ R \quad \text{and} \quad y = \Delta\varphi,$$

with which q is found using:

$$q = \text{arc Sin } [(x\ R)\ /\ \sqrt{x^2 + y^2}]$$

Clearly this corresponds to a Marinos-type projection for the parallel of Mecca – see Fig. 2b in Section 2. Without the emendation of the text, the qibla value would depend on the value used for R. Further confirmation for the emendation, if any were needed, is that this method was actually used by the astronomers of al-Ma'mūn to find the qibla at Baghdad.[27]

[27] See *King* 6.

5. The Standard Approximate Method

We now turn to the method proposed by al-Battānī.[28] This is also approximate, although he makes no reference to the fact. It is difficult to imagine that al-Battānī is the originator of the procedure, not least because a more "sophisticated" version was proposed before his time (see Section 6). However, as we shall see, the leading astronomers of the tenth and eleventh centuries blamed him alone for the method. The following translation is based on the text as published by C. A. Nallino:

"If you want to know the azimuth of Mecca, which is the azimuth of the qibla for prayer, by this method (*bāb*), you extend on (the sundial) a line from the centre of the circle and that line will be the azimuth of the qibla in that locality. Find the latitude of your locality and the latitude of Mecca and determine the general direction (*jiha*) of Mecca, (the City) protected (by God), from that locality, either towards the north or south. Find the longitude of Mecca and the longitude of the city and subtract the lesser from the greater to determine the longitude difference between them and also the general situation of Mecca from that city, either towards the east or the west. If the longitude of Mecca is greater than the longitude of the city recorded in the tables of latitudes and longitudes (in the *Zīj*), then Mecca is east of the city; if it is less, then Mecca is west of the city.

Next place one end of a ruler on the latitude difference measured from the east line in the direction of Mecca with respect to latitude (that is, north or south) and do likewise from the west line in that direction on the circumference of the circle until the side of the ruler falls on the same latitude difference. Then draw a line with the side of the ruler to connect the eastern mark and the western one. Then take the longitude difference and count it on the circumference of the circle from the meridian in the direction of Mecca with respect to longitude in the south on the circumference, and do likewise in the north. Place the side of the ruler on the two points and use it to draw a straight line. Where these two lines intersect represents the position of Mecca with regard to azimuth from that locality. Place the edge of the ruler

[28] See note 9 above.

on the centre of the circle and on the point of intersection and draw a straight line extending it on the sundial to the circumference of the circle in the south [for localities north of Mecca]. This line is the azimuth of the qibla in that locality.

If you want to calculate the value of the azimuth of the qibla, take the Sine (*watar*) of the longitude difference between the two localities and the Sine of their latitude difference. Multiply each of these by itself and add (the squares), then take the square root of the sum. The result is the hypotenuse of the triangle which subtends the right angle, and this is the distance between the centre of the circle and the point of intersection of the longitude and latitude lines on the circumference of the circle, so keep it in mind. Then go back to the Sine of the latitude difference and multiply it by the radius (of the base circle, *i.e.* 60), and divide (the product) by the hypotenuse of the triangle. Take the arc Sine of the result and the arc will be the azimuth of Mecca. Measure it on the circumference from the east point or the west point according to the direction of Mecca for that locality with regard to longitude and latitude. Make a mark on the circumference at that point and draw a straight line from the centre of the circle to that mark: this line represents the azimuth of Mecca from that locality."

Commentary:

The geometric construction is shown in Fig. 7. Draw a circle about a point X in a horizontal plane and mark the cardinal directions N W S E. Make arc EB = $\Delta\varphi$ and draw AB parallel to WE. Make arc SC = ΔL and draw CD parallel to SN. Denote by M the point of intersection of AB and CD. Then XM marks the direction of Mecca. al-Battānī measures the qibla from the east-west line: I denote this by $q' = 90° - q$. Since the distances of M from NS and EW are respectively:

$$x = \text{Sin}\,\Delta L \text{ and } y = \text{Sin}\,\Delta\varphi,$$

this construction may be, and, in medieval treatises, frequently was, represented by the formula

$$q' = \text{arc Sin}\,[R\,\text{Sin}\,\Delta\varphi\;/\;\sqrt{\text{Sin}^2\,\Delta L + \text{Sin}^2\,\Delta\varphi}]$$

which in modern terms is more simply expressed as

$$q' = \text{arc tan}\,(\sin\Delta\varphi\;/\;\sin\Delta L).$$

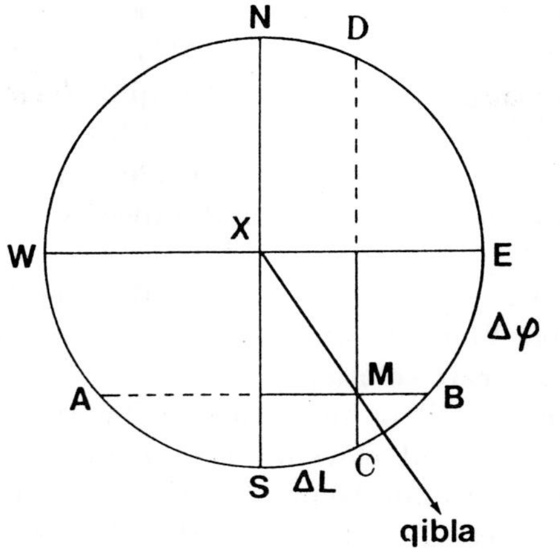

Figure 7:
al-Battānī's construction for finding the qibla,
popular for almost a millenium.

The approximation can be justified in various ways, none satisfactory - see Section 2 and Fig. 2c. It would have been interesting to hear al-Battānī justify it!

In his monumental work on mathematical geography, the *Kitāb Taḥdīd nihāyāt al-amākin*, the celebrated eleventh-century scholar al-Bīrūnī[29] criticized a procedure used by the Indians, and added:

"The authors of this work have made the same assumptions ... which Marinos made in his map of the earth and which al-Battānī made in his determination of the qibla. They have treated the meridian circles as parallel straight lines, and the parallels of latitude as parallel straight lines. Thus they have fallen into this outrageous error."

In his treatise on cartographic projections, the *Kitāb Tasṭīḥ al-ṣuwar* ..., al-Bīrūnī again criticized al-Battānī.[30] This time he adds that some of his predecessors who had written on the qibla had also found fault with al-Battānī's qibla method; he mentions

[29] On al-Bīrūnī, see the article by E. S. Kennedy in *DSB*, and *Sezgin*, V, pp. 375-383, VI, pp. 261-276, and VII, pp. 188-192. For this remark (text, translation and commentary), see *al-Bīrūnī* 1, pp. 233; *Ali*, pp. 198-199; and *Kennedy* 3, p. 151.

[30] See *Berggren* 4, pp. 50-51 (trans.), 69 (comm.), and 91 (text).

specifically al-Sijzī,[31] Abū Naṣr[32] and al-Khujandī.[33] Unfortunately none of these works are extant. Also unfortunate is the fact that some of al-Bīrūnī's writings on the qibla have not survived.[34] In view of his interest in the subject and his familiarity with earlier astronomical works, it is conceivable that he actually commented on the other methods and tables we describe in this paper.

al-Battānī was also criticized by the tenth-century Cairo astronomer Ibn Yūnus,[35] who, like various of his predecessors, presented an accurate solution.

Notwithstanding all of this criticism by the leading astronomers of the tenth and eleventh centuries, al-Battānī's method was widely used in later Islamic astronomy and is advocated in numerous treatises. For example, the early-thirteenth-century Central Asian scholar al-Jaghmīnī in his extremely popular *Mulakhkhaṣ fī 'l-hay'a*, a simple non-technical introduction to astronomy, advocates this method.[36] So does the late-thirteenth-century Yemeni astronomer Muḥammad ibn Abī Bakr al-Fārisī in his *Maʿārij al-fikr*, another popular general work.[37]

As Nallino pointed out, the formula works quite well for

[31] On al-Sijzī, see the article in *DSB* by Y. Dold-Samplonius, and *Sezgin*, V, pp. 329–334, VI, pp. 224–226 (esp. p. 226, no. 7), and VII, pp. 177–182.

[32] On Abū Naṣr, see *Samsó* and the same author's article "Manṣūr . . ." in *DSB*; and *Sezgin*, V, pp. 338–341, and VI, pp. 242–245 (esp. p. 245, no. 4, where it is stated that in the Tehran manuscript of the *Tasṭīḥ al-ṣuwar*, Abū Naṣr's treatise is compared with one bearing the same title by Ḥabash al-Ḥāsib).

[33] On al-Khujandī, see the article in *DSB* by S. Tekeli, and *Sezgin*, V, pp. 307–308, and VI, pp. 220–222 (esp. p. 222, no. 3).

[34] See, for example, *Sezgin*, VI, pp. 274–275, nos. 22–27, for six non-extant works by al-Bīrūnī on the qibla. One of these (no. 25) deals specifically with the qibla at Bust and was thus of the same genre as his *Taḥdīd nihāyāt al-amākin* for Ghazna, if much shorter. Another (no. 27) is a critique of a work entitled *Dalā'il al-qibla* by Ibn al-Qāṣṣ of Tabaristan (on whom, see *Sezgin*, I, pp. 496–497, esp. no. 2, and *Cairo Survey*, no. B27), in which astronomical alignments were advocated for the qibla.

[35] On Ibn Yūnus, see my article in *DSB*, and *Sezgin*, V, pp. 342–343, VI, pp. 228–231, VII, p. 173, and *Cairo Survey*, no. B59. On this remark of his, see already *King* 1, Part III, Section 28.2.

[36] On al-Jaghmīnī, see *Suter* 2, no. 403, and *Cairo Survey*, no. G17. His work is translated into German in *Rudloff & Hochheim*: see especially pp. 271–272.

[37] On al-Fārisī (*Suter* 2, no. 349), see *MAY*, no. 6 (pp. 23–26,) and *Cairo Survey*, no. E7.

Syria.[38] It was popular in Andalusia, but the qibla which can be found using this formula for, say, Cordova is as much as 10° off the true qibla because the longitude difference is so large.[39] According to S. H. Nasr, this method is still in use in Iran, although it is no longer associated with al-Battānī but rather with the seventeenth-century authority Bahā' al-Dīn al-'Āmilī,[40] author of a number of popular summaries of astronomy and mathematics which had very wide circulation in the eighteenth and nineteenth centuries.

6. A Fourth Approximate Method

Yet another computational method is described in both the Tashkent and Istanbul manuscripts. The latter source (fol. 187r) reads:

عمل القبلة أيضاً فى كل عرض تأخذ ما بين عرض مكة وعرض البلد الذى تريد فاجعله جيباً ثم اضربه فى مثله ثم خذ فضل ما بين طول مكة وطول مدينتك فاجعله جيباً واضربه فى جيب تمام عرض مكة واقسمه على قن أبداً فما خرج فهو الجيب المعدل فاضربه فى مثله وزده على المضروب فى مثله الذى حفظته فما بلغ فخذ جذر فاحفظه ثم اضرب الجيب المعدل فى قن واقسمه على الجذر فما بلغ فقوسه واحفظه ثم أدر فى المسجد الذى تريد دائرة هندية ثم اعرف خط الزوال فإن كانت مدينتك غربية عن مكة فعد مثل تلك القوس عن خط الزوال الجنوبى إلى ناحية المشرق فحيث انتهيت فهو موضع القبلة من موضع العود الذى فى وسط الدائرة وإن كانت مدينتك شرقية عن مكة فعد من ناحية المغرب وافعل كما فعلت أولاً

"To find the qibla also for all latitudes, take the difference between the latitude of Mecca and the latitude of the locality (for) which you wish (to determine the qibla), make it a Sine, and then multiply it by the same. Next take the difference between the two longitudes, (that of) Mecca and that of your city, make

[38] *Nallino* 1, I, p. 318.

[39] See further *King* 5, XV, p. 374.

[40] *Nasr*, p. 93. On al-'Āmilī, see *Suter* 2, no. 480, *Cairo Survey*, no. G68, and *King* 4, XVII, p. 214.

Table 2

Recomputed values for a 20x20 qibla table
based on the standard approximate formula

ΔL \ $\Delta\varphi$	1°	2	3	4	5	6	7	8	9	10
1°	45; 0	26;34	18;27	14; 3	11;19	9;29	8; 9	7; 9	6;22	5;44
2	63;26	45; 0	33;42	26;35	21;49	18;28	15;59	14; 5	12;35	11;22
3	71;33	56;18	45; 0	36;53	30;59	26;36	23;14	20;37	18;30	16;46
4	75;57	63;25	53; 7	45; 0	38;40	33;43	29;47	26;37	24; 2	21;53
5	78;41	68;11	59; 1	51;20	45; 0	39;49	35;34	32; 3	29; 7	26;39
6	80;31	71;32	63;24	56;17	50;11	45; 0	40;37	36;55	33;45	31; 3
7	81;51	74; 1	66;46	60;13	54;26	49;23	45; 0	41;12	37;55	35; 4
8	82;51	75;55	69;23	63;23	57;57	53; 5	48;48	45; 0	41;39	38;43
9	83;38	77;25	71;30	65;58	60;53	56;15	52; 5	48;21	45; 0	42; 1
10	84;16	78;38	73;14	68; 7	63;21	58;57	54;56	51;17	47;59	45; 0
11	84;46	79;38	74;40	69;55	65;27	61;17	57;26	53;54	50;39	47;42
12	85;12	80;28	75;52	71;27	67;15	63;19	59;37	56;12	53; 3	50; 8
13	85;34	81;11	76;54	72;46	68;49	65; 5	61;33	58;15	55;11	52;20
14	85;52	81;47	77;48	73;55	70;11	66;38	63;16	60; 5	57; 7	54;20
15	86; 9	82;19	78;34	74;55	71;23	68; 0	64;47	61;44	58;51	56; 8
16	86;23	82;47	79;15	75;48	72;27	69;14	66; 9	63;13	60;25	57;47
17	86;35	83;12	79;51	76;35	73;24	70;20	67;22	64;33	61;51	59;18
18	86;46	83;33	80;23	77;17	74;15	71;19	68;29	65;45	63; 9	60;40
19	86;56	83;53	80;52	77;54	75; 1	72;12	69;29	66;51	64;20	61;56
20	87; 5	84;10	81;18	78;28	75;42	73; 0	70;23	67;51	65;25	63; 5

it a Sine, and multiply by the Sine of the complement of the latitude of Mecca and always divide (the product) by 150. The result will be the 'modified Sine' (*al-jayb al-muʿaddal*): multiply it by the same and add (the product) to the (first) product that you kept in mind. Take the root of the result and keep it in mind. Next multiply the modified Sine by 150 and divide it by the root. Then find the arc corresponding to the result and keep it in mind.

Then draw in the mosque (for) which you want (to find the qibla) an Indian circle and determine the meridian. If your city is west of Mecca measure the arc (that you calculated) from the southern meridian towards the east, and where you end up is the position of the qibla from the stick that is in the middle of the circle. If your city is east of Mecca, measure from the western side and proceed as you did before."

Table 2
(continued)

ΔL \ Δφ	11	12	13	14	15	16	17	18	19	20
1°	5;14	4;48	4;26	4; 8	3;51	3;37	3;25	3;14	3; 4	2;55
2	10;22	9;32	8;49	8;13	7;41	7;13	6;48	6;27	6; 7	5;50
3	15;20	14; 8	13; 6	12;12	11;26	10;45	10; 9	9;37	9; 8	8;42
4	20; 5	18;33	17;14	16; 5	15; 5	14;12	13;25	12;43	12; 6	11;32
5	24;33	22;45	21;11	19;49	18;37	17;33	16;36	15;45	14;59	14;18
6	28;43	26;41	24;55	23;22	22; 0	20;46	19;40	18;41	17;48	17; 0
7	32;34	30;23	28;27	26;44	25;13	23;51	22;38	21;31	20;31	19;37
8	36; 6	33;48	31;45	29;55	28;16	26;47	25;27	24;15	23; 9	22; 9
9	39;21	36;57	34;49	32;53	31; 9	29;35	28; 9	26;51	25;40	24;35
10	42;18	39;52	37;40	35;40	33;52	32;13	30;42	29;20	28; 4	26;55
11	45; 0	42;33	40;18	38;16	36;24	34;42	33; 8	31;42	30;22	29; 9
12	47;27	45; 0	42;45	40;41	38;47	37; 2	35;25	33;56	32;34	31;18
13	49;42	47;15	45; 0	42;55	41; 0	39;13	37;34	36; 3	34;39	33;20
14	51;44	49;19	47; 5	45; 0	43; 4	41;16	39;36	38; 3	36;37	35;16
15	53;36	51;13	49; 0	46;56	45; 0	43;12	41;31	39;57	38;29	37; 7
16	55;18	52;58	50;47	48;44	46;48	45; 0	43;19	41;44	40;15	38;52
17	56;52	54;35	52;26	50;24	48;29	46;41	45; 0	43;25	41;55	40;32
18	58;18	56; 4	53;57	51;57	50; 3	48;16	46;35	45; 0	43;30	42; 6
19	59;38	57;26	55;21	53;23	51;31	49;45	48; 5	46;30	45; 0	43;35
20	60;51	58;42	56;40	54;44	52;53	51; 8	49;28	47;54	46;25	45; 0

Commentary:

The method involves forming the two quantities:

$$x = (\mathrm{Sin}_R\ \Delta L\ \mathrm{Cos}_R\ \varphi_M)\ /\ R \quad \text{and} \quad y = \mathrm{Sin}\ \Delta\varphi,$$

where $R = 150$, and then applying the formula:

$$q = \mathrm{arc\ Sin}\ [(x \cdot R)\ /\ \sqrt{x^2 + y^2}].$$

This seems to be a modification of al-Battānī's method, the factor cos φ_M being introduced to preserve distances along the parallel of Mecca – see Fig. 2d in Section 2. I have not encountered it in any other sources. It could not enjoy the popularity of al-Battānī's method because there was no corresponding trivial geometrical construction for the qibla. Note that the text advocates the use of the "Indian circle" to find the local meridian (see Section 2). This concludes our discussion of the approximate procedures for finding the qibla outlined in our sources.

7. Tables Based on the Standard Approximate Method

In several Egyptian and Syrian astronomical manuscripts, all copied after the fifteenth century, we find isolated qibla tables displaying the function $q\ (\Delta\varphi, \Delta L)$ for various domains of $\Delta\varphi$ and ΔL, with entries computed to degrees and minutes, and based on the formula underlying the method discussed in Section 5.[41]

I make no claim that any of these tables goes back to the Abbasid period, although any Abbasid astronomer worth his salt could have prepared such a simple table. The fact that in one manuscript the table is attributed to al-Khwārizmī[42] is surely not insignificant. However, this copy was made only about a hundred years ago.

The tables are distinguished from other qibla tables by the fact that the entries for $\Delta\varphi = \Delta L$ are all 45;0°. The entries in the various copies of the tables are similar, but the variants are many; indeed they are too many, and the manuscripts are too dispersed to make a "critical edition" of the table feasible. The vast majority of the entries are either correct or in error by $\pm 1'$ or $\pm 2'$. Isolated larger errors occur, and it is these which would enable an eventual classification of the various copies. The accurately computed values are presented in Table 2.

We can distinguish between three different categories of tables of this type. In the first category the domains of the arguments are:

$$\Delta\varphi = 1°, 2°, \ldots, 20° \text{ and } \Delta L = 1°, 2°, \ldots, 20°.$$

Copies of this table that have come to light are six in number:

(a) MS Cairo Muṣṭafā Fāḍil *mīqāt* 167,13, fols. 205v–206r, copied 989 H (= 1581/82), Egyptian provenance.[43a] The table occurs amidst various fifteenth-century Egyptian treatises on timekeeping.

(b) MS Cairo Muṣṭafā Fāḍil *mīqāt* 204,9, fols. 83v–84r, copied 1025 H (= 1642/43), Egyptian provenance.[43b] The table occurs

[41] Five of the eight manuscripts investigated in this section are listed in *King* 4, XIII, p. 122.

[42] *King* 2, pp. 12 and 16 (Fig. 16).

[43a–e] On these Cairo manuscripts, see *Cairo Catalogue*, I, *sub* DM 707, MM 167, MM 204, DM 1030, TR 103, and II, Section 3.3.6, and *Cairo Survey*, Section Z25.

amidst various Abbasid astrological treatises and Mamluk treatises and tables and notes relating to timekeeping.

(c) MS Damascus Ẓāhirīya 7564, fol. 1v, late copy, Syrian provenance.[44a] The table occurs at the beginning of various tables for timekeeping computed for the latitude of Aleppo.

(d) MS Cairo Dār al-Kutub *mīqāt* 1030,2, fols. 3r–4r, copied *ca.* 1200 H (≈ 1785), Egyptian provenance.[43c] The table occurs after a calendrical circle and rules for finding the solar longitude given the date in the Syrian calendar, and is attributed (see below).

(e) MS Cairo Taymūr *riyāḍa* 103,2, pp. 100–104, copied *ca.* 1300 H (≈ 1880), Egyptian provenance.[43d] The table is copied at the end of a copy of the *Zīj* entitled *al-Lum'a* by the fifteenth-century Egyptian astronomer al-Kawm al-Rīshī, and is attributed to al-Khwārizmī (see below).

In the fourth source the table is attributed to Yūsuf al-Damīrī, an individual new to the modern literature (although see the remarks on the Gotha manuscript below).[45] In the fifth source the table is attributed to Abū Ja'far Muḥammad ibn Mūsā al-Khwārizmī, the correct name of the Abbasid astronomer which is often given incorrectly in late Arabic manuscripts. As stated above, this attribution is not without its problems.

The second category of qibla table of this kind is characterized by the fact that the argument domains go up to 40. I have not compared the entries in the two copies which have come to my attention. They are:

(f) MS Cairo Dār al-Kutub *mīqāt* 707, 3 fols., copied *ca.* 1100 H (≈ 1675), Egyptian provenance, incomplete.[43e] Here the argument domains are:

$$\Delta\varphi = 1°, 2°, \ldots, 20° \text{ and } \Delta L = 1°, 2°, \ldots\ldots, 20°, \text{ and}$$
$$\Delta\varphi = 21°, 22°, \ldots, 40° \text{ and } \Delta L = 21°, 22°, \ldots, 40°.$$

(g) MS Damascus Ẓāhirīya 4804, fols. 30r–31r, late copy, Syrian provenance.[44b] Here the first page of tables is missing and entries are given only for:

$$\Delta\varphi = 11°, 12°, \ldots, 20° \text{ and } \Delta L = 1°, 2°, \ldots, 20°, \text{ and}$$
$$\Delta\varphi = 21°, 22°, \ldots, 40° \text{ and } \Delta L = 21°, 22°, \ldots, 40°.$$

[44a–b] On these two Damascus manuscripts, see *Khoury*, pp. 54–55 and 54, where these qibla tables are not mentioned.

[45] See *Cairo Survey*, no. C166.

The third category of this kind of qibla table is in only one source:

(h) MS Gotha Forschungsbibliothek A 1411, fols. 7v–8r, undated (?). Here the table occurs in a fragment of the main Cairo corpus of tables for timekeeping,[46] and this Gotha manuscript is the only known copy of the Cairo corpus which contains such a qibla table. The table is here attributed to Yūsuf ibn al-Damīrī, who must be identical to the Yūsuf al-Damīrī mentioned above. The function tabulated is q ($\Delta\varphi$, ΔL) for each 1° of both $\Delta\varphi$ and ΔL from 1° to 20°, but the table is in fact made up of two distinct parts, although there is no indication of this in the manuscript. The entries for $\Delta\varphi \leq 10°$ are taken from the anonymous Abbasid table described in Section 10 above and those for $\Delta\varphi > 10°$ are based on al-Battānī's formula. A marginal note states that for Cairo $\Delta\varphi = 9°$ and $\Delta L = 12°$, so that from the table $q = 53;0°$.[47] However, the entry for this pair of arguments has been "cooked" to satisfy Egyptians who were used to that value. See further Table 3.

8. The Earliest Exact Method

One method described only in the Istanbul manuscript is distinguished from the others in the treatise by the fact that it is exact. The text (fols. 187r–187v) reads as follows:

باب آخر فى عمل القبلة إذا أردت ذلك فخذ فضل ما بين الطولين فاضربه فى تمام عرض البلد أعنى مكة واقسمه على جملة الجيب فما خرج فاحفظه وهو الأول ثم اجعله قوساً وانقص القوس من ص̄ واجعل ما بقى جيباً وهو الثانى واقسم عليه جيب عرض مكة واضربه فى جملة الجيب وصيره قوساً ثم انقص هذه القوس من عرض بلدك واجعله جيباً مستوياً ومنكوساً ثم انقص المحفوظ الثانى من جملة الجيب ثم انقص منه الجيب المنكوس الأقل من الأكثر ثم اضرب ما بقى فى مثله واضرب المحفوظ الثالث فى مثله ثم اجمعه وخذ جدر ما اجتمع فما كان فنصفه واجعله قوساً فما كان فاضعفه معما اجتمع ثم

[46] On the Cairo corpus, see *King* 1, and also *King* 4, III, pp. 540–542.

[47] On this value of the qibla for Cairo, see *King* 1, Part III, Section 28, and 4, IX, p. 368.

Table 3
Sample entries from the qibla table in MS Gotha A 1411, fols. 7v–8r, attributed to Yūsuf ibn al-Damiri

ΔL \ $\Delta\varphi$	1°	2	3	...	8	9	10	11	...	18	19	20
1°	42;10°	24;19	17;15	...		5;56	5;21	5;13	...	3;17	3; 6	2;56
2	61;46	42;59	31;51			11;45	10;56	10;21		6;29	6; 8	5;50
3	70;19	55;26	43; 0			17;20	15;42	15;20		9;35	9; 6	8;40
.	.											
.	.											
.	.											
9	83;53	76;39	70;26			43;16	40;58	39;19		26;51	25;40	24;35
10	83;55	76;53	72; 2			46;20	43;21	42;18		29;29	28; 4	26;55
11	84;26	79; 8	73;49		52;23	49; 6	47; 4	45; 0		31;42	30;34	29;10
12					53;53	53; 0	48;40					
13					56;12	53;52	50;59					
.	.											
.	.											
.	.											
18	86;36	84;21	80; 7			62;30	59;55	58;18		45; 0	43;28	42;11
19	86;49	83;52	80;43			63;45	61;22	59;26		46;32	45; 0	43;35
20	87; 7	83; 3	81; 8			65; 2	62;40	60;50		47;19	46;25	45; 0

اجعله جيباً مستوياً واقسم عليه المحفوظ الأول فما كان فاضربه فى جملة الجيب فما بلغ
فاجعله قوساً وهى قوس بعد القبلة من خط نصف النهار وجهة القبلة على جهة العمل
الأول

"Another method for finding the qibla. If you wish (to do) this, take [the Sine of] the difference between the two longitudes and multiply it by [the Sine of] the complement of the latitude of the locality, I mean Mecca, and divide (the product) by the total Sine. Keep the result in mind: it is the first (quantity). Next make it an arc and subtract the arc from 90 and make the remainder a Sine: it is the second (quantity). (Now) divide the Sine of the latitude of Mecca by it, multiply (the result) by the total Sine, and make (the product) an arc. Subtract this arc from the latitude of your locality and make it a Sine and a Versed Sine. [Let the Sine be the third quantity.] Then subtract the second quantity from the total Sine and subtract from the Versed Sine, the smaller from the larger, and then multiply the remainder by itself. Multiply the third quantity by itself and add it (to this square). [To the sum add the first quantity multiplied by itself][a] and (then) take the root of the result (*i. e.* of the sum of the three squares). Take half the result and make it an arc., then double the result.[b] Then make it a Sine and divide the first quantity by it. Multiply the result by the total Sine and make the product an arc. It will be the distance of the qibla from the meridian, and the direction of the qibla (can be found in the same way) as the direction in the first method."

[a] this instruction is not in the original.
[b] text: *fa-'ḍ'afhu ma'ma 'jtama'a* (*sic.*). I have translated the text as *fa-'ḍ'af ma 'jtama'a*, although *fa-'ḍ'af mā ḥasala* would be more appropriate since the result was obtained by a square-root operation rather than an addition (*j–m–'*).

Commentary:

The text is obscure and slightly garbled. Some insight into the original meaning was kindly provided by Professor Len Berggren. We are to form three quantities, the first of which is:

$$x_1 = \Delta L \cdot (90° - \varphi_M) / R.$$

The value of R is not specified. In view of the ensuing instructions, this expression for the first quantity must be garbled, and we restore:

$$x_1 = \text{Sin}\ \Delta L \cdot \text{Sin}\ (90° - \varphi_M) / R$$

The second quantity is defined by:

$$x_2 = \text{Sin}\ [90° - \text{arc Sin}\ (x_1)],$$

and using:

$$\alpha = \text{arc Sin}\ [(\text{Sin}\ \varphi_M / x_2) \cdot R\ ],$$

the third quantity is:

$$x_3 = \text{Sin}\ (\varphi - \alpha).$$

We also form:

$$y = \text{Vers}\ \Delta\varphi'.$$

Next we are to form the quantity:

$$z = \sqrt{\ [(R - x_2) \sim y]^2 + x_3{}^2 + x_1{}^2},$$

with which we determine:

$$\beta = \text{arc Sin}\ (1/2\ z).$$

Finally the qibla is to be found using:

$$q = \text{arc Sin}\ (x_1 \cdot R / \text{Sin}\ 2\beta).$$

We explain this procedure as follows with reference to Fig. 4.[48] The two quantities x_1 and x_2 measure ML = d and OL = e, respectively. Then α measures arc SB = φ' ("the modified latitude" – see the next Section), so that $\varphi - \varphi'$, which I label $\Delta\varphi'$ ("the modified latitude difference"), measures arc XS. Thus x_3 measures XT = sin XS = g and y measures ST = h. We now investigate the expression for z. In ΔLTX, right-angled at T, we have:

$$LT = OS - OL - ST = (R - e) \sim h \quad \text{and} \quad XT = g.$$

Thus, by Pythagoras' Theorem:

$$XL^2 = LT^2 + XT^2 = [(R - e) - h]^2 + g^2.$$

Further, ΔXLM is right-angled at L, and so the chord XM (not shown in Fig. 4) is given by:

$$XM^2 = XL^2 + LM^2 = [(R - e) - h]^2 + g^2 + d^2.$$

Thus z measures XM. Then because 2 β measures the arc XM:

$$ZM = \text{Sin } 2\,\beta = i.$$

The qibla can then be found from ΔLZM, right-angled at L, in which also:

$$LM = d \quad \text{and} \quad ZM = i.$$

The compiler of this rather clumsy method deserves maximum credit for pursuing the solution to its conclusion. His procedure is to determine successively ML, OL, and arc XS. The use of the Versed Sine marks the method as belonging to the Indian tradition of determining segments inside the sphere, particularly for time-reckoning.[49] Then he finds XT and ST and XM, whence arc XM and MZ and finally q. We now turn to the "method of the *zījes*", which also involves determining ML, OL, and arc XS, but then finding q by simpler means.

9. On the So-Called "Method of the *Zījes*"

The complicated procedure described in Section 8 is of interest for another reason. Its basic approach to the qibla problem is related to that of the so-called "method of the *zījes*," popular amongst various major astronomers from the mid-ninth to the fif-

[48] See the brief discussion in *Berggren* 3, pp. 4–5.

[49] *Cf.*, for example, *Kennedy & Davidian* (al-Bīrūnī) and *Nadir* (Abu 'l-Wafā').

teenth century. The method is described, for example, by Ḥabash in the mid-ninth century; Ibn Yūnus, Abu 'l-Wafā', Kūshyār, al-Qūhī and al-Bīrūnī in the tenth and eleventh centuries; al-Khāzinī in the twelfth century; and al-Kāshī in the fifteenth century. Their descriptions of this method have been investigated recently for their mathematical content in a series of studies by E. S. Kennedy, M.-Th. Debarnot and J. L. Berggren.[50] Also, A. Dallal of Columbia University has recently investigated the treatise of Ibn al-Haytham on the "method of the *zījes*," in which the author considers 16 different mathematical cases of the problem.[51] Finally, we remark that the same method is one of two advocated by the thirteenth-century Cairo astronomer al-Marrākushī.[52]

In the "method of the *zījes*," where the various stages are achieved by the application of spherical trigonometry, the basic idea is the same: in order to find the qibla – now considered as a problem of spherical astronomy – first compute the angular distance of M from the meridian through X; next compute the azimuth of M on the meridian of X; then compute the zenith distance of M (measured by arc XM); and finally compute the azimuth of M relative to the horizon whose zenith is X (measured by $\angle$ BXM = q).

With this method we first calculate arc MS = $\Delta L'$, the "modified longitude difference," using:

$$\sin \Delta L' = \cos \varphi_M \sin \Delta L = d$$

and then arc SB = φ', the "modified latitude," using:

$$\sin \varphi' = \sin \varphi_M \ / \ \cos \Delta L' = b \ / \ e.$$

(Note that both of these quantities are functions of ΔL.) Then determine arc XS = $\varphi - \varphi' = \Delta\varphi'$, the "modified latitude differ-

[50] See *Kennedy* 3, pp. 211–214; *Debarnot* 1, pp. 260–264; and *Berggren* 1 and 3.

[51] On Ibn al-Haytham, see note 13 above. This treatise is listed in *Sezgin*, V, p. 368, no. 15, and VI, p. 259, no. 18. See *King* 3, p. 317, Fig. 4, for a reference to it by the early-fourteenth-century Syrian astronomer Ibn al-Sarrāj (*Cairo Survey*, no. C26, and *King* 5, IX). Ibn al-Haytham's treatise on finding the qibla by geometric construction (listed in *Sezgin*, V, p. 368, no. 16, and VI, p. 259, no. 19) is analyzed in *Schoy* 2.

[52] On al-Marrākushī and his work, see *King* 4, III, pp. 539–540, and the article in *EI*² (forthcoming). His writings on the qibla are translated in *Sédillot*, pp. 319–323 and 352–359 (where they are expressed in what approaches algebraic notation).

ence," with which arc XM = D, the distance between the two localities (that is, the zenith distance of the zenith of Mecca), is found by:

$$\cos D = \cos \Delta L' \ \cos \Delta\varphi' = e \cdot j = k.$$

Finally, the qibla q (measured from the meridian) is determined from ΔMZL using:

$$\sin q = \sin \Delta L' \ / \ \sin D = d \ / \ i.$$

Clearly, up to the determination of $\Delta\varphi'$, the method described in Section 8 corresponds in principle to the "method of the *zījes*."

As Berggren has shown,[53] the "method of the *zījes*", characterized by the intermediate step of finding coordinates with respect to the meridian, is closely related to a method for coordinate conversion also used in Ptolemy's *Analemma* (which was not translated into Arabic) and in various early Islamic works on sundial theory, notably that of Thābit ibn Qurra, who worked in Baghdad at the end of the ninth century.[54] Nowhere is this connection clearer than in the *Zīj* of Ibn Yūnus, who used mathematically equivalent methods for deriving the hour-angle and azimuth from the solar altitude, the coordinates for marking vertical sundials, and the qibla.[55]

Berggren has argued that the "method of the *zījes*" was in all probability actually derived from sundial theory, and this I would concede. A lost work of Ḥabash on sundial theory may be the missing link.[56] The method of Section 8, on the other hand, I believe predates Ḥabash, and this was certainly independent of traditional sundial theory methods, using terminology quite different from either Ptolemy's *Analemma* or the later Islamic sundial treatises. The question remains whether Ḥabash might have been aware of this method.

For more information on the "method of the *zījes*" the reader is referred to the secondary literature mentioned above. For tables based on this method, see Sections 10 and 12.

[53] See *Berggren* 3.

[54] On Ptolemy's *Analemma*, see note 23 above. The treatise of Thābit is published and translated in *Garbers* with commentary in *Luckey* 2.

[55] On Ibn Yūnus, see note 35 above. On the mathematical methods, see *King* 1, Part III, Sections 15.5(b), 20.2(g), 26.1, and 28.1(c), and *Berggren* 3, pp. 8–10 and 13.

[56] *Berggren* 3, p. 13, *apud Sezgin*, V, p. 276.

10. A Table Based on the "Method of the *Zījes*"

(a) *The Sources*:

Nine copies of the Abbasid qibla table have come to my attention in the manuscript sources.[57] A very corrupt version of the table – so corrupt as to be barely recognizable as the same table – occurs in the various manuscripts of the encyclopaedia of Ḥamdallāh Mustawfī – see Section 12, nos. 4 and 5.

The first of the *bona fide* copies occurs on fol. 188r of the Aya Sofya manuscript, immediately following the methods discussed in Sections 6 and 4. The second occurs on fols. 150v–151v of the Tashkent manuscript, following the method discussed in Section 6.

The other seven copies are not accompanied by this text, and in no case is there any indication that the table was borrowed from an earlier work.

In MS Istanbul H. Hüsnü Ef. 1268,3, fols. 143v–146r, copied in 869 H (= 1464/65), the table occurs at the end of an anonymous Arabic treatise on the qibla with a worked example for Maragha (see further Section 11 below).

Two copies are of Ottoman Turkish provenance. In MS Istanbul Hamidiye 1453, copied in Edirne in 869 H (= 1464/65), the same qibla table occurs on fols. 233v–234r between a set of instructions for using al-Khalīlī's universal auxiliary tables and the tables themselves.[58] In MS Istanbul Esat Efendī 3769, fol. 62r, the qibla table occurs amidst various Turkish treatises on astronomy. In both of these Ottoman manuscripts the last column of entries is garbled.

Three other copies of the table are of Yemeni provenance. The first two are found in the two known copies of the *zīj* called *Taysīr al-maṭālib fī tasyīr al-kawākib* by the late-thirteenth-century astronomer Muḥammad ibn Abī Bakr al-Kawāshī.[59] This work contains some spherical astronomical tables computed for latitudes in the Yemen but otherwise seems to be based on earlier

[57] Five of these were listed in *King* 4, XIII, p. 121.

[58] On al-Khalīlī and his various tables, see my article in *DSB*, XV, Supplement 1, pp. 259–261, and the references there cited, to which add *Cairo Survey*, no. C37. On his auxiliary tables, see *King* 4, XI.

[59] On this work, see *MAY*, II, Section 7 (p. 27).

Egyptian and Iraqi sources. The two copies are MSS London British Library Or. 9116, which I estimate to have been copied in the fourteenth century, and Alexandria Municipal Library 5577C, which was copied in 1142 H (= 1730). The third Yemeni copy of the qibla table is contained in a manuscript preserved in the private collection of Sayyid Aḥmad ʿAbd al-Qādir al-Ahdal of Zabīd, Yemen, which is a unique copy of a Yemeni *zīj* entitled *Zād al-musāfir* by the sixteenth-century astronomer Muḥammad al-Daylamī.[60] This manuscript was copied in Ibb in 1091 H (= 1680), and by the side of the table it is stated that the table was taken from the *Taysīr al-maṭālib*. In each of these three Yemeni copies the columns for horizontal arguments 12°, 13°, and 14° are in the order 13°, 14°, 12°.

On the last copy of the table, see Section 7 (h).

(b) *The Table*:

The horizontal argument in the qibla table is labelled *al-ʿarḍ*, "latitude," and the vertical argument *al-ṭūl*, "longitude," and from the instructions it is clear that "latitude difference" and "longitude difference" would be more appropriate (though see the commentary below). The entries in the table are given in degrees and minutes and are written in the standard Arabic alphabetical notation. The values are displayed in Table 4, edited from four copies, hereafter referred to by the appropiate sigla:

AS: MS Istanbul Aya Sofya 4830, fol. 188r.
H: MS Istanbul Hamidiye 1453, fols. 233v–234r.
L: MS London British Library Or. 9116, fols. 132v–133r.
T: MS Tashkent Oriental Institute 177,3, fols. 150v–151v.

In Table 4, an asterisk indicates that the values in the sources are at variance, but the variants can easily be explained in terms of errors arising from the alphanumerical (*abjad*) notation. An exclamation mark indicates that there are variants which cannot be thus explained. Errors in the minutes are displayed in parentheses – see below on the underlying formula. Values reconstructed by me and not attested in any of the sources are shown in italics.

In general, AS and T constitute the most reliable sources for a

[60] See *MAY*, II, Section 28 (p. 42) on al-Daylamī's *Zīj*.

Table 4

The anonymous Abbasid qibla table

ΔL \ Δφ	1°	2	3	4	5	6	7	8	9	10
1°	42;58°!	24;59*	17;15	13; 7	10;34	8;50*	7;36	6;40	5;56	5;21*
2	61;46 −1	42;59	31;51 −1	25; 0	20;28	17;17*	*14;57**	13; 9!	11;45*	10;35! −2
3	70;19*	54;26*	43; 0	34;59*	29;15	25; 2*	21;49 −1	19;19 −1	17;20	15;42*
4	75; 2* +2	61;48 −1	51;12* −1	43; 2	36;46	31;55	28; 7*	25; 4	22;36	20;33
5	77;54*	66;48	57;17*	49;25	43; 4*	37;56*	33;45	30;20*	27;31 +1	25; 8
6	79;53	70;22	61;51	54;31	48;18	43; *7** +1	38;46 +1	35; *7** +1	32; 2 +1	29;24*
7	81;17 −2	73; 1	65;23	58;35* −1	52;40	47;32* −1	43; 9	39;23	36; 9	33;21*
8	82;22 −2	75; 4 +1	68;21 +10	61;55	56;18* −1	51;21	47; 1	43;12 −1	39;53*	36;59
9	83;13* −2	76;39* −1	70;26*	64;40* +1	59;42! +19	54;36 −2	50;23 −1	46;37	43;16* −1	40;18*
10	83;55	77;*58** −1	72;26!* +8	66;*56*! −1	62; 0*	57;12* −16	53;21 −1	49;40	*46*;20!	43;21
11	84;26* −3	79; 3 −1	73;49 −1	68;53!	64;14*	59;56*	55;59	52;23	49; 6	46; 4! −4
12	84;51 −6	79;51 −7	75;11* +2	70;*32**	66;10 −1	62; 6*	58;*19**	54;49!	51;36	48;40
13	85;18* −3	80;47 +2	76;16 −1	71;58	67;52!	64; 0	59;56! −26	57; 0	53;52	50;59*
14	85;38 −3	81;23 −2	77;15*	73;14! +1	69;20 −1	65;40 −1	62;13*	58;58	55;56	53; 6
15	85;59!	82; 1 +1	78; 6!	74;19	70;10! −30	67;12! +1	63;52!	60;43 −1	57;48!	55; 3*
16	86; 9! −5	82;29* −2	78;51*	75;*18*! +1	71;50!	68;31!	65;21!	62;21	59;30!	56;50!
17	86;27* −1	82;58*	79;30 −1	76; 9	72;50 −3	69;*43* −1	66;42!	63;49!	61; 4	58;28*
18	86;36! −5	83;21 −2	80; 7	76;56	73;49! −1	70;49*	67;56*	65; 9!	62;32* +2	59;59*
19	86;49 −3	83;42 −3	*81;43*! +3	77;37* −1	74;42! +1	71;49	69; 3!	66;22! −1	63;49!	61;24! +1
20	87; 2	84; 4	81; 8 −1	78;15 −1	75;28	72;40 −3	70; 4	67;30	65; 2!	62;40* −1

Table 4
(continued)

ΔL \ Δφ	11	12	13	14	15	16	17	18	19	20
1°	4;52*	4;28*	4; 8!	3;51	3;36	3;23*	3;11*	3; 1	2;52*	2;43
2	9;40	8;53*	8;14	7;39!	7;10	6;44	6;21	6; 0! −1	5;42	5;27 +1
3	14;21	13;13*	12;14 −1	11;24	10;41*	10; 2!	9;29	8;59	8;32!	8; 7!
4	18;51	17;23	16; 9*	15; 4*	14; 7*	13;17*	12;33	11;54	11;19	10;47
5	23; 7*	21;24	19;54*	18;36	17;28	16;28	15;34	14;46	14; 3	13;24!!!
6	27; 9	25;12	23;31* +1	22; 1	20;43! +1	19;33 +1	18;30!	17;34	16;45 +1	15;58*
7	30;55*	28;47	26;56*	25;17	23;49	22;31	21;21	20;17*	19;20	18;28!
8	34;25*	32;10*	30;10	28;23	26;48	25;22 −1	24; 6	22;56*	21;53!!	20;55*
9	37;40	35;19	33;13	31;20	29;38* −1	28; 7!	26;44	25;29	24;20 −1	23;18*
10	40;40	38;15	36; 5	34; 7*	32;21	30;44*	29;16* −1	27;57*	26;43	25;36
11	43;25 −1	40;59	38;44 −2	36;44 −1	34;54*	33;14*	31;42	30;18	29; 0 −1	27;50*
12	45;58	43;31	41;16*	39;13!	37;20	35;36 −1	34; 2!	32;34!	31;14	30; 0!
13	48;19	45;52*	43;36 −1	41;32	39;37*	37;52	36;14	34;44*	33;21	32; 4*
14	50;28! −1	48; 3*	45;48	43;43	41;47	40; 0!	38;21	36;49	35;23!	34; 4*
15	52;28	50; 4	47;50	45;45	43;49*	42; 2!	40;21!	38;48	37;20 −1	35;59*
16	54;18 −1	51;57	49;44 −1	47;40! −1	45;45	43;57	42;15* −1	40;41	39;13*	37;50!
17	56; 1	53;41 −1	51;31	49;28 −1	47;35! +1	45;45 −1	44; 4!	42;29	41; 0!	39;36* −1
18	57;35	55;19 −1	53;11!	51;10	49;16*	47;28 −1	45;47*	44;12*	42;43	41;19!
19	59; 3*	56;50 −1	54;44! −1	52;45 −1	50;53	49; 6	47;25! −1	45;50* −1	44;21	42;56*
20	60;24 −1	58;15 −1	56;12*	54;15	52;24	50;38* −1	48;59	47;24*	45;55*	44;30*

Critical Apparatus to Table 4:

(1,1) H: 52;10! (2,1) H: 24;29 (6,1) L,T: 8;7 (10,1) AS: 5;24 (11,1) L: 4;12 (12,1) H: 4;23 (13,1) H: 4;5 (16,1) H: 3;28 (17,1) H: 3;51 (19,1) T: 2;12

(6,2) AS,H: 16;14 L: 17;14 (7,2) AS,H,L: 15;17 T: 14;17 (8,2) AS,H,L: 13;19 T: 13;9! (9,2) T: 14;45 (10,2) AS, H: 10;36!! (12,2) T: 8;13 (14,2) H: 7;19 T: 7;?? (text has *ykh* for minutes) (18,2) H: 6;12 T: 6;9!!

(1,3) T: 70;59 (2,3) AS,H: 54;57 L: 56;26 T: 54;26 (4,3) AS,H: 34;19 L: 32;19 T: 34;59 (6,3) T: 24;12 (10,3) AS,H: 15;44 L: 15;12 T: 15;42 (12,3) L: 13;33 (13,3) T: 42;34 (15,3) T: 10;9 (16,3) T: 10;12 (19,3) AS,H,L: 8;30 T: 8;32! (20,3) AS,H, L: 8;4 T: 8;50!

(1,4) L: 75;7 (3,4) H: 51;15 T: 51;52 (7,4) AS,H,L: 28;50 T: 28;7 (13,4) T: 16;19 (14,4) T: 15:3 (15,4) T: 14;50 (16,4) L: 13;57

(1,5) AS,H: 77;14 L: 77;12 T: 77;54 (3,5) AS,H,L: 57;36 T: 57;17 (5,5) H: 48;4 T: 43;14 (6,5) L: 37;16 (8,5) T: 30;9 (11,5) L: 23;50 (13,5) L: 19;14 (17,5) L: 15;32 (20,5) H: 35;58!!!! L: 13;27

(6,6) AS,H,L,T: 43;50 (8,6) AS, H,L,T: 35;50 (10,6) L: 29;26 T: 21;24 (11,6) T: 27;20 (13,6) L: 24;31 (15,6) T: 22;43! (17,6) T: 18;34 (20,6) H: 18;25!!

(4,7) T: 58;59 (6,7) T: 47;22 (10,7) AS,H: 33;41 L: 23;41 T: 33;21 (11,7) T: 30;15 (13,7) L: 26;16 (18,7) T: 20;16 (20,7) AS,L: 18;25 H: 20;55 T: 18;28!!

(2,8) AS,H,L: 75;5 T: 75;4 (5,8) T: 57;18 (9,8) L: 39;13 (11,8) L: 34;27 (12,8) H: 37;10 (14,8) T: 28;3 (18,8) T: 22;16 (19,8) T: 21;20 (20,8) H: 23;18!! L: 20;15

(1,9) L: 83;53 (2,9) AS,H: 76;19 (3,9) T: 68;26 (4,9) AS,T: 65;39 L: 64;39 (5,9) H: 29;41! (6,9) T: 54;37 (9,9) H: 43;17 (10,9) L: 40;28 (15,9) AS,H: 29;28 L: 24;18 T: 29;38!! (16,9) AS,H,L: 28;30 T: 28;50! (20,9) H: 25;36!!

(2,10) AS,H,L: 77;53 T: 77;33 (3,10) T: 74;2! (4,10) AS,H: 66;16 L: 66;36 T: 68;58 (5,10) T:62;5 (6,10) T: 57;52 (9,10) AS,H,L: 47;20 T: 46;25 (14,10) AS,H: 34;6 L: 34;7 T: 34;50 (16,10) L: 32;44 (17,10) H: 29;15 (18,10) AS,H: 27;56 L: 27;16 T: 27;17 (20,10) H: 27;4 T: 24;36!!

(1,11) L: 83;27 (4,11) AS,H: 68;13 L: 68;23 T: 70;53! (5,11) T: 64;17 (6,11) AS,H,L: 59;16 T: 59;56 (10,11) H: 46;20 T: 46;50! (12,11) AS,H,L: 40;19 T: 40;59 (13,11) T: 38;45 (15,11) L: 34;44 (16,11) AS,H: 33;17 L: 33;54 T: 33;14 (20,11) H: 30;0!!

(3,12) H: 75;41 (4,12) AS,H,L,T: 70;22 (6,12) H: 62;5 L: 62;17 (7,12) AS,H,T: 58;59 (8,12) H: 54;50! (13,12) AS,H,L: 41;36 T: 41;16 (14,12) L: 38;53! (17,12) AS,H,L: 34;4 T: 34;2! (18,12) T: 34;32 (20,12) H: 32;4 T: 30;5!!

(1,13) T: 85;58 (2,13) T: 80;44 (5,13) H: 67;51! (7,13) H: 59;16 L: 59;53 T: 54;56!! (10,13) AS,H: 50;59 (12,13) H,L,T: 45;12 (15,13) H,L: 39;36 (18,13) AS,H: 34;45 (20,13) H: 34;4!!

(3,14) L,T: 77;14 (4,14) L: 73;54 T: 73;52! (7,14) AS,H: 62;53 (11,14) H: 50;8! (12,14) T: 48;8 (16,14) T: 47;0! (19,14) H: 35;21! (20,14) H: 35;59!!

(1,15) T: 85;58 (3,15) L: 78;17! (4,15) AS,H,L: 74;39 T: 74;19 (5,15) T: 70;59! (6,15) H: 67;11 T: 67;10! (7,15) AS,L: 63;44 H: 63;43 T: 63;52!! (8,15) T: 60;44 (9,15) AS,H,L: 57;3 T: 57;48 (10,15) T: 55;8 (15,15) H,L: 43;59 (16,15) AS,H,L: 42;7 T: 42;2! (17,15) H: 40;20 (20,15) H: 47;50 T: 35;19!!

(1,16) T: 86;20! (2,16) T: 82;59 (3,16) T: 78;11 (4,16) AS,H,L: 75;50 T: 75;58!! (5,16) AS,H: 71;31 L: 71;0 T: 71;50! (6,16) AS,H,L: 68;21 T: 68;31! (7,16) H: 65;20! (8,16) AS,H,L: 62;30 T: 64;21! (9,16) AS,H,L: 59;7 T: 59;30! (10,16) AS,H,L: 57;7 T: 56;50 (14,16) T: 47;17 (17,16) H: 40;20! (19,16) AS,H,L: 39;33 T: 39;13 (20,16) AS: 47;50!? H: 39;36

(1,17) T: 86;24 (2,17) T: 82;18 (5,17) AS,H,L: 72;7 T: 72;50 (6,17) AS,H,L: 69;21 T: 69;48! (7,17) AS,L: 66;49 H: 66;50 T: 66;42! (8,17) AS,L: 63;42 H: 63;41 T: 63;49! (10,17) H: 58;23 (15,17) T: 47;1!! (17,17) T: 44;44! (19,17) T: 20;5!!!! (20,17) H: 41;29

(1,18) L: 86;37 T: 86;34! (5,18) AS,H,L: 73;29 T: 73;49! (6,18) AS,H: 70;19 (7,18) AS,H,L: 67;16 T: 67;56 (8,18) AS: 65;39 H: 65;40 L: 65;41 T: 65;9!!!! recomputed value: 65;9 (9,18) L: 62;12 (10,18) L: 59;19 (12,18) AS,H,L: 55;49 T: 55;19! (13,18) AS: 53;51 H: 53;31! (15,18) L: 49;36 (17,18) L: 46;17 (18,18) L: 44;52 (20,18) AS,L: 41;29 H: 42;16 T: 41;19!!

(3,19) AS,H,L: 81;8 T: 80;43!! (4,19) AS: 77;34 (5,19) H: 74;41 (7,19) H: 69;1! (8,19) H: 66;21! (9,19) H: 63;50 (10,19) AS: 61;22 H: 61;21!! (11,19) T: 59;8 (13,19) T: 54;42! (17,19) L: 48;25 (18,19) T: 45;7 (20,19) AS,L: 42;16 H: 44;30 T: 42;56!!

(3,20) AS,H,L: 81;58 T: 81;8 (9,20) H: 65;1! (10,20) H: 64;40 (13,20) L: 56;52 (14,20) T: 54;14 (16,20) L: 50;33 (18,20) AS,H,L: 47;25 T: 47;24 (19,20) H: 42;55 (20,20) H repeats 44;30.

reconstruction of the table. Several errors common to AS, H and L, especially in rows 15–18, were caused by a copyist who muddled the values for the minutes. There are additional copyist's mistakes in L, and in H (as in all the Ottoman copies) the last column of entries is garbled as a result of combining the degrees of one entry with the minutes of the entry below. As noted above, the Yemeni copies, here represented by L, have three columns in the wrong order. In T columns have been drawn for a 30x20 table, and the last ten have no entries.

(c) *The Problem of the Table*:

I confess to have been puzzled for some time about this table. Fortunately, however, working in the history of Islamic science one has colleagues to whom one can turn for inspiration. In this case, Dr. Richard Lorch pointed out to me the fact that the entries in the Abbasid table correspond very closely to recomputation with yet another approximate formula using $\varphi_M{}' = 21;20°$. The formula derived by Lorch is the following:

$$\tan q = \frac{\sin \Delta L \cos \varphi_M}{\sin \Delta\varphi \sqrt{1 - (\sin \Delta L \cos \varphi_M)^2}} = d \,/\, f \sqrt{1 - d^2} = d \,/\, (fe).$$

I have not come across this formula in any medieval text, but it is clearly the one underlying the Abbasid table. Lorch has shown the way in which this formula can be derived by spherical trigonometry.[61] The following derivation by solid geometry is inspired by the procedure outlined in the accurate method discussed in Section 6.

In Fig. 4, we have in ΔZLO (see Section 4):

$$\text{OL} = e \quad \text{and} \quad \angle \text{Z} = 90°,$$

and we introduce the approximation:

[61] To derive the required approximation, Lorch (1, p. 264) reasonably suggests that the parallel of latitude MZ be considered as a great circle yet perpendicular to both meridians through X and M – see Fig. 8. Then by the Sine Theorem applied to ΔPZM, we have:

$$\sin \text{ZM} = \cos \varphi_M \sin \Delta L,$$

and by the Tangent Theorem applied to ΔMXZ, we have:

$$\tan q = \tan \text{ZM} \,/\, \sin \Delta\varphi,$$

from which Lorch's formula follows immediately.

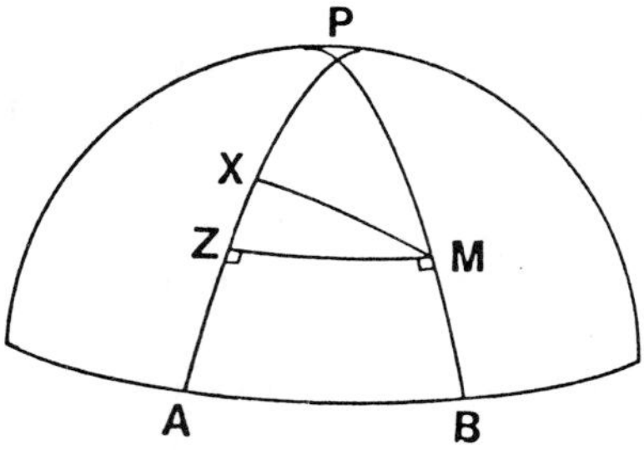

Figure 8:
Figure to derive Lorch's formula (see note 61).

$$\angle O \approx \Delta\varphi.$$

Then:

$$ZL \approx e \sin \Delta\varphi = e \cdot f.$$

Now in ΔMZL we have:

$$ML = d, \quad ZL \approx e \cdot f, \quad \text{and} \quad \angle Z = q,$$

whence:

$$\tan q \approx d \,/\, (ef), \quad \text{as required.}$$

I wrote the above commentary in 1984 but could not understand why the values in the table were so carefully computed using such a curious approximate formula. In March 1986, whilst trying to recompute the table in the *Zīj* of al-Wābkanawī – see Section 12, no. 7 – I quite by chance generated the Abbasid table with my computer. There emerged the following conclusion (which I thought surprising at the time): the Abbasid table is based on the *accurate* formula corresponding to the "method of the *zīj*es" if one considers the horizontal argument to be $\Delta\varphi' = \varphi - \varphi'$ rather than $\Delta\varphi$. (This revelation is less surprising when one recalls that the "approximate" formula can be justified by assuming $\Delta\varphi \approx \Delta\varphi'$.)

The most probable explanation is that the compiler of the table, who was clearly well aware of the "method of the *zīj*es", realized that for the range of arguments of the table $\Delta\varphi$ is quite close to $\Delta\varphi'$, so he simply advocated the use of $\Delta\varphi$ as the horizontal argument. A less likely explanation is that the original instructions for the table were not understood and a new set of instructions, stipulating $\Delta\varphi$ for the horizontal argument, was prepared. In this case, alongside the original instructions there should have been an auxiliary table of $\varphi'(\Delta L)$, as we find together with the table preserved in the *Zīj* of al-Wābkanawī.

It is appropriate to consider the nature of the trigonometric tables which the compiler might have had at his disposal. It is possible that he might have had a table of sines to base $R = 150$ with values for each degree of argument to three sexagesimal digits and a table of cotangents to base $R = 12$ with values for each degree of argument to two sexagesimal digits. Such were the tables of al-Khwārizmī,[62] and such are the anonymous trigonometric tables on the folio (188v) after the qibla table in the Aya Sofya manuscript. But already in the mid-ninth century, more sophisticated tables were available, such as those of Ḥabash, with sines for each 0;15° stretched to four sexagesimal digits and tangents for each 0;30° to three digits.[63] I have not attempted to use such tables to generate qibla values according to any of the medieval formulae, but this could be a worthwhile undertaking.

(d) *The Coordinates for Baghdad*:

According to the medieval geographical coordinates data-processed by E. S. Kennedy and his former colleagues in Beirut,[64] the only early source which gives $\varphi_M = 21;20°$ is the *Zīj* of al-Bīrūnī.[65] For Mecca and Baghdad his coordinates are:

$$\varphi_M = 21;20° \quad L_M = 67;\ 0°$$
$$\varphi_B = 33;25° \quad L_B = 70;\ 0°$$

On fols. 145v–147v of the Tashkent manuscript there is also a set of geographical tables, stated to be by al-Bīrūnī, and actually an abridgement of the tables in his *Zīj*. Here the values for Baghdad and Mecca are as shown above.

Unfortunately al-Bīrūnī in his treatise on mathematical geography does not mention the value 21;20° for φ'. (Those values which he does discuss are 21;0° and 21;40°.) According to the two reports of the observations of the astronomers commissioned

[62] On al-Khwārizmī's trigonometric tables, see *Neugebauer* 1, pp. 104 and 105; *Neugebauer & Schmidt*, p. 227; and *King* 2, p. 2 and note 12 on p. 34.

[63] See *Kennedy* 1, pp. 151 and 153, and *Debarnot* 2, *ad* Tables 24 and 47.

[64] See note 17 above.

[65] On al-Bīrūnī, see note 29 above. These coordinates are found in *al-Bīrūnī* 2, pp. 551 and 558.

by al-Ma'mūn, the results were 21° (?) and 21;42°.[66] Note that the mean of these is about 21;20°, and one can hope that this was not the origin of the parameter underlying the table.

In both the Tashkent and Istanbul manuscripts there are instructions on the use of the table. Text and translation of the Istanbul version (fols. 187v–188r) follow:

معرفة سمت القبلة بالجدول فإذا أردت ذلك فخذ فضل ما بين طول المدينة التى تريد سمت قبلتها وطول مكة فاحفظه ثم خذ فضل ما بين عرض مكة وعرض المدينة فاحفظه ثم أدخل ما حفظت مما بين الطولين فى السطر الموقع عليه الطول واعرف أى بيت وافيت من العدد ثم أدخل ما حفظت من فضل ما بين العرضين فى السطر الموقع عليه العرض فاعرفه أيضا ثم مر فيه حتى تحاذى البيت الأول الذى حفظت فأى عدد وافيت فهو بعد ما بين خط الجنوب إلى سمت القبلة فإن كانت مكة شرقية عن البلد فهو بعدها من الجنوب إلى ناحية المشرق وإن كانت غربية فهو بعدها من الجنوب إلى ناحية المغرب واعلم أن أكثرها يلوه فهو إلى المشرق أقرب ومثال ذلك إن فضل ما بين طول مكة وطول بغداد ثلثة أجزاء وفضل ما بين عرض مكة وعرض بغداد اثنا عشر جزءًا فتدخل ثلثة اجزاء فى سطر الطول وتدخل اثنا عشر جزءًا فى سطر العرض وتعمل كما وصفت [لك] فتوافى بيتاً فيه ثلثة عشر درجة وثلثة عشر دقيقة فذلك سمت قبلة بغداد مما يلى المغرب من خط الجنوب لأن بغداد أكثر طولاً من مكة فإن استوا طول المدينة وطول مكة فالصلوة على خط نصف النهار وإن استوا عرض المدينة وعرض مكة فالصلوة على خط المشرق والمغرب وهو الخط القاطع لخط نصف النهار على زوايا قائمة فإن أردت أن تعلم مكة شرقية أم غربية عن البلد الذى أنت فيه فانظر فإن كان طول مكة أكثر من طول البلد الذى تريد فمكة شرقية وإن كان طولها أقل فمكة غربية عن البلد

"Finding the azimuth of the qibla using the table. If you want (to do) that, take the difference between the longitude of the city whose qibla you want (to find) and the longitude of Mecca and keep it in mind. Then take the difference between the latitude of Mecca and the latitude of the city and keep it in mind. Next

[66] *Langermann*, p. 109, and *King* 6.

enter the difference between the two longitudes that you kept in mind on the scale along which (the word) "longitude" is written and find which argument number you reach. Then enter the difference between the two latitudes that you kept in mind on the scale along which (the word) "latitude" is written, and know (which argument number it corresponds to) also. Then pass along (the corresponding column) until you get level with the argument box (corresponding to the) first (quantity) that you kept in mind. Whatever number you reach will be the distance between the south line and the azimuth of the qibla. If Mecca is east of the locality it is the distance (of the qibla) from the south towards the east, and if (Mecca) is west it is the distance (of the qibla) from the south towards the west, *wa-'lam anna aktharahā yalūhu fa-huwa ila 'l-mashriq aqrab* [this phrase is not in the Tashkent manuscript and its meaning is not clear].

"*Example.* The difference [Tashkent: *we have found* the difference] between the longitude of Mecca and the longitude of Baghdad is three degrees and the difference between the latitude of Mecca and the latitude of Baghdad is twelve degrees. You enter three degrees in the longitude scale and twelve degrees in the latitude scale, and do as I described to you: you will find the entry thirteen degrees and thirteen minutes, and that is the azimuth of the qibla of Baghdad, westwards from the south line because Baghdad has a greater longitude than Mecca. If the longitude of the city and the longitude of Mecca are equal, the (direction of) prayer is along the meridian, and if the latitude of the city and the latitude of Mecca are equal, the (direction of) prayer is along the east-west line which is the line that cuts the meridian at right angles. If you want to know whether Mecca is east or west of the locality where you are, investigate whether the longitude of Mecca is greater than the longitude of the locality (whose qibla) you want, (in which case) Mecca is east, or whether its longitude is less, (in which case) Mecca is west of the locality."

Commentary:

The most significant part of these instructions is the numerical example for Baghdad. The parameters used in the text are:

$$\Delta L = 3^\circ \quad \text{and} \quad \Delta\varphi = 12^\circ.$$

The value $\Delta L = 3°$ was one result of the observations commissioned by al-Ma'mūn. The other result was:

$$\Delta\varphi = 11;45° \quad (\varphi_M = 21;42°)$$

according to the report of Ḥabash, and:

$$\Delta\varphi = 12;15° \quad (\varphi_M = 21;0\ (?)°)$$

according to the judge entrusted by the Caliph to oversee the observations.[67] Notice that 21;20° is the mean of these results. More surprising is that the judge was not relieved of his post, for al-Bīrūnī confirms the value 21;40°. But al-Bīrūnī was not convinced. His own coordinates (see above) correspond to:

$$\Delta L = 3° \text{ and } \Delta\varphi = 12;5° \approx 12° \text{ with } \varphi_M = 21;20°.$$

It seems probable that our author was using this set of coordinates.

In passing we observe that the coordinates for Baghdad and Mecca given in an anonymous set of geographical tables on fols. 194v–196v of the Aya Sofya manuscript, as in many other Islamic sources, are:

$$\varphi_B = 33;0° \quad L_B = 70;0°$$
$$\varphi_M = 21;0° \quad L_M = 67;0°$$

which correspond precisely to the values of ΔL and $\Delta\varphi$ given in the instructions. Nevertheless, we should bear in mind that our table is based on $\varphi_M = 21;20°$. I have discussed various other medieval sets of coordinates for Baghdad elsewhere.[68]

We should also mention that the author's remark that the qibla is east or west when $\Delta\varphi = 0$ is incorrect,[69] because the qibla is defined by the great circle between the two localities and not by the small circle which is the parallel of latitude through Mecca.

[67] *ibid.*

[68] See *King* 6.

[69] Even Ibn Yūnus (on whom, see note 35 above) makes this error: see further *King* 1, Part III, Section 28.3(a).

11. References to Early Qibla Tables in Medieval Sources

I have noted three isolated references to 20x20 qibla tables in the Islamic astronomical literature. In all cases, it is not completely clear which qibla table is refered to.

The first reference occurs amidst some miscellaneous notes on spherical astronomy preserved in MS Paris B. N. ar. 2506, fols. 42r–42v, copied *ca.* 600 H (≈ 1200) in Cairo:

... ومعرفة الانحراف من الجدول العشرينى الذى وضعه علماء أهل هذا الشأن واعلم أن أكثرهما طولاً أقربهما من المشرق لأن الطول من المغرب ومثال ذلك أن فضل ما بين طول مكة وبغداد ٣ اجزاء وفضل عرضيهما ١٢ فتدخل بثلاثة اجزاء فى سطر الطول واثنى عشر فى سطر العرض وتدخل بهما فتوافى يجـ درجة و يجـ دقيقة فذلك سمت قبلة بغداد من خط الجنوب إلى ما يلى المغرب لأن بغداد أكثرهما طولاً وكذلك اعمل بكل سمت تريد معرفة بعده لعمل القبلة إن شاء الله ...

The anonymous author states that to find the qibla one can use a *jadwal 'ishrīnī*, and shows how to use such a table, giving the same example for Baghdad found in the original instructions which accompany the table in the Aya Sofya manuscript. The following folios (43v–44r) contain some sundial tables for Cairo probably by the same author: here the qibla at Cairo is given as 37° (south of east). A value of q which rounds to 53° was derived by Ibn Yūnus[70] using coordinates:

$$\varphi_C = 30;0° \quad L_C = 55;0°$$
$$\varphi_M{}' = 21;0° \quad L_M = 67;0°.$$

If we enter in the table based on the standard approximate method with:

$$\Delta\varphi = 9° \text{ and } \Delta L = 12°$$

we obtain q = 53;3°. The corresponding value in the Abbasid table is 51;36°.

Another reference to the *jadwal 'ishrīnī* occurs in an anonymous Egyptian treatise on astronomy compiled in 723 H (= 1323), a small part of which is extant in MS Leiden 468 [192 fols.,

[70] On this value of the qibla at Cairo, see note 47 above.

ca. 800 H (≈ 1400)]. This work relies heavily on the *Kitāb Jāmiʿ al-mabādiʾ wa-'l-ghāyāt* of Abū ʿAlī al-Marrākushī, compiled in Cairo about forty years earlier, and well known as a major Islamic work on spherical astronomy.[71] The anonymous author mentions the determination of the qibla at Cairo (fol. 189r), quoting numerical examples from al-Marrākushī, and then introduces the *jadwal ʿishrīnī*, which he says was compiled by "one of the scholars in the field":

... وهذا جدول يسمى العشرينى حسبه بعض علماء هذا الفن يستخرج منه سمت أى بلد شئت فى بلد آخر بشرط أن لا يزيد فضل ما بين طوليهما أو فضل ما بين عرضيهما على سَ ...

He goes on to say that the table is for finding the direction of any locality from any other, and that it can only be used if ΔL and $\Delta\varphi$ are both less than 60°, which is a curious remark. The table itself, promised for the next page, has not been copied at all. After this blank page (fol. 189v), the author continues (fols. 190r–191r) with a discussion of how to use a magnetic compass (there is no mention of magnetic declination) and gives a list of coordinates and qibla values for various localities.

A third reference to a *jadwal ʿishrīnī* occurs in some notes appended to the treatises on astrolabes, sundials, and the magnetic compass by the late-thirteenth-century Yemeni Sultan al-Ashraf, preserved in MS Cairo Taymūr *riyāḍa* 105 (149 fols., autograph?).[72] al-Ashraf himself in the treatise on the compass states that the qibla for the central Yemen (Aden and Taiz) is 27° east of north, without giving any details about the geographical coordinates with which this value was derived, but implying that it was taken from a *jadwal ʿishrīnī*. However, some notes at the end of this treatise (fol. 143r) state that the qibla is 27° "and a fraction" east of north, and that this value is for the Yemen with $L = 63;30°$ and Mecca with $L = 60;0°$. Now, if we assume that $\varphi = 14;30°$, which is one of the values used by al-Ashraf for the Yemen, and enter in the table with arguments $\Delta\varphi = 6;30°$ and $\Delta L = 3;30°$, then using linear interpolation between the values:

[71] On al-Marrākushī and his work, see note 52 above.

[72] On al-Ashraf and his works, see *MAY*, II, Section 8 (pp. 27–29); *Cairo Survey*, no. E8; and *King* 5, II.

ΔL \ $\Delta\varphi$	6°	7°
3°	25; 2°	21;49
4°	31;55	28;50

we obtain $q = 26;54° \approx 27°$. But the author states that the qibla is 27° "and a fraction," and alsewhere in a passage that is hardly intelligible, he seems to imply that $\Delta\varphi = 5°$ and $\Delta L = 5°$. Notice that for arguments $\Delta\varphi = 9°$ and $\Delta L = 5°$, the entry in the Abbasid table is $q = 27;31°$. However, for al-Ashraf the value $\Delta\varphi = 9°$ would imply $\varphi = 12°$ for the Yemen and he used $\varphi = 13°$ for Aden!

It seems to me probable that al-Ashraf's qibla value for the Yemen was somehow derived using the Abbasid qibla table, although this was devised for $\varphi > \varphi_M$. As we have seen in Section 8, the qibla table occurs in the contemporary Yemeni *Zīj* of al-Kawāshī. al-Ashraf gives the following coordinates of localities in the Yemen and the Hejaz in his treatise on astrology (MS Oxford Bodleian Hunt. 233, fol. 153v):

	L	φ
Medina	65;20°	25; 0°
Mecca	67; 0*/60; 0	21; 0
Sanaa	63; 0	14;30
Aden	65;30	13; 0
Taiz	66;30	13;43
Zabid	62; 0	14; 0
Saada	64;25	15; 0
Zofar	78; 0!	15; 0
Marib	63; 0	15;15
Hadramawt	71; 0	12;30

* After the first value the text has *wa-bi-'l-rasā'il sīn wa-huwa 'l-aṣaḥḥ*, meaning "in some treatises 60° and this is more correct."

There is no mathematical discussion of the qibla in this work.

Finally, we should discuss the qibla value for Maragha in MS Istanbul H. Hüsnü Ef. 1268,3 [see Section 10(a)] which is presented as if it had been derived from the accompanying table. The parameters are $\Delta L = 5°$ and $\Delta\varphi = 15;40°$, and the result is $q = 16;38°$. But using linear interpolation in the table, I obtain 16;48°. I have no explanation for this divergence. The value underlying the table for finding the qibla at Maragha by means of the sun which is found in the thirteenth-century *Zīj* of Muḥyi 'l-Dīn al-Maghribī – see, for example, MS Dublin Chester Beatty 3665, fols. 96v–97r – is 16;49°. The value stated to underlie the table is 16;59°, which is the result of a calculation reproduced by al-Maghribī elsewhere,[73] as well as in the *Zīj* of the fourteenth-century Persian scholar al-Wābkanawī: see MS Istanbul Aya Sofya 2694, fols. 85v–86v.[74]

12. A Revised Survey of Medieval Qibla Tables

Some nine different qibla tables are now known to have been compiled in the medieval period. At least four of these are based on accurate formulae. In 1975 I wrote that al-Khalīlī's table was "the most sophisticated trigonometric table known to me from the entire medieval period." Now I can claim only that it is one of several highly sophisticated trigonometric tables compiled by Muslim scientists. Qibla tables currently known to me include:

(1) The anonymous Abbasid qibla table discussed in Section 8, extant in both copies of the Abbasid treatise and in seven more copies occurring in later works. See also nos. 4 and 5 below.

(2) The anonymous qibla tables in three versions based on the standard approximate method, extant in at least ten copies and discussed in Section 10.

(3) The celebrated eleventh-century scholar Ibn al-Haytham mentions in his autobiography, recently published by A. Heinen, that he himself compiled a table for finding the qibla.[75] He refers to the work as: "A treatise on the determination of the azimuth of the qibla in all of the inhabited world by tables which I compiled . . . . . ."[76] In view of his exhaustive treatment of the "method of the *zīj*es" in his second treatise on the qibla (see Section 9), his qibla table was probably computed using the same

[73] *Saliba*, p. 396.

[74] On al-Wābkanawī and his *Zīj*, see *Kennedy* 1, no. 35, and *Storey*, p. 65.

[75] On Ibn al-Haytham, see notes 13 and 51 above. On his autobiography, see *Heinen*, esp. p. 263, lines 12–13.

[76] The text adds the curious remark: "and I did not present any proof for this."

Table 5a

The 20x20 qibla table attributed to al-Khāzinī
by Ḥamdallāh Mustawfī, as published by Le Strange

ΔL \ $\Delta\varphi$	1°	2	3	4	5	6	7	8	9	10
1°	41;18	26;59	17;15	12; 2	10;14	8; 4	7;56	7;40	7;36	7;21
2	69;17	42; 5	31;11	25; 8	20;40	15; 9	14; 7	13;11	11;32	10;20
3	70;11	54; 9	43; 7	34;18	29;30	26;10	21; 4	19;20	16;19	15;30
4	75;14	61;37	51;17	43;10	36;20	31;16	28;14	25;41	22;11	20;30
5	77; 4	67;18	56;42	49;15	43;15	36;40	33; 7	30; 4	26;17	25;36
6	79; 7	74;18	60;14	54;18	48;16	43;20	38;14	36;16	32;11	29;40
7	81; 4	75;34	72; 7	58;41	52;14	46;56	43;24	39;22	36;23	33;32
8	82; 5	78;14	74; 9	61;14	56; 9	51; 7	46;50	43;31	39;42	36;46
9	83; 9	80;19	76;11	64; 7	59;49	54;34	50;14	46;45	43;36	40;30
10	84;11	82;23	77;34	66;17	62;36	56;45	53;19	49;35	46;25	43;39
11	85;19	83;42	79;16	68;12	64;14	59;19	55;17	52;12	49;15	46;15
12	86;21	85;51	79;42	71; 6	66; 9	62; 6	58;41	54;49	51;49	48;58
13	87;25	86; 8	82;31	73; 9	67;17	64;14	59; 2	56;17	53;34	50;43
14	88;19	87;11	82;17	74;14	69;40	65;29	62;14	58;18	55;44	53;14
15	89; 4	88;16	82;43	75;13	70; 4	68;13	63; 9	60;42	56;58	55; 4
16	89;44	88;32	82;36	76;49	71;36	68;46	65;11	62;17	59;56	56;52
17	90; 0	89;14	82; 4	76;14	72;29	69;17	66; 3	63;19	60;17	58;12
18	90; 0	90; 0	83;18	76;51	73;42	70;30	67; 6	65;30	62;14	59;59
19	90; 0	90; 0	83;31	77; 7	74;30	71;21	69; 4	66;46	63;18	61;14
20	90; 0	90; 0	84; 9	83;17	75;17	72;34	70; 7	67;23	65;29	62;17

method. Unfortunately we do not know which value he advocated for φ_M: in neither of his two treatises on the determination of the qibla is this parameter mentioned.

(4) A table displaying q ($\Delta\varphi$, ΔL) for each degree of both arguments from 1° to 20° is attributed to the early-twelfth-century Seljuq astronomer ʿAbd al-Raḥmān al-Khāzinī[77] by the fourteenth-century author Ḥamdallāh Mustawfī (Qazwīnī). The table is reproduced in the geographical part of Mustawfī's encyclopaedia entitled *Nuzhat al-qulūb*.[78] The text and an English translation of this section of the treatise was published by G. Le

[77] On al-Khāzinī, see the article by R. E. Hall in *DSB*. See also note 80 below.

[78] On the author and his work, see *Storey*, pp. 129–131. This treatise is available in numerous copies and a defective edition was published in Bombay in 1894.

Table 5a
(continued)

ΔL \ Δφ	11	12	13	14	15	16	17	18	19	20
1°	6;32	6;28	4; 8	3;51	3;34	3;33	3;31	3;21	3;17	2;28
2	9;44	6; 4	8;30	7;20	7;10	6;39	6;15	6; 4	5;30	5;26
3	14;34	14;14	14; 4	11;14	10;31	10; 3	9;40	8;31	8;17	8; 7
4	18;14	17;11	16; 7	15;21	15;19	13;17	12;36	11;41	11;30	10;17
5	23;47	21;36	19;42	18;24	17;16	16;44	15;14	14;47	14; 4	13;34
6	27;43	25;57	23;51	22;46	22;34	19;43	18;19	17;42	16;47	15;50
7	32;28	28;14	26;18	25;34	23;36	22;32	21;48	20;19	19;12	18;23
8	35; 4	32;33	32;42	28;36	26;19	25;14	24;23	21;30	21;54	20;19
9	37;30	35;23	33;50	33;33	29; 7	28;32	26;14	25;36	25;49	23;38
10	40;41	38; 2	36; 8	34;17	34;15	31;15	29;32	27;14	26;16	25;36
11	43;43	40;50	38;12	36;18	35;36	34;20	31;45	31;30	29;42	26;50
12	45;54	43;45	41;29	39; 4	36;32	36;12	34;24	32;46	31;16	31;28
13	48;48	45;48	43;47	41;39	39;44	36;40	36;20	34;44	33;41	32; 4
14	50;23	48;38	45;44	43;49	41;49	40; 4	38;36	36;29	35;12	34; 4
15	51;46	50;13	48;28	45;40	45;32	42;16	40;24	38;50	36;34	35;49
16	53;33	51;36	49;46	47;45	45;35	43;58	42;26	40;44	39;12	36;50
17	55;42	53;23	51;25	49;36	47;35	45;30	44;23	42;36	41; 2	39;56
18	56;44	55;32	53;13	51; 5	49;26	47;25	45;25	44;32	42;46	41;19
19	59;39	56;14	54;22	52;14	50;38	49;16	47;15	45;20	44;41	42;56
20	60;14	58; 7	56; 4	54;12	52; 4	50;18	48;15	47; 5	45;15	44;50

Strange in 1919,[79] and as he pointed put – much credit to him! – the qibla table – see Table 5a – is contained in neither the Vatican nor the London manuscripts of al-Khāzinī's *Sanjarī Zīj*.[80]

Now Mustawfī specifically states that al-Khāzinī compiled the table at the request of the Seljuq Sultan Sanjar, to whom al-Khāzinī's *Zīj* is dedicated. However, as R. Lorch – see below – has already pointed out, perhaps this remark should be taken as refering to the *Zīj* rather than the table. Now a qibla table is actually mentioned in the text of the *Sanjarī Zīj* (MS Vatican ar. 761, fol. 47r and MS London B. L. 6669, fol. 25r):

... وعلى هذا عملنا جدولاً ثلثين فى ستين لما بين العرضين والطولين ثم استخرجنا
منه سمت قبلة كل بلد وأثبتناه بإزائه فى الكتاب وما بين طولى مرو ومكة حرسها الله

[79] *Le Strange*, pp. 24–31, especially pp. 30–31.

[80] On the *Sanjarī Zīj*, see *Kennedy* 1, no. 27.

تسعة عشر جزءًا ونصف وما بين عرضيهما سنة عشر جزءًا وثلث وسمت قبلتها عن خط الزوال نحو المغرب أحد وخمسون جزءًا وثلث

Here al-Khāzinī actually states that he compiled a qibla table for arguments $\Delta\varphi$ from 1° to 30° and ΔL from 1° to 60°. Thus his table was of the same format as the one in the *Zīj* of al-Wābkanawī (see no. 7 below), although one of the arguments in that table is $\Delta\varphi'$ rather than $\Delta\varphi$.

al-Khāzinī also states that he used his table to compute the qibla at Marw (with $\Delta L = 19\ 1/2°$ and $\Delta\varphi = 16\ 1/3°$) as 51 1/3°. (Inevitably these values are inconsistent with the entries in the tables of geographical coordinates in his *Zīj*!) If we use linear interpolation in the Abbasid table, we obtain from the above values of $\Delta\varphi$ and ΔL *ca.* 49 1/2° for the qibla at Marw, which proves – if proof were needed – that this table was not computed by al-Khāzinī. Using al-Khalīlī's table, accurately computed for $\varphi_M = 21;30°$ (which is the value given in al-Khāzinī's geographical tables), I obtain 51;24°. This proves that al-Khāzinī produced a 30x60 table of q ($\Delta\varphi$, ΔL) based on an accurate formula. This table is not known to have survived in the manuscript sources.

(5) Mustawfī's table, as published by Le Strange, is reproduced in Table 5a. It was not listed in my previous survey (in spite of the fact that it was the only qibla table that was then published!!).[81] The table as edited by Le Strange has been investigated by R. Lorch.[82] His conclusion regarding this table was that it was probably computed according to the approximate formula discussed in Section 10(c), and this I can confirm. Given the state of the table, Lorch's achievement in cracking the underlying formula and parameter is most remarkable. But there is more that can be said.

At first sight, the table appears as if it might be based on an accurate formula: notice the values of 90° for small $\Delta\varphi$ and large ΔL (the qibla is north of the east-west line for such arguments). One might suspect that a copyist simply dropped the values of the minutes. But closer inspection reveals that the values presented in the table are *extremely* corrupt. Even if we bear in mind that in the *abjad* system, careless copying can lead to a 13

[81] The qibla table is mentioned in *Sarton*, III², p. 632.
[82] See *Lorch* 1.

Table 5b
Selected values from the qibla table in MS Istanbul Fatih 3417
of the *Encyclopaedia* of Ḥamdallāh Mustawfī

ΔL \ $\Delta\varphi$	1°	2	3	4	5	6	7	8	9	10 ...
1°	62;18°	27;39	17;15	18; 2	10;34	8; 7	7;16	6;40	5;56	5;21
2	69;43	42;30	31;11	25; 0	20;28	15;17	14;17	13;20	11;44	10;16
3	70;18	44;26	13; 0	34;18	29;14	25;12	21;43	19;18	17;20	15;41
4	75; 7	10;18	51;41	48; 2	36;51	31; 0	28; 7	25; 9	22; 7	20;33
5	77;14	66;40	16;53	49;27	15; 4	7;16	13;14	4;20	27;31	25;70,1 (!)
6	79;18	70;11	10;51	57;31	48;18	43; 7	18;47	35; 7	32;18	29;28
7	81; 4	73; 1	65;28	18;35	12;40	46;31	48; 9	39;38	36;30	13;21
8	82;22	75; 5	63;21	10;55	16;38	11;21	46; 1	13;51	19;18	16;18
9	48;18	76;40	70;26	67;38	19;51	14;33	55; 8	46;33	48;17	40; 4
10	18;55	77;13	72; 2	77;46	62;74 (!)	57;11	38;21	49;40	46;21	18;21

being mistaken for a 58, a 21 for a 9, and even a 17 for a 40, many of the entries for the minutes are inexplicable. Notice that there are several places where isolated entries correspond exactly to those in the Abbasid table, *e.g.* 64;14° at (5,11); 62;6° at (6,12); 59;59° at (10,19); 3;51° at (14,1); and 47;35° at (15,17). Then there are pairs of entries which show switching of the minutes: compare, *e.g.* 45;40° at (14,15) and 47;45° at (14,16) in Table 5a and 45;45° and 47;40° in the Abbasid table. By considering first differences between adjacent entries horizontally, vertically and diagonally, we can reconstruct a few more. Also, there is an uncanny resemblance between the entries in the two tables for $\Delta\varphi = 20°$. In fact, the minutes in general start to make more sense if we consider the possibility that they have been shuffled by some incompetent; the column for $\Delta\varphi = 20°$ is the only one which has been spared this indignity. Nevertheless, I would be the first to admit that the minutes entries in extensive sections of the table bear no obvious relation to those in the Abbasid table.

I have examined only one manuscript of Mustawfī's treatise, namely the elegant copy MS Istanbul Fatih 4517. Before I realized the significance of the table, I recorded an array of 10x10 entries – see Table 5b. My notes on the table state: "total disaster – not worth copying," but the entries confirm my suspicions about Table 5a. Also I recall that the minutes were written above the degrees in an unusual way: part of the confusion in

Table 6

Selected values from the 30x30 qibla table in the *Ashrafī Zīj*

ΔL \ $\Delta\varphi$	1°	5	10	15	20	25	30
1°	42;18°	12; 1	5;54	3;52	2;46	2;13	1;52
5	77; 4	44;15	26;51	17;18	13;24	10;58	9;18
10	80;59	64;22	44;48	31;32	25;42	21;28	18; 8
15	85;19	73;12	56;15	45;12	36;28	31; 5	26;26
20	89;30	78;19	65;59	54;41	46;41	39;52	34;57
25	87;45	83;28	71;11	62; 3	54;12	47;27	42;26
30	85;10	87;45	76;54	68;22	60;59	54;33	49;12

this table and in the one recorded by Le Strange can probably be attributed to some "cute" way of writing the minutes.

In brief, then, Mustawfī's table has nothing to do with al-Khāzinī and is simply a hopelessly corrupt version of the Abbasid table.

(6) A table displaying the qibla for each degree of two arguments from 1° to 30° is preserved in the fourteenth-century Persian *Zīj-i Ashrafī* by Muḥammad ibn Abī ʿAbd Allāh Sanjar al-Kamālī, extant in the unique copy MS Paris B. N. supp. pers. 1488, fols. 130v–131r.[83] See Table 6 for sample entries. The instructions in the margin (in Arabic) indicate that the arguments are $\Delta\varphi$ and ΔL. In the accompanying text (fol. 104v), the author shows how the table yields 53;31° for the qibla at Shiraz. The *Ashrafī* qibla table clearly contains numerous copyists' mistakes and appears to be based on an approximate formula, which I have been as yet unable to determine.

(7) Another qibla table is preserved in the early-fourteenth-century Persian *Zīj* of Shams al-Munajjim al-Wābkanawī, of which I have examined MS Istanbul Aya Sofya 2694, fols. 85v–86v (text) and 165r–166v (table). Other copies of this work are preserved in Tehran and Bombay.[84]

The table consists of two parts. In the first, smaller subtable a function labelled [*al*-]*ʿarḍ al-muʿaddal*, "modified latitude," is tabulated for each degree of *faḍl mā bayn al-ṭūlayn*, "longitude difference," from 1° to 60°. In the main table, the horizontal argument is again the longitude difference and the vertical argu-

[83] On the *Zīj-i Ashrafī*, see *Kennedy* 1, no. 4, and *Storey*, pp, 64–65.

[84] On al-Wābkanawī, see note 74 above.

ment is called *al-maḥfūẓa*, "that which was kept in mind (after an arithmetical operation described in the text)." It is clear from these arguments and from the accompanying text that the compiler used the "method of the *zījes*."

In the first subtable $\Delta\varphi'$ is tabulated as a function of ΔL. In the main table q is tabulated as a function of ΔL and $\Delta\varphi'$. The underlying value of φ' appears to be 21;20°. It is not clear to me how the table was computed because the entries in the main table, especially for arguments less than about 20°, show divergences from the entries in the Abbasid table. Sample entries from the Istanbul copy are displayed in Table 7.

Finally we note that a disadvantage of the table is that the argument $\Delta\varphi'$ is invariably non-integer, and so one must interpolate in the table to find the qibla for integer coordinates. In al-Khalīlī's table, to which we now turn, there is no such problem.

(8) A table displaying q (φ, ΔL), with values based on the accurate formula for each degree of φ from 10° to 50° and of ΔL from 1° to 60°, was computed by the fourteenth-century Damascus astronomer Shams al-Dīn al-Khalīlī.[85] This is the most practical and sophisticated qibla table. It is extant in at least three copies, namely:

MS Paris B.N. ar. 2558, fols 53v-61r, copied 811 H (= 1408/09);
MS Berlin Ahlwardt 5757, fols 47v–55r, copied *ca.* 1500; and
MS Paris B.N. ar. 2560,11, fols 47v–55r, copied *ca.* 1600.

The table has been published using the first two of these copies. Attempts to establish precisely the way in which it was compiled have been inconclusive.[86]

(9) Finally, a qibla table is preserved in MS Cairo Dār al-Kutub *mīqāt* 832, fols 90v–91r, copied 832 H (= 1428/29), of a treatise on timekeeping compiled by the mid-fourteenth-century astronomer 'Alā' al-Dīn Abu' l-Ḥasan 'Alī ibn Ṭaybughā, *muwaqqit* at the Umayyad Mosque in Aleppo.[87] This table (fol. 90v) displays the function q (ΔL, φ) for domains somewhat less extensive than those of al-Khalīlī, namely:

[85] On al-Khalīlī and his works, see note 58 above.

[86] See *King* 4, XIII.

[87] This author and his works are new to the modern literature: see *Cairo Survey*, no. C54.

Table 7
Selected values from the 20x20 qibla table
in the *Zīj* of al-Wābkanawī

ΔL	φ'
1°	21;20°
2	21;21
3	21;22
4	21;23
10	21;38
20	22;34
30	24;16
40	27; 1
50	31;17
60	38; 0

$\Delta\varphi'$ \ ΔL	1°	2	3	4	5	10	15	20
1°	42;27°	61;39	68;40	74;26	77;39	83;38	85;13	87;13
2	22;13	44;21	59;38	62;31	66;48	76; 8	82; 9	83;32
3	18; 6	31;48	44;20	54; 6	58; 0	73; 5	78;21	81;49
4	13;30	25; 4	35;38	45;23	49;20	67; 5	74;20	78; 6
5	10;45	20;28	29;29	35;55	42;36	62; 3	70;42	75; 5
10	5;22	10;35	15;43	20;34	25;23	43;16	54;50	62;20
15	3;36	7; 9	10;40	14; 8	17;39	32;16	43;43	52;22
20	2;43	5;26	8; 7	10;56	13;25	25;35	35;58	44;30
25	2;32	4;24	6;35	8;48	10;55	21;12	30;26	38;30
30	1;52	3;43	5;35	7;25	9;16	18; 9	26;26	33;14

$\Delta\varphi'$ \ ΔL	25	30	35	40	45	50	55	60
1°	87;26	87;34	87;51	88;52	88;30	88;52	88;55	88;58
2	84;47	86; 9	86;48	87;18	87;34	88; 8	88;31	88;57
3	83; 0	84;26	85; 8	86;15	86;40	87; 3	87;28	87;57
4	80;30	82;31	83;42	85;11	85; 8	86;59	86;22	87;26
5	78;21	80;32	81;35	83;39	84;24	85;18	85;56	86;30
10	67;51	71;39	74;24	76;55	78;47	80;21	81;31	82;14
15	58;45	63;48	67;36	70;13	73;27	75;43	77;34	79;13
20	51;27	57; 0	61;33	65;24	68;37	71;25	73;52	75;58
25	45;23	51;17	56;11	60;59	64;14	67;37	70;15	72;47
30	40;36	46;30	51;38	57; 5	60;14	63;51	67; 2	69;39

$$\Delta L = 1°, 2°, \ldots, 12° \text{ and}$$
$$\varphi = 16°, 17°, \ldots, 21°, 21;30° \text{ (Mecca)}, 22°, 23°, \ldots, 44°.$$

The entries in this table are accurately computed according to an exact formula. They show small variations from the corresponding entries in al-Khalīlī's more extensive table and hence were independently computed. Sample entries are shown in Table 8.

Table 8

Sample entries from Ibn Ṭaybughā's qibla table

φ \ ΔL	1°	6	12
16°	9;36°	44;48	62;16
30	6;16	33;43	54;58
44	2;26	14;23	27;42

Ibn Ṭaybughā's qibla table is accompanied by another with the same format displaying the corresponding distance from Mecca $D(\Delta L, \varphi)$. Values are again accurately computed according to an exact formula. Sample entries are shown in Table 9. I know of no other tables of this kind in the Islamic sources. The simplest formula, which follows from the use of the sine rule for spherical ΔPXM in Fig. **4**, is:

Table 9

Sample entries from Ibn Ṭaybughā's table of distance from Mecca

φ \ ΔL	1°	6	12
16°	5;36°	7;57	12;37
30	8;33	10; 6	13;44
44	22;31	23; 3	24;36

$$D = \text{arc sin } (\sin \Delta L \cos \varphi_M / \sin q)$$
$$= \text{arc sin } (d / \sin q).$$

But more probably the table of D was computed first and corresponding values of q were generated from it.

Interest in the distance of localities from Mecca goes back to the ninth and tenth centuries,[88] but this is the only table of that distance known to me from the medieval sources.

[88] See, for example, *Kennedy* 6, *Langermann*, and *King* 6.

Acknowledgements:

My research on the sacred direction in Islam was supported by grants from the U. S. National Science Foundation (1980–1982), which support is gratefully acknowledged.

It is a particular pleasure to record my gratitude to Mr. Muammer Ülker, Director of the Süleymaniye Library in Istanbul, and to Dr. Hayrulaev, Director of the Tashkent Oriental Institute, and to their respective staffs, for the privilege of working in these two libraries. I should also like to express my gratitude to the Directors of the Municipal Library in Alexandria, the Egyptian National Library in Cairo, the Ẓāhirīya Library in Damascus, the Chester Beatty Library in Dublin, the Forschungsbibliothek in Gotha, the British Library in London, the Bodleian Library in Oxford, the Bibliothèque Nationale in Paris, the Special Collections Library of Princeton University, and the Biblioteca Vaticana for the generous facilities afforded me in each of these libraries. Sayyid ʿAbd al-Qādir al-Ahdal of Zabid kindly allowed me to inspect the manuscript in his private collection.

It is also a pleasure to thank my friends Professor J. Len Berggren (Simon Fraser University), Dr. J. P. Hogendijk (Mathematical Institute, Utrecht, formerly Institut für Geschichte der Naturwissenschaften, Johann Wolfgang Goethe-Universität, Frankfurt), Prof. E. S. Kennedy (American University of Beirut, currently visiting the Institut für Geschichte der Arabisch-Islamischen Wissenschaften, Frankfurt), and Dr. Richard Lorch (Bayerische Akademie der Wissenschaften, Munich) for their comments on earlier drafts of this paper.

List of Manuscripts Consulted

Alexandria Municipal Library 5577C

Berlin Deutsche Staatsbibliothek Ahlwardt 5757

Cairo Dār al-Kutub *mīqāt* 707, 832, 1030
  Muṣṭafā Fāḍil *mīqāt* 167, 204
  Taymūr *riyāḍa* 103, 105

Damascus Ẓāhirīya 4804, 7564

Dublin Chester Beatty 3665

Gotha Forschungsbibliothek A1411

Istanbul (Süleymaniye) Aya Sofya 2694, 4830
Esat Ef. 3769
Fatih 4517
H. Hüsnü Ef. 1268
Hamidiye 1453

Leiden Universiteitsbibliotheek 143, 468, 1068,15 (see *Berggren* 4)

London British Library Or. 6669, 9116

Oxford Bodleian Hunt. 233, 331

Paris Bibliothèque Nationale ar. 2506, 2558, 2560
supp. pers. 1488

Tashkent Oriental Institute 177

Vatican ar. 761

Zabid, manuscript in the private collection of Sayyid ʿAbd al-Qādir al-Ahdal (microfilm now available from the Institute of Arabic Manuscripts, Cairo)

Bibliography and Bibliographical Abbreviations

Aaboe Festschrift — J. L. Berggren and B. R. Goldstein, eds., *From Babylon to Bohr: Studies on the History of Science in Honor of Asger Aaboe, Acta historica scientiarum naturalium et medicinalium* (in press).

Ali — J. Ali, *The Determination of the Coordinates of Cities: al-Bīrūnī's Taḥdīd [nihāyāt] al-amākin*, Beirut: American University of Beirut, 1966.

Berggren 1 — J. L. Berggren, "On al-Bīrūnī's 'Method of the Zījes' for the Qibla," *Proceedings of the XVIth International Congress for the History of Science*, (Bucharest, 1981), pp. 237–245.

Berggren 2 — –, "A Comparison of Four Analemmas for Determining the Azimuth of the Qibla," *Journal for the History of Arabic Science* 4 (1980), pp. 69–80.

Berggren 3 — –, "The Origins of al-Bīrūnī's "Method of the Zījes" in the Theory of Sundials," *Centaurus* 28 (1985), pp. 1–16.

Berggren 4 — –, "al-Bīrūnī on Plane Maps of the Sphere," *Journal for the History of Arabic Science* 6 (1982), pp. 47–96.

Berggren 5 — –, "History of Mathematics in the Islamic World: The Present State of the Art," *Bulletin of the Middle East*

Studies Association of North America 19 (1985), pp. 9–33.

al-Bīrūnī 1 — al-Bīrūnī, *Kitāb Taḥdīd nihāyāt al-amākin*, ed. P. Bulgakov, *Majallat Ma'had al-Makhṭūṭāt al-'Arabīya*, vol. 8 (1962). See also *Ali* and *Kennedy* 3.

al-Bīrūnī 2 — –, *al-Qānūn al-Mas'ūdī*, 3 vols., Hyderabad, Deccan: Osmania Oriental Publications Bureau, 1955.

al-Bīrūnī 3 — –, *Kitāb al-Tafhīm* . . .: R. R. Wright, trans., *The Book of Instruction in . . . Astrology* . . ., London: Luzac & Co., 1934, repr. Baghdad, n. d.
See also *Berggren* 4.

Brockelmann — C. Brockelmann, *Geschichte der arabischen Litteratur*, 2 vols., 2nd ed., Leiden: E. J. Brill, 1943–49, and Supplementbände, 3 vols., Leiden: E. J. Brill, 1937–42.

Cairo Catalogue — D. A. King, *A Catalogue of the Scientific Manuscripts in the Egyptian National Library* (in Arabic), 2 vols., Cairo: General Egyptian Book Organization, 1981–1986.

Cairo Survey — D. A. King, *A Survey of the Scientific Manuscripts in the Egyptian National Library*, (Publications of the American Research Center in Egypt), Winona Lake, Ind.: Eisenbrauns, 1986.

Carandell — J. Carandell, "An Analemma for the Determination of the Azimuth of the Qibla in the *Risāla fī 'ilm al-ẓilāl* of Ibn al-Raqqām," *Zeitschrift für Geschichte der Arabisch-Islamischen Wissenschaften* 1 (1984), pp. 61–72.

Debarnot 1 — M.-Th. Debarnot, *al-Bīrūnī; Kitāb Maqālīd 'Ilm al-Hay'a. La trigonométrie sphérique chez les Arabes de l'est à la fin du Xe siècle*, Damascus: Institut Français de Damas (Publication no. 114), 1985.

Debarnot 2 — –, "The *Zīj* of Ḥabash al-Ḥāsib: A Survey of MS Istanbul Yeni Cami 784/2," in *Kennedy Festschrift*.

DSB — *Dictionary of Scientific Biography*, 16 vols., New York: Charles Scribner's Sons, 1970–80.

*EI*² — *Encyclopaedia of Islam*, 2nd ed., 5 vols. to date, Leiden: E. J. Brill, 1960 to present.

Garbers — K. Garbers, "Ein Werk Thābit b. Qurra's über ebene Sonnenuhren," *Quellen und Studien zur Geschichte der Mathematik, Astronomie und Physik*, Abt. A. 4 (1936).

Heinen — A. Heinen, "Ibn al-Haiṭam's Autobiographie in einer Handschrift aus dem Jahr 556 H./1161 A. D.," in U. Haarmann and P. Bachmann, eds., *Die islamische Welt zwischen Mittelalter und Neuzeit: Festschrift für Hans Robert Roemer zum 65. Geburtstag*, Beirut, in commission for Franz Steiner Verlag, Wiesbaden, 1979, pp. 254–277.

Hogendijk J.-P. Hogendijk, "The Width of the Lunar Crescent according to Ya'qūb ibn Ṭāriq and al-Battānī," to appear.

Irani R. A. K. Irani, "Arabic Numeral Forms," *Centaurus* 4 (1955), pp. 1–12, repr. in *Kennedy et al.*, pp. 710–721.

Juschkewitsch A. P. Juschkewitsch, *Geschichte der Mathematik im Mittelalter* (translated from the 1961 Russian edition), Leipzig: B. G. Teubner, 1964. See also *Youschkevitch.*

Kennedy 1 E. S. Kennedy, "A Survey of Islamic Astronomical Tables," *Transactions of the American Philosophical Society,* N. S., 46 (1956), pp. 123–177.

Kennedy 2 –, "The Lunar Visibility Theory of Ya'qūb ibn Ṭāriq," *Journal of Near Eastern Studies* 27 (1968), pp. 126–132, repr. in *Kennedy et al.*, pp. 157–163.

Kennedy 3 –, *A Commentary upon al-Bīrūnī's Kitāb Taḥdīd [nihāyāt] al-amākin,* Beirut: American University of Beirut Press, 1973.

Kennedy 4 –, *The Exhaustive Treatise on Shadows by Abū al-Rayḥān ... al-Bīrūnī: Translation and Commentary,* 2 vols., Aleppo: Institute for the History of Arabic Science, 1976.

Kennedy 5 –, "A History of Trigonometry," published 1969, reprinted in *Kennedy et al.*, pp. 3–29.

Kennedy 6 –, "Applied Mathematics in the Tenth Century: Abu'l-Wafā' Calculates the Distance Baghdad–Mecca," *Historia Mathematica* 11 (1984), pp. 193–206.

Kennedy 7 –, "Spherical Astronomy in Kāshī's Khāqānī Zīj," *Zeitschrift für Geschichte der Arabisch-Islamischen Wissenschaften* 2 (1985), pp. 1–46.

Kennedy 8 –, "Suhrāb and Ma'mūn's World Map," to appear in *Aaboe Festschrift.*

Kennedy 9 –, "Mathematical Geography," to appear in a volume of studies on different aspects of Islamic science edited by R. Rashed.

Kennedy & Davidian E. S. Kennedy and M. L. Davidian, "al-Bīrūnī on the Time of Day from Shadow Lengths," *Journal of the American Oriental Society* 80 (1960), pp. 330–335, reprinted in *Kennedy et al.*, pp. 274–279.

Kennedy & Haddad E. S. Kennedy and F. I. Haddad, "Geographical Tables of Medieval Islam," *Al-Abḥāth* 24 (1971), pp. 87–103, reprinted in *Kennedy et al.*, pp. 636–651.

Kennedy & Id E. S. Kennedy and Y. Id, "A Letter of al-Bīrūnī: Ḥabash al-Ḥāsib's Analemma for the Qibla," *Historia Mathematica* 1 (1974), pp. 3–11, reprinted in *Kennedy et al.*, pp. 621–629.

Kennedy & Regier E. S. Kennedy and M. H. Regier, "Prime Meridians in Medieval Islamic Astronomy," *Vistas in Astronomy* 28 (1985), pp. 29–32.

Kennedy et al. E. S. Kennedy, Colleagues and Former Students, *Studies in the Islamic Exact Sciences*, Beirut: American University of Beirut Press, 1983.

Kennedy Festschrift D. A. King and G. Saliba, eds., *From Deferent to Equant: A Volume of Studies in the History of Science in the Near East Dedicated to E. S. Kennedy, Annals of the New York Academy of Science*, 500 (1987).

Khoury I. Khoury, *Fihris Makhṭūṭāt Dār al-Kutub al-Ẓāhirīya: 'Ilm al-hay'a wa-mulḥaqātuhu*, Damascus: Arabic Language Academy, 1969.

King 1 –, *Spherical Astronomy in Medieval Islam: the* Ḥākimī Zīj *of Ibn Yūnus*, to appear (Frankfurt).

King 2 –, "al-Khwārizmī and New Trends in Mathematical Astronomy in the Ninth Century," *Occasional Papers on the Near East* (New York University: Hagop Kevorkian Center for Near Eastern Studies) 2 (1983).

King 3 –, "The Sacred Direction in Islam: A Study of the Interaction of Religion and Science in the Middle Ages," *Interdisciplinary Science Reviews* 10 (1985), pp. 315–328.

King 4 –, *Islamic Mathematical Astronomy*, London: Variorum Reprints, 1986.

King 5 –, *Islamic Astronomical Instruments*, London: Variorum Reprints, 1987.

King 6 –, "The Earliest Muslim Longitude Measurements" (to appear).

King 7 –, "Some Medieval Islamic Qibla-Maps" (to appear).

King 8 –, "Universal Solutions in Medieval Islamic Astronomy," to appear in *Aaboe Festschrift*.

Krause M. Krause, "Stambuler Handschriften islamischer Mathematiker," *Quellen und Studien zur Geschichte der Mathematik, Astronomie und Physik*, Abt. B. 3:4 (1936), pp. 437–532.

Langermann Y. Z. Langermann, " 'The Book of Bodies and Distances' of Ḥabash al-Ḥāsib," *Centaurus* 28 (1985), pp. 108–128.

Le Strange G. Le Strange, tr., *The Geographical Part of Nuzhat-al-Qulūb Composed by Ḥamd-Allāh Mustawfī of Qazwin in 740 (1340)*, Leiden: E. J. Brill and London: Luzac & Co., 1919.

Lorch 1 R. Lorch, "The Qibla-Table Attributed to al-Khāzinī," *Journal for the History of Arabic Science* 4 (1980), pp. 259–264.

Lorch 2 –, "Naṣr b. ʿAbdallāh's Instrument for Finding the Qibla," *Journal for the History of Arabic Science* 6 (1982), pp. 123–131.

Luckey 1 P. Luckey, "Das Analemma von Ptolemäus," *Astronomische Nachrichten* 230 (1927), no. 5498, cols. 17–46.

Luckey 2 –, "Ṯābit b. Qurra's Buch über die ebenen Sonnenuhren," *Quellen und Studien zur Geschichte der Mathematik, Astronomie und Physik*, Abt. B, Bd. 4, (1937–38), pp. 95–148.

Matvievskaya & Tllashev G. P. Matvievskaya and Kh. Tllashev, *Matematicheskie i astronomicheskie rukopisi uchen'ikh Srednei Azii X–XVIII vv.* (= *Mathematical and Astronomical Manuscripts of Scholars of Central Asia, 10th–15th Centuries*), Tashkent: Fan, 1981.

MAY D. A. King, *Mathematical Astronomy in Medieval Yemen* ......, (Publications of the American Research Center in Egypt), Malibu, Ca.: Undena Publications, 1983.

von Mžik H. von Mžik, *Das Kitāb Ṣūrat al-Arḍ des Abū Ǧaʿfar Muḥammad ibn Mūsā al-Ḫuwārizmī*, Leipzig: Otto Harrassowitz, 1926.

Nadir N. Nadir, "Abū al-Wafāʾ on the Solar Altitude," *The Mathematics Teacher* 53 (1960), pp. 460–463, reprinted in *Kennedy et al.*, pp. 380–383.

Nallino 1 C. A. Nallino, *al-Battani sive Albatenii opus astronomicum*, (*Pubblicazioni del Reale Osservatorio di Brera in Milano*, XL), 3 vols., Milan and Rome, 1899–1907, repr. in 1 vol., Frankfurt: Minerva G.m.b.H., 1969.

Nallino 2 –, *Raccolta di scritti editi e inediti*, vol. V: *Astrologia – Astronomia – Geografia*, Roma: Istituto per l'Oriente, 1944.

Nasr S. H. Nasr, *Islamic Science: An Illustrated Study*, London: World of Islam Festival Publishing Co. Ltd., 1976.

Neugebauer 1 O. Neugebauer, "The Astronomical Tables of al-Khwārizmī," *Kgl. Danske Vidensk. hist.-fil. Skrifter*, 4: 2 (1962).

Neugebauer 2 –, *The Exact Sciences in Antiquity*, 2nd ed., New York: Dover Publications, Inc., 1969.

Neugebauer 3 –, *A History of Ancient Mathematical Astronomy*, 3 pts., Berlin – Heidelberg – New York: Springer Verlag, 1975.

Neugebauer 4 –, *Astronomy and History: Selected Essays*, New York, *etc.*: Springer-Verlag, 1983.

Neugebauer & Schmidt O. Neugebauer and O. Schmidt, "Hindu Astronomy at Newminster in 1428," *Annals of Science* 8 (1952), pp. 221–226, reprinted in *Neugebauer* 4, pp. 425–432.

Pingree 1 — D. Pingree, "Indian Influence on Sasanian and Early Islamic Astronomy and Astrology," *The Journal of Oriental Research* (Madras) 34–35 (1964–66/1973), pp. 118–126.

Pingree 2 — –, "The Greek Influence on Early Islamic Mathematical Astronomy," *Journal of the American Oriental Society* 93 (1973), pp. 32–43.

Pingree 3 — –, "The Fragments of the Works of Yaʿqūb ibn Ṭāriq," *Journal of Near Eastern Studies* 27 (1968), pp. 97–125.

Pingree 4 — –, "The Fragments of the Works of al-Fazārī," *Journal of Near Eastern Studies* 29 (1970), pp. 103–123.

Rudloff & Hochheim — G. Rudloff and A. Hochheim, "Die Astronomie des Maḥmūd ibn Muḥammad ibn ʿOmar al-Ǧaġmīnī," *Zeitschrift der Deutschen Morgenländischen Gesellschaft* 47 (1893), pp. 213–275.

Saliba — G. Saliba, "An Observational Notebook of a Thirteenth-Century Astronomer," *Isis* 74 (1983), pp. 388–401.

Samsó — J. Samsó Moya, *Estudios sobre Abū Naṣr Manṣūr b. ʿAlī b. ʿIrāq*, Barcelona: Asociación para la Historia de la Ciencia Española, 1969.

Sarton — G. Sarton, *An Introduction to the History of Science*, 5 vols. in 3, Baltimore: Williams and Wilkins Co., 1927, for the Carnegie Institution of Washington.

Schoy 1 — K. Schoy, "Längenbestimmung und Zentralmeridian bei den älteren Völkern", *Mitteilungen der Kaiserlich-Königlichen Geographischen Gesellschaft* 58 (1915), pp. 27–62.

Schoy 2 — –, "Abhandlung des al-Ḥasan ibn al-Ḥasan ibn al-Haiṯam (Alhazen) über die Bestimmung der Richtung der Qibla", *Zeitschrift der Deutschen Morgenländischen Gesellschaft* 75 (1921), pp. 242–253.

Schoy 3 — –, "Abhandlung von al-Faḍl b. Ḥātim al-Nayrīzī über die Richtung der Qibla", *Sitzungsberichte der math.-phys. Klasse der Bayerischen Akademie der Wissenschaften zu München*, 1922, pp. 55–68.

Schoy 4 — –, *Gnomonik der Araber*, in E. von Bassermann-Jordan, ed., *Die Geschichte der Zeitmessung und der Uhren*, Band 1F, Berlin – Leipzig: Heinz Lafaire, 1923.

Schoy 5 — –, "The Geography of the Moslems of the Middle Ages," *The Geographical Review* 14 (1924), pp. 257–269.

Sédillot — J. J. Sédillot, *Traité des instruments astronomiques des Arabes composé au treizième siècle par Aboul Hhassan Ali de Maroc* . . ., 2 vols., Paris: Imprimerie Royale, 1834, reprinted Frankfurt: IGAIW, 1985.

Sezgin	F. Sezgin, *Geschichte des arabischen Schrifttums*, 9 vols. to date, Leiden: E. J. Brill, 1967 onwards.
Storey	C. A. Storey, *Persian Literature: A Biobibliographical Survey*, Vol. II, Part 1 (Mathematics, Astronomy, *etc.*), London: Luzac & Co., 1972.
Suter 1	H. Suter, "Die astronomischen Tafeln des Muḥammed ibn Mūsā al-Khwārizmī in der Bearbeitung des Maslama ibn Aḥmed al-Madjrīṭī und der latein. Übersetzung des Adelard von Bath," *Kgl. Danske Vidensk. Skrifter*, 7. R., *Hist. og filos. Afd.* 3: 1 (1914).
Suter 2	–, "Die Mathematiker und Astronomen der Araber und ihre Werke," *Abhandlungen zur Geschichte der mathematischen Wissenschaften* 10 (1900), and "Nachträge und Berichtigungen," *ibid.* 14 (1902), pp. 157–185 (reprinted Amsterdam: Oriental Press, 1982).
Toomer	G. J. Toomer, "A Survey of the Toledan Tables," *Osiris* 15 (1968), pp. 5–174.
Wiedemann	E. Wiedemann, "Über den indischen Kreis," *Mitteilungen zur Geschichte der Medizin und Naturwissenschaften* 11 (1912), pp. 252–255, repr. in *idem*, *Gesammelte Schriften zur arabisch-islamischen Wissenschaftsgeschichte*, (3 vols., Frankfurt: IGAIW, 1984–85), II, pp. 666–669.
Youschkevitch	A. P. Youschkevitch, *Les mathématiques arabes* (*VIII*e–*XV*e *siècles*), trans. by M. Cazenave and K. Jaouiche, Paris: J. Vrin, 1976.

ADDENDA AND CORRIGENDA

Bibliographical abbreviations:

ASI: refers to the present volume.

ANYAS = *Annals of the New York Academy of Sciences.*

Awraq = *Awraq* (Revista editada por el Instituto Hispano-Árabe de Cultura, Madrid).

Cairo Survey = D. A. King, *A Survey of the Scientific Manuscripts in the Egyptian National Library*, (Publications of the American Research Center in Egypt, Catalogs, vol. 5), Winona Lake, Ind.: Eisenbrauns, 1986.

DSB = *Dictionary of Scientific Biography*, 16 vols., New York: Charles Scribner's Sons, 1970-80.

EI ^{2}= *The Encyclopaedia of Islam*, 2nd ed., 6 vols. to date, Leiden: E. J. Brill, 1960 to present.

Frankfurt Preprints = Preprint series of the Institut für Geschichte der Naturwissenschaften, Johann Wolfgang Goethe-Universität, Frankfurt am Main.

Goldstein, *Studies* = B. R. Goldstein, *Theory and Observation in Ancient and Medieval Astronomy*, London: Variorum, 1985.

IMA = D. A. King, *Islamic Mathematical Astronomy*, 1st. ed., London: Variorum, 1986, 2nd. ed., 1993.

IAI = D. A. King, *Islamic Astronomical Instruments*, London: Variorum, 1987.

JAOS = *Journal of the American Oriental Society.*

JHAS = *Journal for the History of Arabic Science.*

JNES = *Journal of Near Eastern Studies.*

Kennedy Festschrift = D. A. King and G. Saliba, eds., *From Deferent to Equant: A Volume of Studies in the History of Science in the Ancient and Medieval Near East in Honor of E. S. Kennedy, Annals of the New York Academy of Sciences* 500 (1987).

Kennedy, *Studies* = E. S. Kennedy, Colleagues, and Former Students, *Studies in the Islamic Exact Sciences*, Beirut: American University of Beirut Press, 1983.

La Palma Catalogue = [J. Vernet, J. Samsó, *et al.*], *Instrumentos astronómicos en la España medieval: Su influencia en Europa*, Santa Cruz de la Palma, 1985.

Madrid 1992 Catalogue = J. Vernet and J. Samsó, eds., *El Legado Científico Andalusí*, Madrid: Ministerio de Cultura, 1992.

MAY = D. A. King, *Mathematical Astronomy in Medieval Yemen*, (Publications of the American Research Center in Egypt, Catalogs, vol. 4), Malibu, Ca.: Undena, 1983.
Samsó, *Ciencias en al-Andalus* = Julio Samsó, *Las ciencias de los antiguos en al-Andalus*, Madrid: Editorial Mapfre, 1992.
SAMI = D. A. King, *Spherical Astronomy in Medieval Islam: The Ḥākimī Zīj of Ibn Yūnus*, to appear.
SATMI = D. A. King, *Studies in Astronomical Timekeeping in Medieval Islam*, to appear.
ZGAIW = *Zeitschrift für Geschichte der Arabisch-Islamischen Wissenschaften*.

IMA = *Islamic Mathematical Astronomy*

General remarks: These addenda have been incorporated in the second edition (1993). The numerous papers of E. S. Kennedy cited throughout my own are now reprinted in Kennedy, *Studies*. A bibliography of studies either by Kennedy or inspired by him is contained in *Kennedy Festschrift*.

I: See now my paper "Some Remarks on Islamic Scientific Manuscripts and Instruments, and Past, Present, and Future Research," in John Cooper, ed., *The Significance of Islamic Manuscripts*, London: Al-Furqan Islamic Foundation, 1992, pp. 115-143 (also available in Arabic). On astronomy and other sciences in Andalusia see now the masterful study: Samsó, *Ciencias en al-Andalus* (and also *Madrid 1992 Catalogue*).

II: It should be stressed again that the tables described here are generally not found in *zīj*es and so were not mentioned by E. S. Kennedy in his *zīj* survey - see p. 40. This paper has been updated in "Towards a Classification of Medieval Islamic Astronomical Tables," paper presented at the XVIIIth International Congress of History of Science, Hamburg & Munich, 1st.-9th. August 1989, available in *Frankfurt Preprints* 25 (1990). See also J.-P. Hogendijk, "The Mathematical Structure of Two Islamic Astrological Tables for 'Casting the Rays'," *Centaurus* 32 (1989), pp. 171-202, *idem* and E. S. Kennedy, "Two Tables from an Arabic Astronomical Handbook for the Mongul Viceroy of Tibet," in E. Leichty *et al.*, eds., *A Scientific Humanist. Studies in Memory of Abraham Sachs*, Philadelphia, 1988, pp. 233-242, and the addenda to *ASI* II. • **Pp. 40-41:** For a survey of multiplication tables, see now XIV and and XV. • **Pp. 41-44:** For a more detailed survey of Islamic trigonometric tables, see *SAMI*, Part II, Section C, and "Trigonometric Tables in the Medieval Islamic World", to appear in A. Djebbar and K. Jaouiche, eds., Actes du